Autorenkollektiv

Photochemistry and Organic Synthesis

Recent Photochemistry of α-Diketones

M. B. Rubin

Photochemistry in Micelles

N. J. Turro, G. S. Cox and M. A. Paczkowski

Arylated Phenols, Aroxyl Radicals and Aryloxenium Ions Syntheses and Properties

K. Dimroth

Natural Polyamines-Linked Cyclophosphazenes Attempts at the Production of More Selective Antitumorals

J.-F. Labarre

AKADEMIE-VERLAG · BERLIN

Autorenkollektiv

Photochemistry and Organic Synthesis

Autorenkollektiv

Photochemistry and Organic Synthesis

With 91 Figures and 50 Tables

AKADEMIE-VERLAG · BERLIN
1985

Die Originalausgabe erscheint im Springer-Verlag Berlin · Heidelberg · New York · Tokyo
in der Schriftenreihe „Topics in Current Chemistry“,
Volume 129

Vertrieb ausschließlich für die DDR und die sozialistischen Länder

Erschienen im Akademie-Verlag Berlin, DDR-1086 Berlin, Leipziger Straße 3–4
Lizenznummer: 202 · 100/514/85
Printed in the German Democratic Republic
Gesamtherstellung: VEB Druckerei „Thomas Müntzer“, 5820 Bad Langensalza
Umschlaggestaltung: Karl Salzbrunn
LSV 1275
Bestellnummer: 763 492 4 (3072/129)
13800

Table of Contents

Recent Photochemistry of α-Diketones

Mordecai B. Rubin

Department of Chemistry, Technion-Israel Institute of Technology, Haifa, Israel

Table of Contents

Activity in the photochemistry of α-diketones has continued unabated in the past decade. In addition to special attention to absorption and emission spectra, photoelectron spectroscopy has been applied widely. Areas of recent emphasis include (1) cyclobutene- and (2) cyclobutanediones, (3) bridged cyclohexenediones, and (4) reactions in the presence of oxygen, particularly epoxidation of olefins. The more venerable aspects such as inter- and intramolecular hydrogen atom abstraction reactions and additions to multiple bonds continue to receive attention. A considerable number of examples of synthetic applications have accumulated in recent years. In parallel, mechanistic understanding has broadened considerably.

I Introduction

The photochemistry of α-diketones has been a subject of interest for about a century. Since the appearance of comprehensive review articles [1] in 1969 and 1971, activity in this area has continued with investigation of a number of new systems, particularly unsaturated diketones and diketones incorporated in a four-membered ring. New types of chemistry of synthetic and mechanistic interest have been revealed. The purpose of this review is to summarize these newer developments with briefer reference to some significant developments in the older reactions.

The Scheme below summarizes the "classical" photochemistry of saturated and aryl diketones. These undergo efficient intersystem crossing (very weak fluorescence, strong phosphorescence) to the chemically reactive triplet state (n_+, π^*) which may (inter- or intramolecularly) abstract a hydrogen atom of a wide variety of types or add to a multiple bond; in both cases two new radical centers are formed. The resulting radical pair or 1,4-biradical will proceed to product(s) or revert to starting material(s) by appropriate free radical processes which are of considerable intrinsic interest but whose only relation to photochemistry may be the multiplicity deriving from the excited state precursor. In addition to the two very common reaction types mentioned above, two possible α-cleavages, as illustrated in Scheme I, might occur and will be recognizable by fragmentation or loss of carbon monoxide. These cleavages are generally of negligible importance [2] and are observed mainly with

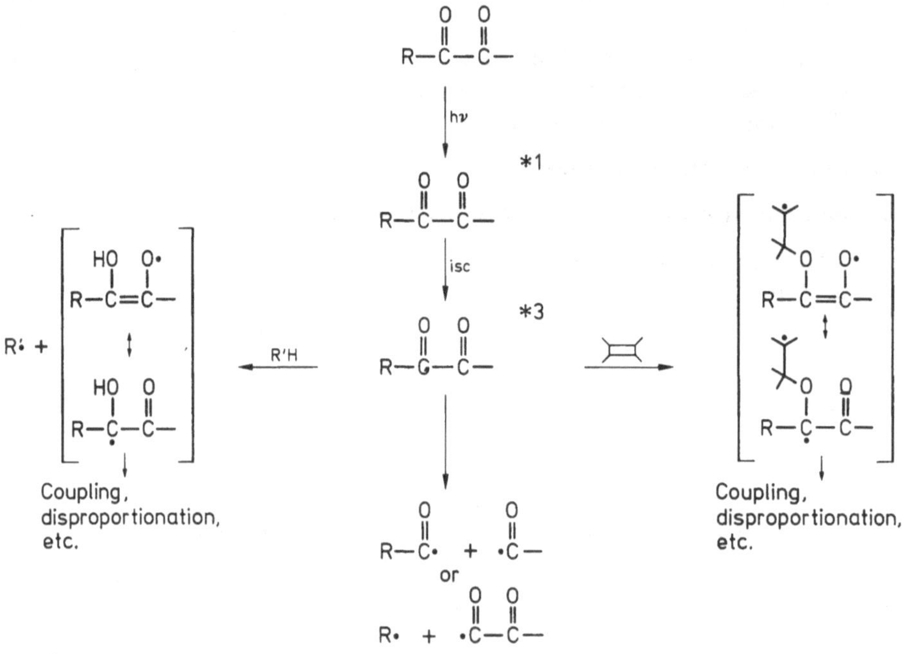

Scheme I Photochemistry of Saturated and Aryl Diketones

seven-membered cyclic systems, particularly those containing a heteroatom, or in irradiations in inert medium.

Most of the newer photochemistry, with the exception of open-chain unsaturated diones and recent results with the older reactions, is very different from that described in Scheme I. Reactions occur from the singlet state; bond cleavage, either with rearrangement or loss of carbon monoxide, is the major process.

II Spectroscopy

On the basis of extended Huckel and CNDO/2 calculations, Swenson and Hoffmann [3] proposed in 1970 that through-bond interaction between non-bonding orbitals n_1 and n_2 of the two carbonyl groups of α-diketones would result in two molecular orbitals n_+ and n_- with clearly split orbital energies (rather than two orbitals of identical energies as had been assumed previously). The effect of through-space interactions was estimated to be negligible. Experimental confirmation was forthcoming one year later [4] from vertical ionization potentials (IP) determined by photoelectron (PE) spectroscopy. PE spectra of many dicarbonyl compounds have been measured since; representative results are presented in Table I together with long wavelength absorption maxima. Assignments were based on theoretical calculation and analogy.

As can be seen in the Table, the splitting of n_+ and n_- orbitals lies in the range 1.5–2.1 eV for a large number of dicarbonyl compounds of differing ground state conformations. Typical are planar biacetyl (entry 1, λ_{max} 440 nm, ΔIP 1.84 eV) and tetramethylcyclobutanedione (entry 5, 492 nm, 2.08 eV) on the one hand and skewed di-t-butyldiketone (entry 2, 362 nm, 1.99 eV) and tetramethylcyclooctanedione (entry 7, 348 nm, 2.08 eV) on the other. Introduction of homoallylic conjugation in cyclic systems (entries 12, 15) results in a small hypsochromic shift of the absorption maximum and a much larger splitting of n_+ and n_- energies. This has been attributed to through-bond effects. Much smaller effects are observed with more remote double bonds. The combination of PE, absorption, and emission spectra provides a powerful tool for detailed characterization of excited states.

Turning to absorption spectra, the long-standing generalization [14] that long wavelength (n_+, π^*) absorption maxima of α-diketones vary as a function of torsion angle (maximum values for 0° (~500 nm) and 180° (~450 nm), minimum for 90° (~330 nm)) continues to receive support. This generalization was originally based on absorption spectra in ethanol solution of the Leonard series of α,α,α′,α′-tetramethyldiones of varying ring size where the ring provides a conformational constraint. Repetition of these measurements [8,9] in cyclohexane solution confirmed the earlier results but with higher extinction coefficients due to the absence of a perennial problem with α-diketones, hemiketal (or hydrate) formation. The four methyl groups introduced to prevent enolization apparently lead to conformational complications in larger rings since the value of 384 nm for a tetramethyl substituted 16-membered ring is much lower than the range 442–448 nm observed for the maxima of four compounds of similar ring size lacking methyl substitution [15,16]. Long wavelength absorption maxima (band of highest intensity) will be included wherever possible in the sections to follow.

Table 1. Long Wavelength Absorption Maxima[a] and Vertical Ionization Potentials[b] of Selected α-Diketones

Entry	Compound	λ_{max}(nm)	I.P. (eV) n_+	n_-	π	Ref.
1	Biacetyl	440	9.57	11.41		4)
2	Di-t-butyldiketone	362	8.66	10.65		5)
3	Benzil	370	9.1	11.1		6)
4	Cyclobutanedione	489	9.61	11.71	12.83	7)
5	Tetramethylcyclobutanedione	492	8.79	10.87		8,9)
6	3,3,7,7-Tetramethylcycloheptanedione	337	8.67	10.55		5,8,9)
7	3,3,8,8-Tetramethylcyclooctanedione	348	8.61	10.59		9)
8	Cyclobutenedione	340	9.79	11.87	11.55, 13.61	7)
9	Benzocyclobutenedione	420	9.23	11.23	10.14, 10.43	10)
10	Camphorquinone	470	8.80	10.40		4)
11	Bicyclo[2.2.1]heptanedione	484	9.0	10.5		11,12)
12	Bicyclo[2.2.1]heptenedione	460	8.7	11.1	10.6	11,12)
13	7-Oxabicyclo[2.2.1]heptenedione	486	8.9	11.7	10.8	11)
14		460	8.9	11.5	10.3, 10.7[c]	11)
15		450	8.7	11.8	10.3, 10.5[c], 10.9[c]	11)
16		526	8.85	10.65		13)
17	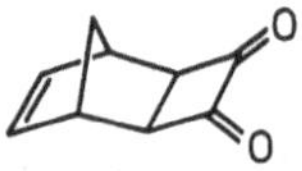	546	8.65	10.80		13)

[a] In hydrocarbon solvent. Values are for the most intense maximum.
[b] Vertical ionization potentials obtained from He(I) photoelectron spectra.
[c] IP assigned to Walsh orbital of cyclopropane ring.

The intriguing observation [17] that absorption maxima of [4,4,2]-propellanediones *1*, *2*, and *3* depended markedly on the presence of remote unsaturation has stimulated considerable activity. As summarized below, the saturated compound *1* had an absorption band of Gaussian shape with maximum at 461 nm, the most intense maximum of the diene *3* was at 537 nm with fine structure at shorter wavelengths, and the monoene *2* gave a composite spectrum. This remarkable effect was originally

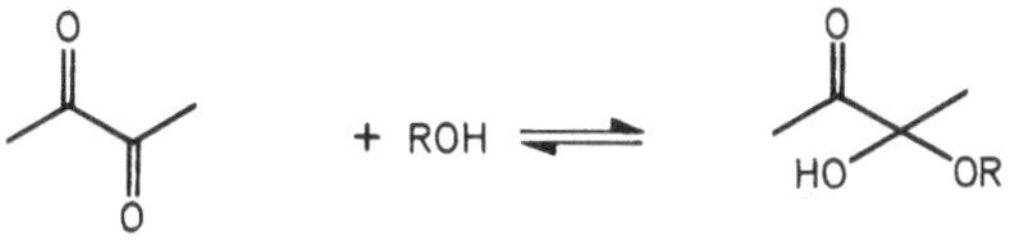

attributed [17] to through-space interaction of the π-electrons with the dione moiety and subsequently [18] to through-bond interaction. The present view [19,20], based on low-temperature absorption, PE, and fluorescence spectra and on calculation, is simply that the shift in absorption in going from *1* to *3* reflects conformational factors, specifically increasing rigidity of the system with introduction of double bonds. The questions posed by the spectra of *1*, *2*, and *3* prompted synthesis of the diketones *4–6* (and a cyclopropane analogue) in which conformational flexibility is not a problem [19]. As can be seen below, the presence of unsaturation results in a small hypsochromic shift of the absorption maximum while PE spectra were similar. Calculated values for λ_{max} in this series were in good agreement with observed values. Quite good agreement between observed and calculated absorption has also been obtained with a number of bicyclic diketones [21]. Additional examples [22] pertinent to

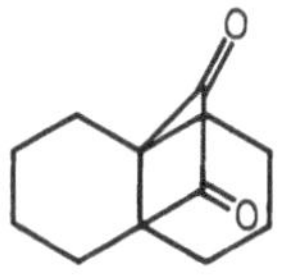

1

λ_{max} 461 (73)
IP 8.65, 10.4

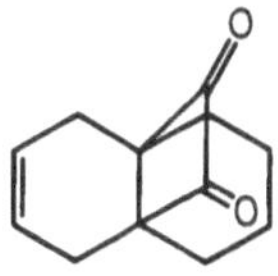

2

460—464 (39), 532—535 (34)
8.60, 9.5, 10.5

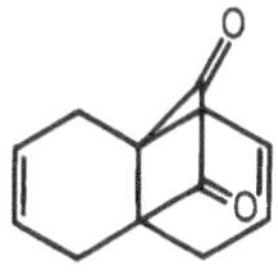

3

537.5 (72)
8.7, 9.35, 10.0, 10.7

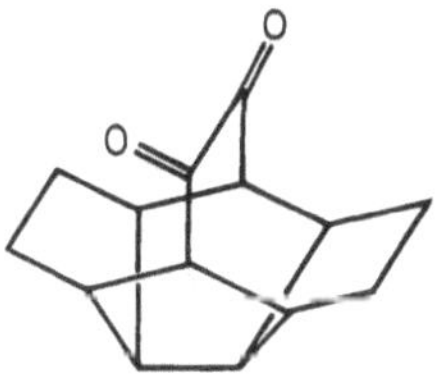

4

λ_{max} 427
IP 8.82, 10.26

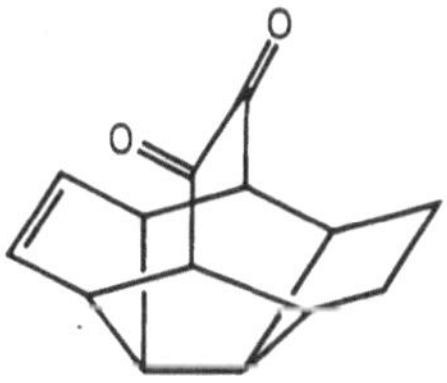

5

421 (22)
8.80, 9.45, 10.4

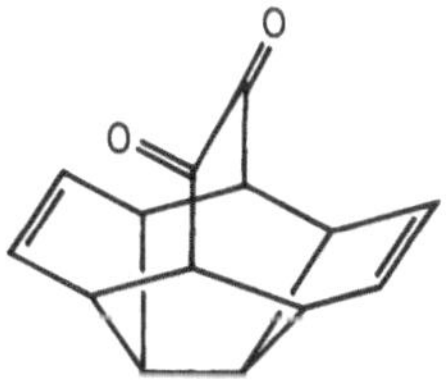

6

408
8.85, 9.95, 10.0, 10.3

the question of long-range interactions are the bis-α-diketones *7* and *10*. Comparison of *7*, where the two diketo-chromophores are approximately orthogonal, with *8* and *9* possessing a single diketone function, shows little difference in absorption spectra but a considerable one in ionization potentials. On the other hand, the biscyclobutanedione *10*, with the two chromophores approximately parallel, shows opposite behaviour when compared to *11*. The geometry of *10* is considered to be particularly favorable for through-bond interactions. In general, absorption maxima of cyclobutanediones exhibit large variation as a result of relatively minor structural change as can be seen from the examples cited. Another illustration is the pair of stereoisomers *12* and *13* shown below, both of which appear to possess very rigid structures [23]. While cyclobutanedione itself has been shown to have a planar structure in the gas phase [24,25], the small deviations from planarity possible as a result of substituent effects do not seem sufficient to account for the considerable variation in absorption spectra in solution.

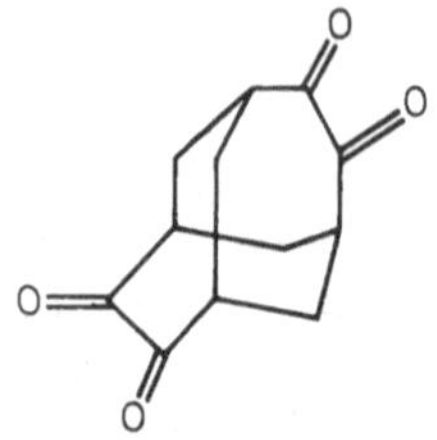

7
λ_{max} 408 (61)
IP 9.23

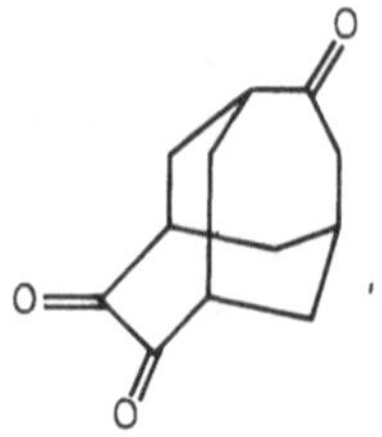

8
418 (23)

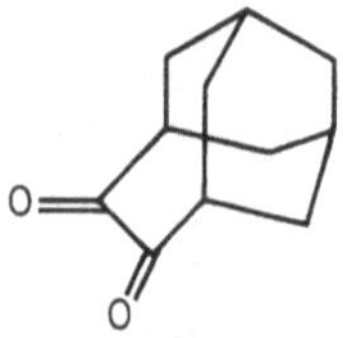

9
418
8.82

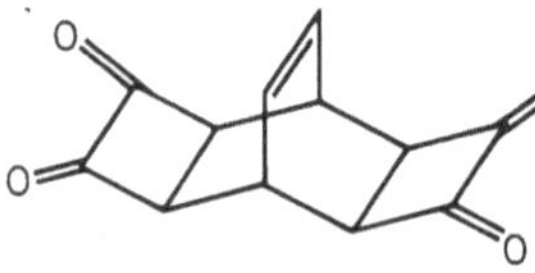

10
λ_{max} 540 (260)
IP 9.08, 9.80, 10.32

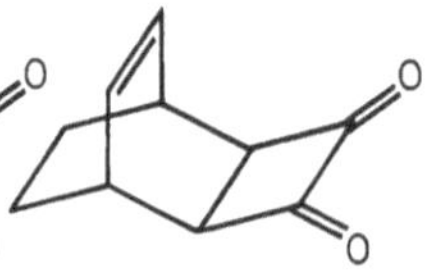

11
507 (26)
9.02, 9.77

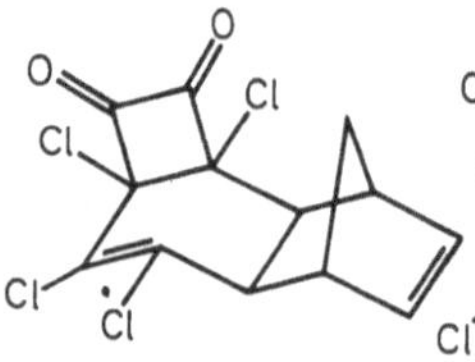

12
517

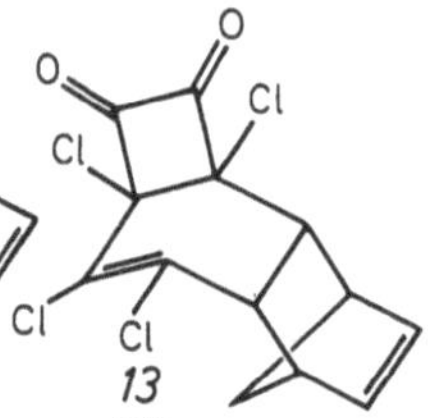

13
495

α-Diketones exhibit weak fluorescence and strong phosphorescence. In addition to their usefulness in characterizing excited states, these properties have frequently been exploited, particularly using biacetyl, as mechanistic probes in many types of photoreactions using the diketone either as a sensitizer or as a quencher [1,2]. An interesting application is the macrocyclic compound *14a* incorporating separated phenanthrene and diketone chromophores [26a]. Excitation of the aromatic moiety resulted in dual fluorescence (and phosphorescence) arising from partial energy transfer. The rate of singlet energy transfer was relatively slow when excitation was in the O—O band of the phenanthrene moiety and much faster upon shorter wavelength excitation. The direction of energy transfer could be reversed by two-photon excitation. This work has been extended [26b] to series of analogous compounds such as *14b* where the efficiency of singlet energy transfer from the aromatic moiety to the dione depended markedly on the length of the polymethylene chains separating the chromophores. A successful theoretical treatment of this Dexter type of energy transfer has been achieved [26c]. Comparison of the solvent-dependence of emission spectra of the two steroidal diketones shown has been interpreted in terms of a long range interaction between ring A and the dione chromophore [27].

Conformational factors can play an important role in emission properties. The earlier view that, no matter what their ground state conformations, excited α-diketones assume a coplanar (s-trans if possible) conformation is generally accepted [6,8,9,28–34]. This is based in part on observations that diketones having skewed ground states show marked lack of mirror image symmetry between absorption and fluorescence

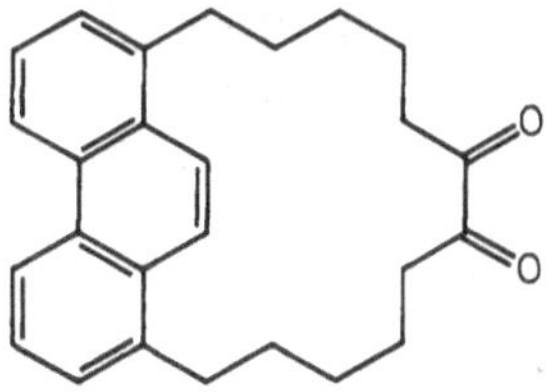

14a

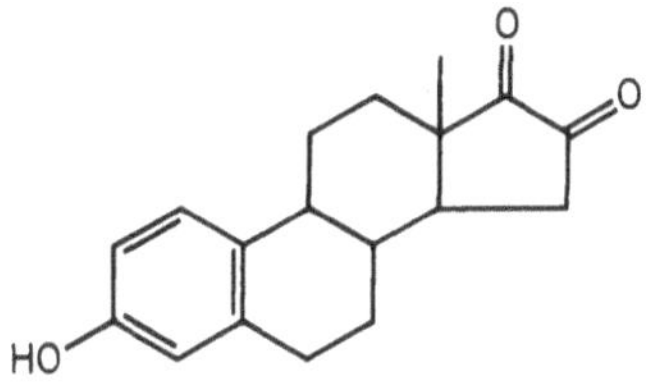

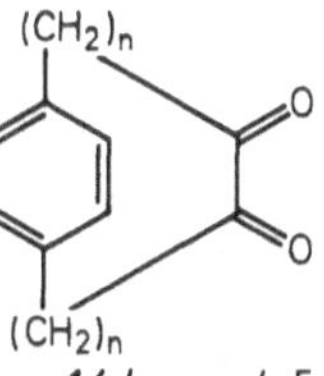

14b n = 4, 5, 6

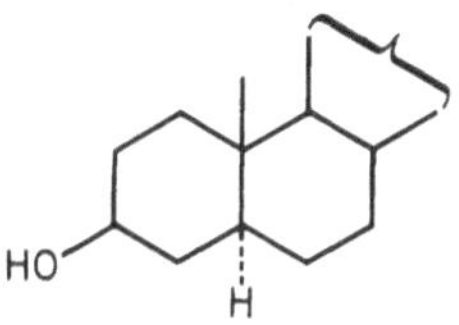

spectra with large differences in energy between O—O states but fairly constant differences between fluorescence and phosphorescence transition energies. The comparison between benzil, whose ground state consists of two nearly planar benzoyl groups in an approximately orthogonal relationship, and mesitil [29], which consists of an s-trans coplanar dicarbonyl system with orthogonal aromatic rings, is an instructive one. Benzil has a broad absorption band with a maximum at 370 nm and a

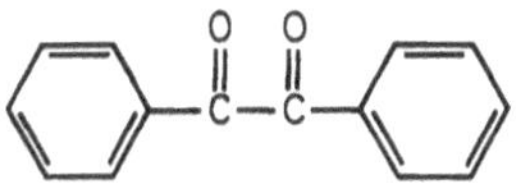

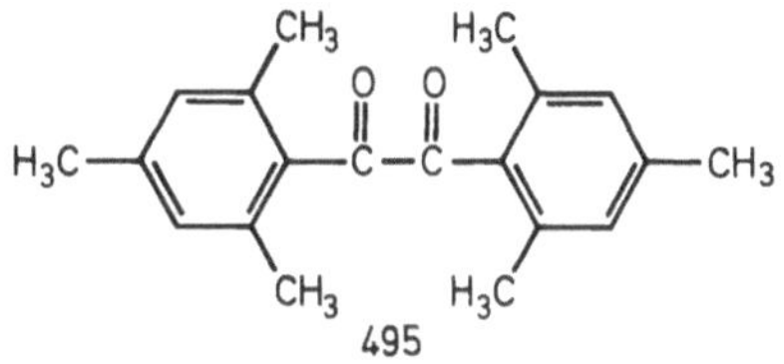

	Benzil	Mesitil
λ_{max}	370	495
λ_{fl}	505	505
λ_{phos}	562	577

separation of about 5500 cm^{-1} between absorption and fluorescence while mesitil has a structured absorption spectrum with maximum at 495 nm and a separation of 400 cm^{-1} between O—O bands of absorption and fluorescence. Both compounds have similar separation ($\sim$2300 cm^{-1}) between fluorescence and phosphorescence maxima. This has been interpreted in the following way [30]. Excitation of skewed ground-state benzil results in formation of skewed singlet which then relaxes to the lower energy singlet having an s-trans planar dione system. This may require rotation of phenyl groups out of the plane of the dione. The relaxed singlet may emit or undergo intersystem crossing to a triplet of the same conformation which is again the most stable one for that state. It has been pointed out [30] that the energy required to promote stable, skewed benzil to its (higher energy) skewed triplet state will be higher than that released by transformation of the relaxed planar (lower energy) triplet to planar (higher energy) ground state benzil. In other words, ground state benzil acting as a triplet quencher will have a higher apparent triplet energy than triplet benzil acting as a sensitizer.

Differences in low temperature emission spectra of benzil in methylcyclohexane and in isopentane have been ascribed to inhibition of the conformational changes involved in the skewed to planar relaxation in isopentane [33]. Emission spectra were identical in both solvents at temperatures above the glass-forming temperature. Preference for an s-cis conformation in ethylene glycol solution has been suggested [33b] to account for anomalous emission spectra of benzil in that medium. Other aspects of benzil emission have been examined [33c].

In the Leonard series of four- to eight-membered tetramethyl-cycloalkane diones mentioned earlier in connection with the angular dependence of λ_{max}, the four-membered compound gave no emission, five- and six-membered showed only fluorescence, and the two larger ring members exhibited both fluorescence and phosphorescence [9]. The separation between absorption and fluorescence varied as expected from the assumption of planar emitting and non-planar absorbing species.

We note that cyclobutanediones and unsaturated diketones which form the major part of this review show only fluorescence, often with very low yield.

Circular dichroism provides an additional spectroscopic tool for characterization of excited states [35]. Considerable interest has also been extended to esr-spectra of anion radicals of α-diketones [36]. Circular polarization of the phosphorescence of camphorquinone has been determined [151]. Biacetyl has been the subject of a CIDNP study [152], of fluorescence quenching by a variety of substrates [153], and of steric effects in quenching of triplet states of alkylbenzenes [154].

III Ethylenedione

Ethylenedione, the dimer of carbon monoxide, has attracted chemists' interest since at least 1913 when its synthesis by dechlorination of oxalyl chloride was unsuccessfully attempted [37]. This substance, which lies between carbon dioxide and carbon suboxide in the series of oxycumulenes, is the simplest possible unsaturated diketone. It can be represented by a number of canonical structures as shown below:

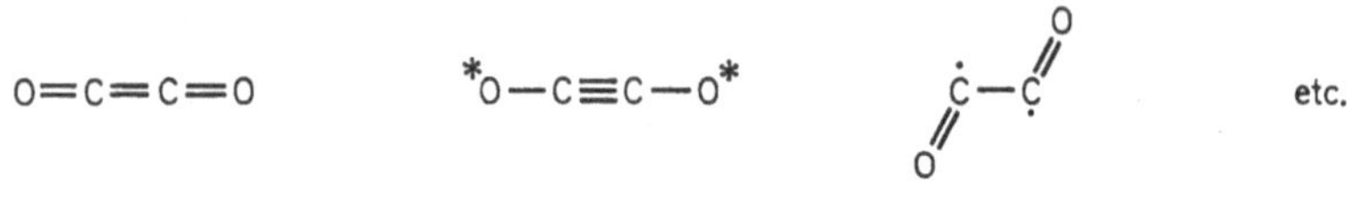

Interest in the possibility of detecting C_2O_2 received special impetus as a result of the photobisdecarbonylation reactions of bridged cyclohexenediones and of cyclobutanediones to be discussed later. Concerted cycloelimination to give ethylenedione and olefin was envisaged as a reaction pathway. Observation of a fragment corresponding to $C_2O_2^{+\cdot}$ in mass spectra of such diketones was taken as an indication that similar fragmentation might occur from electronically excited states. As a result, a considerable

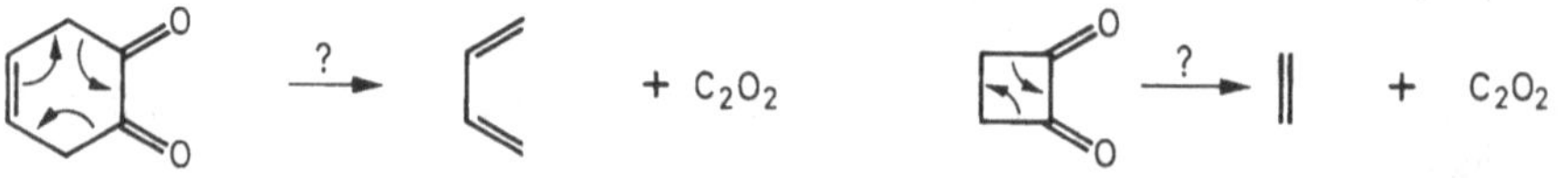

number of papers have appeared [38] presenting results of calculations of the structure and properties of C_2O_2. In the first detailed theoretical paper [38a] it was concluded that "ethylenedione is kinetically (singlet) and thermodynamically (singlet and triplet) unstable with respect to two molecules of carbon monoxide". The latest treatment [38b] concludes that, because of spin restrictions, the triplet ground state is a minimum on the potential energy hypersurface and that metastable C_2O_2 "should be detectable in a carefully designed experiment". The analogous ethylenedithione (C_2S_2) was predicted to be of significantly greater stability.

On the experimental side, C_2O_2 has never been observed by physical methods nor has it been possible to trap it (e.g. by reaction with dienes or with chlorine [39]). Its existence remains an interesting question.

IV Cyclobutenediones

In view of the destabilization resulting from incorporation of four trigonal carbon atoms in a four-membered ring and the juxtaposition of dipolar carbonyl groups in an *s-cis* arrangement, it is not surprising that all of the known photochemistry of 1,2-cyclobutenediones involves unimolecular reactions with ring cleavage or ring enlargement. The major primary process is ring opening to bis-ketenes (*15*). Formation of *15* is supported by observation of ketene bands in the infrared upon photolysis of dimethyl- [40] (*16a*) and diphenylcyclobutenedione [40,41] (*16b*) (and a monoimine derivative [41]) at 77 K. Intermediacy of *15* had been proposed earlier on the basis of isolation of the derived succinnic diester from irradiation of phenylcyclobutenedione

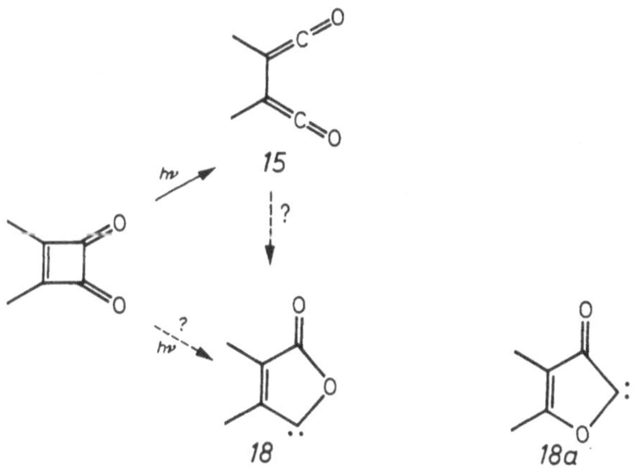

in alcohol solution and of a Diels-Alder adduct when benzocyclobutenedione (*17*) was irradiated in the presence of dienophile. A number of additional examples of isolation of substituted succinnic esters from photolyses in the presence of alcohols are summarized below, as well as an intramolecular case [42].

Additional examples of trapping of the bisketene from benzocyclobutenedione (in low yield) by Diels-Alder reactions have also been reported [44].

In addition to products derived from bis-ketenes, reactions of *17* [45] and of diethyl

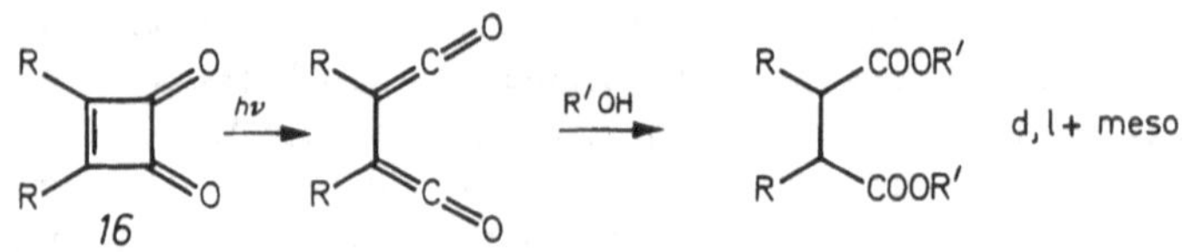

16, *a* R = CH_3 [40]; *b* R = C_6H_5 [40,41]; *c* R = $(C_6H_5)_2CH$ [42]; *d* R = C_2H_5O [43]

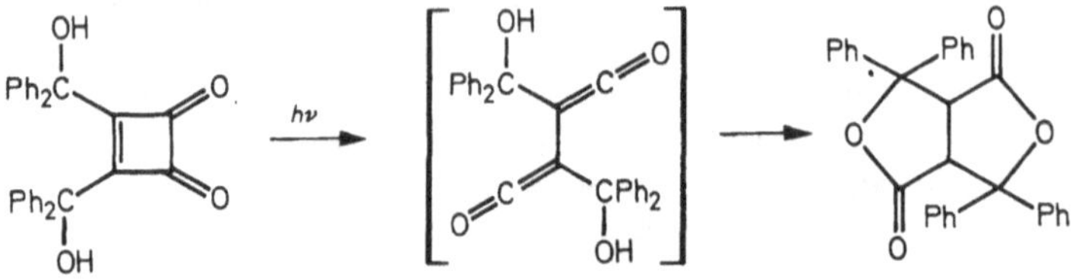

squarate [43] (*16d*) gave products derived from lactocarbenes *18*. This ring enlargement, analogous to a well-known reaction of cyclobutanones, could arise from a second, competing primary process or from thermal or photochemical isomerization of *15*. No products derived from the isomeric oxaketocarbene *18a* have been observed. Interestingly, infrared bands assigned to dimers of *18* and ultraviolet absorption attributed to biphenylene [46] were observed in photolyses of *17* at 77 K but the bisketene *19*, in photoequilibrium with *17*, was observed [47] in an Argon matrix at 10 K in addition to benzyne and, "under special conditions", benzocyclopropenone (*20*).

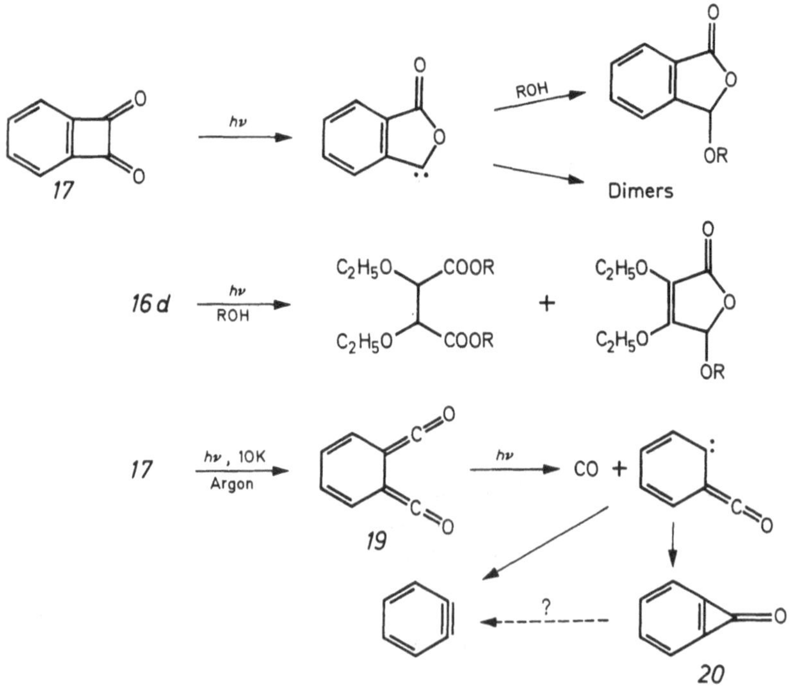

One of the reasons for confusion in this area is the fact that cyclobutenediones absorb at much shorter wavelengths than their saturated counterparts, prompting the use of short wavelength light in photolyses; this can result in formation of secondary photoproducts. It seems likely that formation of cyclopropenones and of acetylenes is the result of photochemical reaction of bisketenes, the primary products of irradiation. Acetylenes could also be formed from cyclopropenones.

The most dramatic synthetic achievement with cyclobutenediones is the synthesis of deltic acid derivatives. As illustrated below, photolysis [43] of diethyl squarate (*16d*) yielded, among other products, diethyl deltate (*21*). The bis-trimethylsilyl ester (*16e*) of squaric acid similarly furnished a deltate ester which was successfully hydrolysed to deltic acid itself (*22*), the lowest member of the series of oxocarbon acids [48].

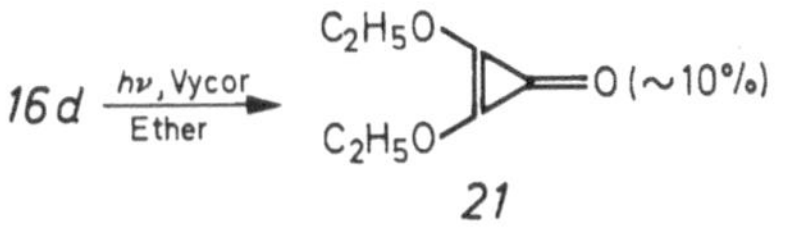

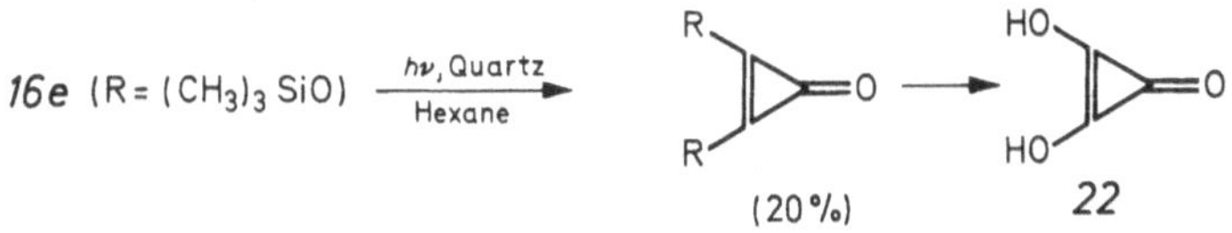

Formation of the dianilide of acetylenedicarboxylic acid from still another reaction of *16d*, photolysis in ether containing aniline, may involve prior reaction [49] of aniline with *16d*. The bis-anhydride (*16f*) of squaric acid also afforded an acetylene dicarb-

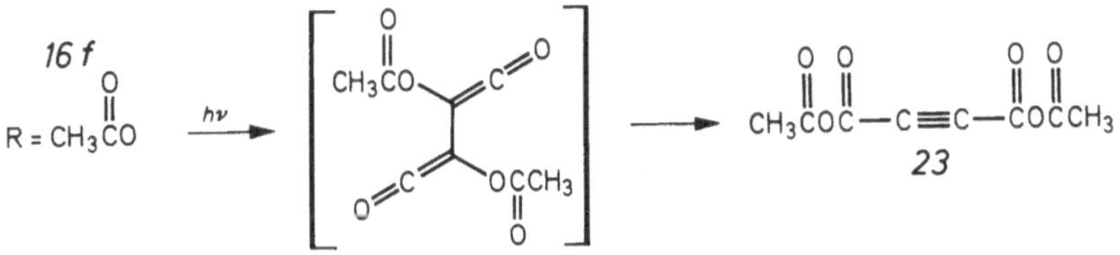

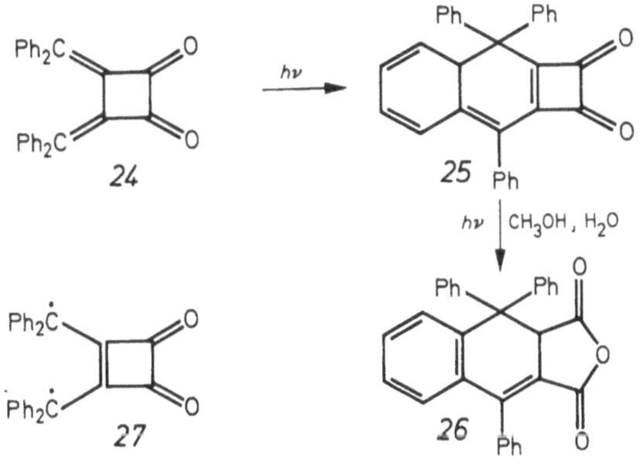

oxylic acid derivative, the mixed anhydride *23*, upon photolysis [50]. In the latter case, rearrangement of intermediate ketene was proposed to account for the observed result.

The photolysis of bis-benzhydrylidenecyclobutanedione *24* to cyclic anhydride *26* in aqueous methanol may involve prior photocyclization to cyclobutenedione *25* followed by photolysis to bisketene and reaction with solvent rather than formation of biradical *27* and subsequent reactions as originally suggested [51].

With the exception of the low temperature studies mentioned earlier, mechanistic details of cyclobutenedione reactions have not yet been investigated. The products clearly indicate that clevage of the intercarbonyl bond is the major reaction of excited state(s). Interestingly, there is no evidence for a primary process involving formation of a triple bond and carbon monoxide (or C_2O_2).

V Cyclobutanediones

Cyclobutanediones, once exotic compounds represented by a few perhalo derivatives, have become readily available as a result of new synthetic developments in recent years. These include the modified acyloin condensation [52] in which the intermediate enediolate is trapped as bis-trimethylsilyl ether (*28*) which can be converted to cyclobutanedione by reaction with bromine or hydrolyzed to acyloin and oxidized in a separate step. In addition to this efficient and general method, bi- or polycyclic unsaturated cyclobutanediones (*30*) have become available from photolysis of bridged cyclohexenediones (*29*) to be discussed in the following section. Photocycloaddition of dichlorovinylene carbonate (DCVC) to olefins [53] promises to provide a third route if the problems associated with hydrolysis of the photoadducts (*31*) can be overcome.

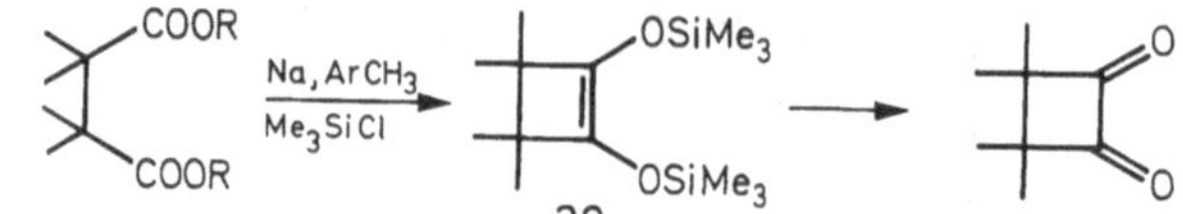

28

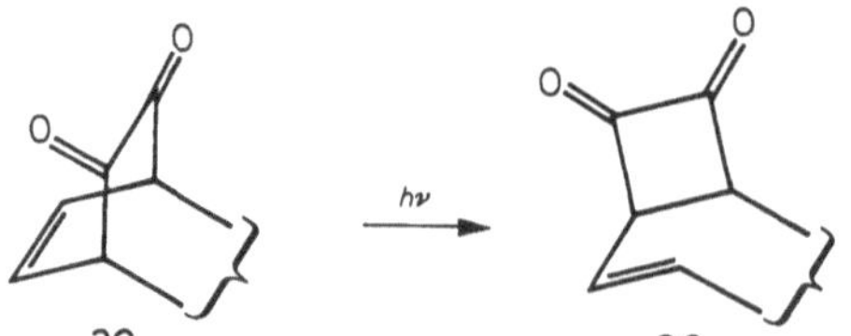

29 30

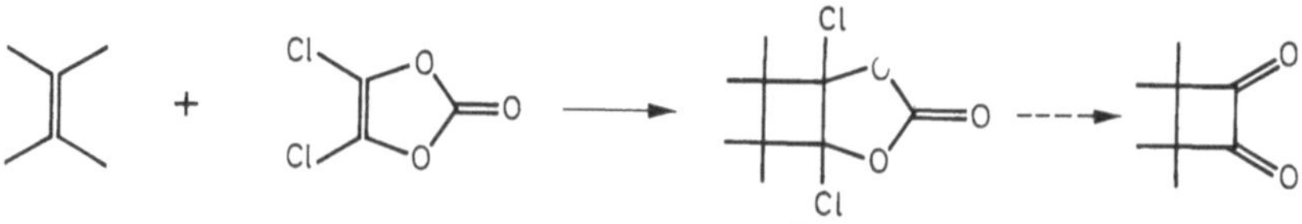

DCVC 31

The major difficulty in working with cyclobutanediones is their extreme sensitivity to moisture or protic solvents. Very rapid hydrate or hemiketal formation is followed by uncatalyzed benzilic acid type rearrangement to give ring-contracted products as illustrated below [54]. The usual effects on hydrate equilibria, such as enhancement by electron-withdrawing substituents and inhibition by bulky groups, are observed. The use of carefully dried, aprotic solvents is essential in working with these compounds.

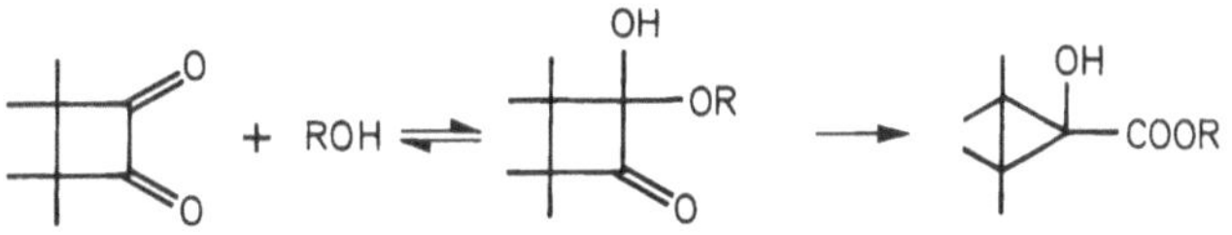

The n_+,π^* absorption of cyclobutanediones is observed at longer wavelengths than for any other α-dicarbonyl compounds (except 2,2,5,5-tetramethyltetrahydrofuran-3,4-dione, λ_{max} 559 nm). As noted in the section on spectroscopy, values range from 461–546 nm (extinction coefficients of 50–100) with considerable fine structure in some cases. Vinylcyclobutanediones of general structure *30* exhibit maxima of Gaussian shape in the range 485–530 nm (extinction coefficients 200–500) and maxima at 300–350 nm (extinction coefficients of several thousand). Very limited information on emission from cyclobutanediones is available. Attempts [55] to observe phosphorescence from *trans*-di-t-butylcyclobutanedione (*32*), propelladienedione *3*, and a number of unsaturated compounds of type *30* gave negative results; triplet energies are not known. The latter compounds also did not give detectable fluorescence while very weak fluorescence ($\phi_f < 0.01$) was detected [56] from *3* and *32*. Cyclobutanediones are also characterized by two high frequency carbonyl bands (~1760 and 1780 cm^{-1}) in the infrared.

The same factors, ring strain and unfavorable dipolar interactions, operating in cyclobutenediones also obtain in their saturated analogues. Although nearly all of the results to date have appeared in the form of preliminary communications with minimal experimental detail, almost all reactions appear to involve singlet states (no effect of oxygen or triplet quenchers on quantum yield) with cleavage of the ring. A priori, concerted fragmentation to two molecules of ketene (A below) or one molecule each of alkene and ethylenedione (B) or cleavage of a single bond as illustrated in C and D are possible as primary processes. No hint of formation of ketenes has

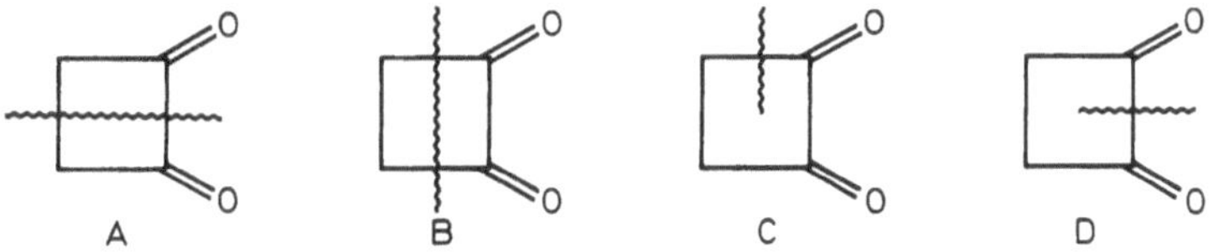

A B C D

been reported to date but only very limited evidence is available to allow any distinction among the remaining possibilities.

The most common reaction, with the exceptions to be noted later, is photobisdecarbonylation to alkene plus two molecules of carbon monoxide (C_2O_2?). This is observed with tetramethylcyclobutanedione [57] (*33*), the propellanedione *3* (in a variety of solvents)[58], and a considerable number of vinylcyclobutanediones[23,59] of type *30* (summarized in Table II of section VI A), all in nearly quantitative yield. Irradiations of the latter type of compound have been performed using visible light and gave quantum yields of 0.1–0.4 except for tetrachloro-derivatives which exhibited lower values in some cases. Triplet-sensitized reactions[23,55] also produced dienes with quantum yields approaching unity. Synthetic aspects of these reactions will be discussed separately.

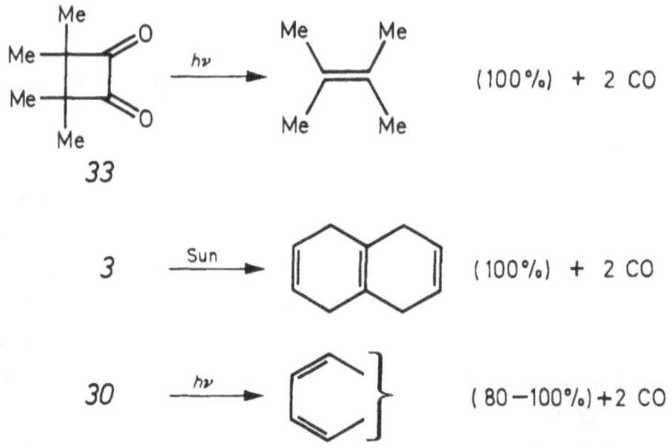

As noted earlier, compounds of type *30* are obtained by photolysis of *29*. Irradiations of *30* at $\lambda > 500$ nm where *29* have no absorption (cf. Fig. 1 for typical spectra) allow examination of the possible occurrence, even to a small extent, of the reverse photoisomerization 30 → *29*. No such isomerization has been observed except for the overcrowded tetrasubstituted substance *34* where a "trace" of such reversal has been detected[61] together with a small amount of monodecarbonylation in addition to the major product, the diene *35*.

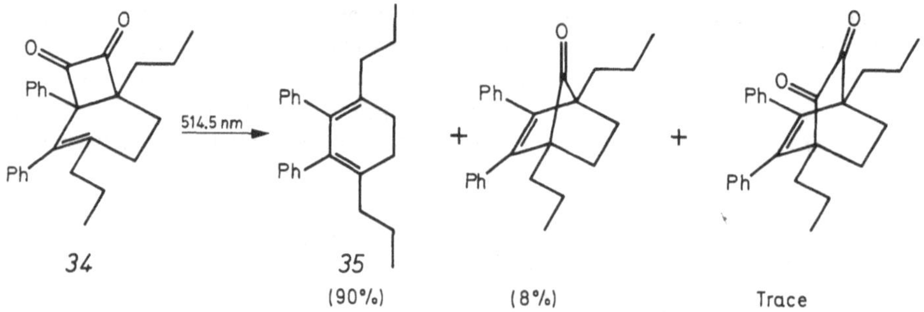

The contrast between thermal and photochemical reactions of vinyl cyclobutanedione *36* is of interest[60]. Photolysis of *36* proceeded quantitatively with bisdecarbonylation to tetrachlorodiene *37* while thermolysis (70°) of *36* resulted in quantitative isomerization to isomeric diketone *39* (the photochemical precursor of *36*). Similar effects have been observed with related compounds. The most reasonable mechanism for isomerization *36* → *39* is homolysis to biradical *38* which can either revert to *36* or

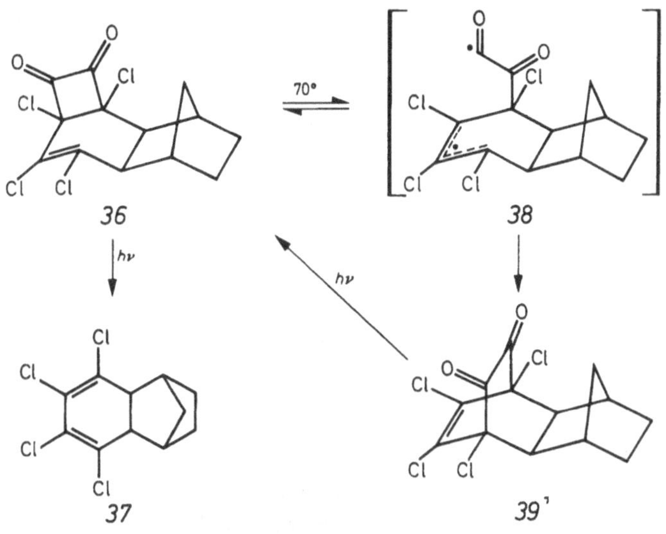

yield *39* (thermally stable to above 150°). If this interpretation is correct, biradical *38* cannot be an intermediate in photolysis of *36*. This leaves concerted elimination of C_2O_2 or intercarbonyl bond homolysis as possible pathways in photolysis of *36*.

Attempts to observe intermediate(s) in photolysis of *3* by irradiation at 77 K were thwarted by the extremely low quantum yield for reaction at this temperature [23]. No such marked temperature dependence was observed with compounds of type *30* where quantum yields at 77 K were only slightly lower than room temperature values [23,55,62]. Results obtained with tricyclic compounds which are of special interest in

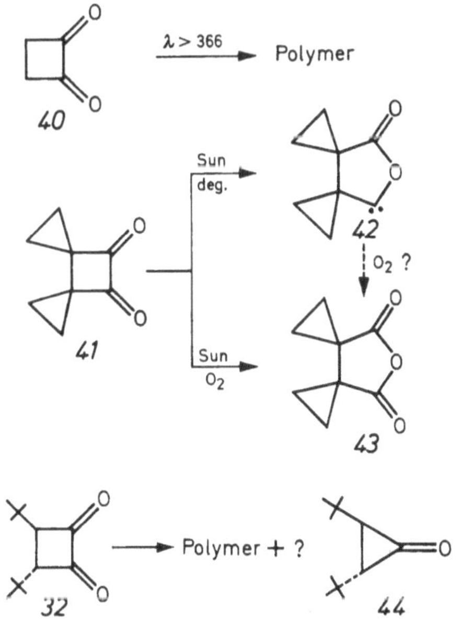

connection with the synthesis of norcaradiene will be discussed in the section (VI D) on synthetic applications.

The three cyclobutanediones which did not undergo bisdecarbonylation were cyclobutanedione (*40*) itself, the dispiro compound *41*, and trans-di-t-butylcyclobutanedione (*32*). Irradiation [63] of *40* with visible light afforded a carbonyl containing (ν_{max} = 1715 cm^{-1}) polymer at room temperature or −75 °C. Formation of lactocarbene *42* from *41* was inferred [64] from its trapping by methanol and insertion into starting material as illustrated above. It is not clear if formation of anhydride *43* from irradiation in the presence of oxygen proceeds via *42* or is another example (cf. Section VII) of reaction of triplet diones with oxygen. The third compound, *32*, exhibited behaviour characteristic of a triplet state, namely quenching by oxygen and by anthracene ($E_T \sim 42$ kcal/mole). Nmr-monitoring [55] of photolysis of *32* in degassed deuteriochloroform at $\lambda > 390$ nm indicated that a complex product mixture was formed. It has not been possible to reproduce the original report [65] that *trans*-di-t-butylcyclopropanone *44* is formed from *44*.

In summary, the photochemistry of cyclobutanediones shows a marked dependence on substitution. While intercarbonyl bond cleavage (path D) appears to be a reasonable candidate for the primary process, additional investigation is required to clarify reaction mechanisms.

VI Bridged Cyclohexenediones

A Mono-enes

Unlike the complex photochemistry of four-membered ring diketones discussed in the previous two sections, 3,6-bridged 4-cyclohexene-1,2-diones (*29*) undergo a single photoreaction upon excitation with visible light [23, 59, 66]. This is isomerization via a 1,3-acyl migration to unsaturated cyclobutanediones (*30*). Subsequent irradiation of *30* gives 1,3-dienes and two molecules of carbon monoxide (or C_2O_2) as described in the previous section. Since *30* have absorption maxima (485–530 nm) at considerably longer wavelength than *29* (430–470 nm), the change is often a visible one in

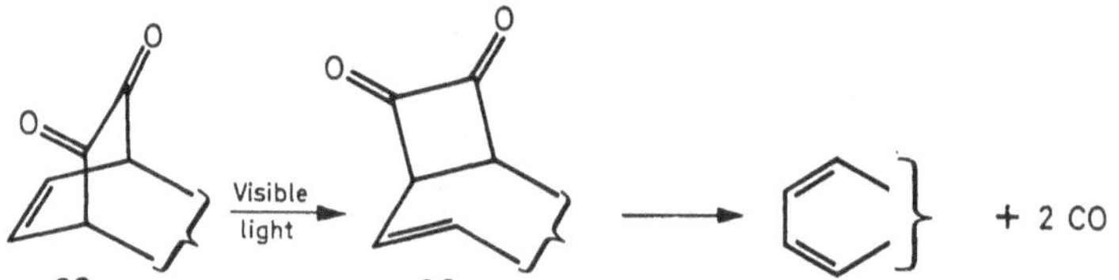

29 30

which the originally yellow solution turns pink or red as if orange juice had been transformed into rosé wine. Absorption spectra in the visible region provide a convenient tool for quantitative study.

One of the best examples of this type of transformation is illustrated in Fig. 1 showing spectra determined after intervals of irradiation of *45* at 404 nm [55]; the new maximum at 515 nm due to *46* appears with isosbestic points (*381*, 463 nm, two

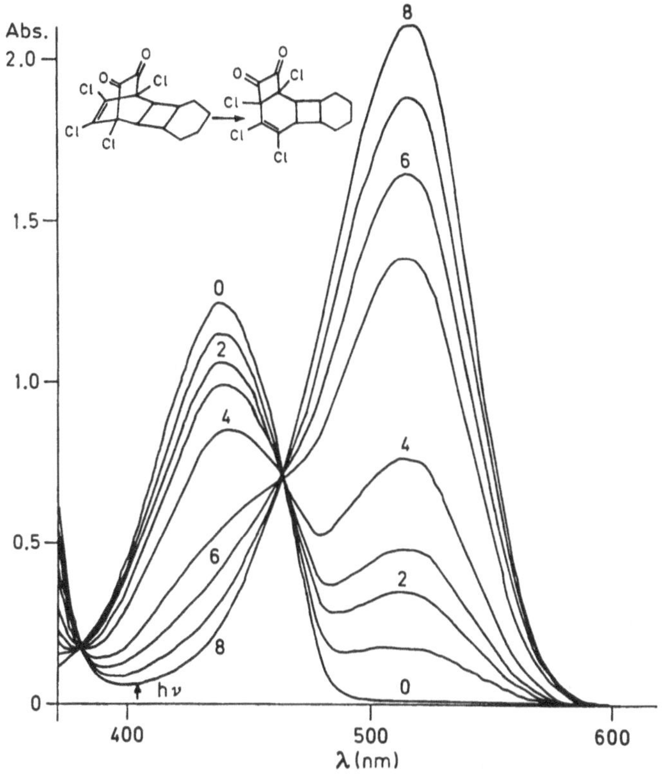

Fig. 1. Photoisomerization of *45* upon irradiation at 404 nm

additional isosbestic points are observed in the ultraviolet when a more dilute solution is used) persisting throughout the reaction. Subsequent bisdecarbonylation to *47* can be achieved by irradiation of *46* at shorter or longer wavelengths or with a broad spectrum of light. The key to the lovely conversion *45* → *46* is a combination

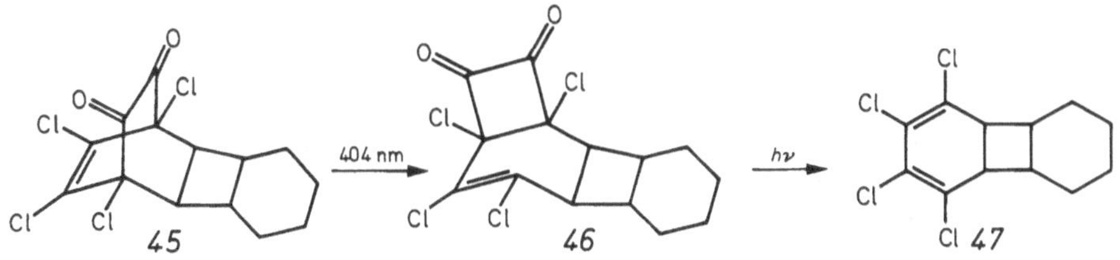

of two factors. First, as can be seen in Fig. 1, the irradiating wavelength (404 nm) coincides with a minimum in the absorption spectrum of *45* so that light absorption by the product is minimal. Secondly, the quantum yield for subsequent reaction of *46* is much less than the quantum yield for its formation ($\phi_{45\rightarrow46} = 0.4$, $\phi_{46\rightarrow47} = 0.03$). Since the rate of a photochemical reaction is given by the product of quantum yield and rate of light absorption, the subsequent reaction is particularly slow in this case.

A more typical example is shown in Fig. 2 which presents the results[62] of 404 nm

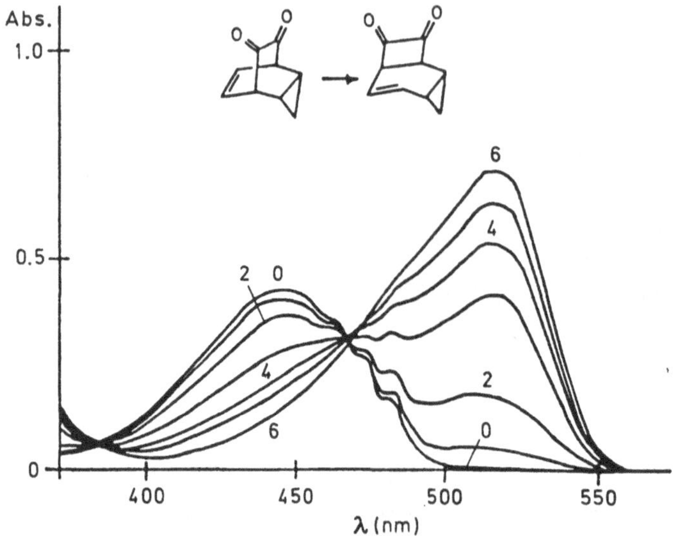

Fig. 2. Photoisomerization of *48* upon irradiation at 404 nm

irradiation of tricyclic enedione *48*. While the rearrangement product *49* again has a minimum at agout 404 nm, the quantum yield for its bisdecarbonylation is 0.3. This is reflected in spoiling of the isosbestic points due to some conversion of *49* into cycloheptatriene in the late stages of irradiation.

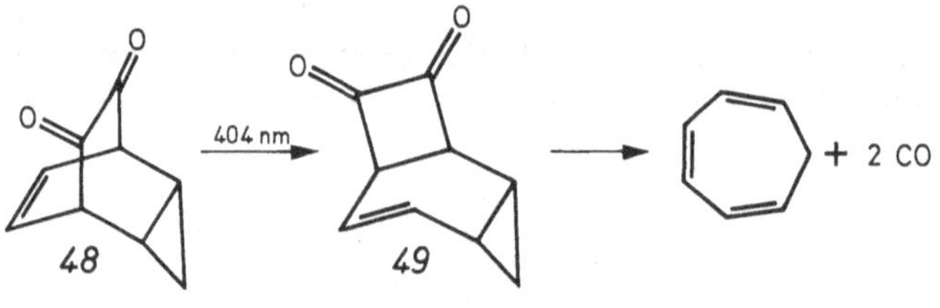

The first bridged cyclohexadiene shown to undergo 1,3-acyl migration was bicyclo-[2.2.1]heptenedione (*50*) and its photochemistry has been investigated extensively. The results obtained are quite general and will be discussed in some detail. Irradiation[66] of *50* at 404 or 436 nm in benzene solution at room temperature to high

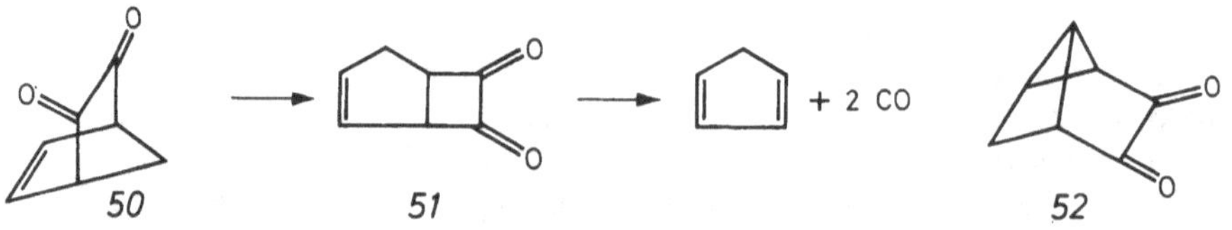

conversion produced bicyclo[3.2.0]heptene-6,7-dione (*51*) contaminated with cyclopentadiene from which pure *51* could be isolated by preparative-scale gas chromatography as a pink solid. Its structure was established by spectroscopic properties and chemical reactions. Isosbestic behaviour was observed during the early stages of photolysis; quantum yields for disappearance of *50* and for formation of *51* were identical ($\phi = 0.21$) at low conversions. Gas-chromatographic analysis confirmed the fact that isomerization *50* → *51* was a quantitative reaction.

An accumulation of evidence [23] points to the singlet state of *50* as the reactive species. The quantum yield for *50* → *51* was independent of the presence or absence of oxygen, of solvent (benzene, cyclohexane, methylene chloride, deuteriochloroform, toluene, 2-methyltetrahydrofuran) or of added anthracene ($E_T = 42$ kcal/mole). The fluorescence of anthracene (366 nm excitation) was quenched by *50*; Stern-Vollmer treatment gave a quenching constant of 1×10^{10} l mole^{-1} sec^{-1} indicating diffusion-controlled singlet energy transfer; the product was *51*. Results obtained with triplet sensitizers are particularly convincing. Quantum yields for *disappearance* of *50* using m-methoxyacetophenone ($E_T = 72.5$ kcal/mole, λ_{irr} 313 nm) or benzil ($E_T \cong 54$ kcal/mole, λ_{irr} 366 nm) were unity and 0.8 respectively, but the quantum yields for *formation* of *51* were significantly smaller, namely 0.3 and 0.2. Thus the triplet state of *50* partitions between isomerization to *51* and direct bisdecarbonylation to cyclopentadiene in contrast to the quantitative isomerization of the excited singlet state. The triplet energy of *50* can be estimated from the above to be slightly less than that of benzil. No phosphorescence has been detected from *50* and attempts to observe the $S_0 \rightarrow T_1$ transition using the oxygen perturbation technique gave negative results [23]. It should be noted that the oxa-di-π-methane reaction which is of major importance in the triplet chemistry of β,γ-unsaturated monoketones [67] has not been observed with diketones. This reaction of *50* would produce *52*; no trace of high field signals characteristic of cyclopropyl protons could be detected when irradiation of an acetone solution of *50* at 313 nm was monitored by nmr-spectroscopy [23].

In addition to chemical reaction, weak fluorescence was detected from *50* at room temperature (λ_{exc} 460 nm, λ_{em} 552 nm, $\phi_f = 0.04$). Temperature effects on reaction and fluorescence from 77–310 K have been studied [68]. A steady decrease in quantum yield for reaction (ϕ_r) and a complementary increase in fluorescence quantum yield (ϕ_f) were observed down to about 150 K where a sharp increase in ϕ_f occurred. Photochemical reaction was negligible at 77 K (436 nm). The fluorescence lifetime at 77 K was a few nanoseconds and the estimated value at room temperature is on the order of 60 ps. Detailed analysis of the data showed that two thermally-activated processes are involved: (1) chemical reaction of the singlet state with an Arrhenius activation energy of 1.5 kcal/mol and (2) radiationless decay of the singlet with $E_{act} = 1.1$ kcal/mol. Both processes would appear to be associated with certain vibrational modes of the excited state which become progressively less populated with decreasing temperature.

These temperature effects were also shown to be influenced by wavelength. Compound *48* showed temperature dependent ϕ_r (436 nm) and ϕ_f (450 nm) analogous to the results described above for *50*. However, from a comparable study at 404 nm it was found [69] that for chemical reaction at 300 K $\phi_r^{404}/\phi_r^{436} = 1$ but at 150 K $\phi_r^{404}/\phi_r^{436} \simeq 30$ while for fluorescence $\phi_f^{406}/\phi_f^{450} \simeq 0.6$ and at 150 K this ratio was

Table 2. Photoisomerization[a] of Bridged Cyclohexenediones to Unsaturated Cyclobutanediones

Entry	R_1	R_2	R_3	R_4	}	λ_{max}(nm)	ϕ_{isom}	λ_{max}(product)	Ref.
1	H	H	H	H	$>CH_2$	460	0.21	505	66)
2	H	H	H	H		446	0.47	515	62)
3	t-bu	H	t-bu	H		447	0.21	530[b]	55)
4	Cl	Cl	Cl	Cl		431	0.66	517	55)
5	n-Pr	Ph	Ph	n-Pr		452		513	61)
6	n-Pr	H	H	n-Pr		450		516	61)
7	H	H	H	H		456		513	55,75)
8	Cl	Cl	Cl	Cl		438		508	55)
9	t-bu	H	t-bu	H		452	0.17	510[b]	55,76)
10	Cl	Cl	Cl	Cl	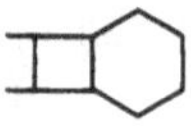	438		515	55)
11	Cl	Cl	Cl	Cl					55)
12	Cl	Cl	Cl	Cl	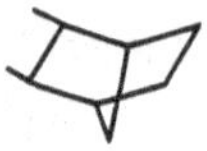				55)
13	H	H	H	H		454	0.09	497	23)
14	Cl	Cl	Cl	Cl		432	0.17	517	23)
15	Br	Br	Br	Br	Do	428		510	23)
16	Cl	Cl	Cl	Cl		445	0.11	495	23)
17	Br	Br	Br	Br	Do	442	0.07	493	23)
18	n-Pr	Ph	Ph	n-Pr					61)

[a] In saturated hydrocarbon or benzene solution.
[b] Mixture of position isomers.

61)

R_1	R_2	R_3	Ref.
H	H	Ph	
H	H	$COCH_3$	
H	H	$COOC_2H_5$	
H	H	OC_2H_5	
H	$COOC_2H_5$		
H	$COOC_2H_5$	$COOC_2H_5$ (trans)	
H	$COOCH_3$	$COOCH_3$ (cis)	
H	R_2R_3 =	$-(CH_2)_5-$	
$COOCH_3$	H	H	
Ph	H	H	

reduced to 0.3. This is interpreted to indicate that at 404 nm, still in the S_1 band of *48*, molecules excited to vibrational levels above the minimum required for chemical reaction can proceed more efficiently to the product so that temperature-dependent thermal equilibration among the vibrational levels of the S_1 state becomes less important. Substituents also can have an important effect as shown by the observation [55] that ϕ_r for a tetrachloroderivative of *48* is not reduced markedly in going from 300 to 77 K and fluorescence becomes very weak at all temperatures.

The bridged cyclohexenediones whose photoisomerization to cyclobutanediones have been observed are summarized in Table II where it can be seen that a wide variety of compounds undergo this reaction. In all cases where the point has been checked, quantum yields for disappearance of starting material and for formation of product were identical. The series of compounds, entries 13–17, were of interest since they involve unsubstituted (entry 13), tetrachloro (14, 16) and tetrabromo (15, 17) isomers where heavy atom effects on intersystem crossing rates might be manifested in quantum yield variations. As can be seen in the Table, the results do not permit such an interpretation.

Isomerizations occurring from the appropriate vibrational levels of the singlet state could be concerted reactions involving a photochemically allowed, suprafacial 1,3-acyl migration with a transition state as represented by E below. Alternatively, homolysis could lead to a short-lived singlet biradical F which could rebond either to product or starting material. The former possibility seems more likely but experimental evidence is lacking. Results of irradiation of unsymmetrically substituted compounds might provide some information. In the case of triplet-sensitized reactions, triplet biradical *G* might be an intermediate in which intersystem crossing to F competes with loss of carbon monoxide, or formation of G which terminates as cyclobutanedione might compete with cleavage to diacyl radical H which undergoes bisdecarbonylation. Concerted isomerization from triplet state of starting material to triplet state of product cannot be excluded. Resolution of these mechanistic questions awaits further experimentation. It should be noted that these reactions proceeded with identical results in a relatively inert solvent such as benzene and in the good H-atom donating solvents toluene or 2-methyltetrahydrofuran.

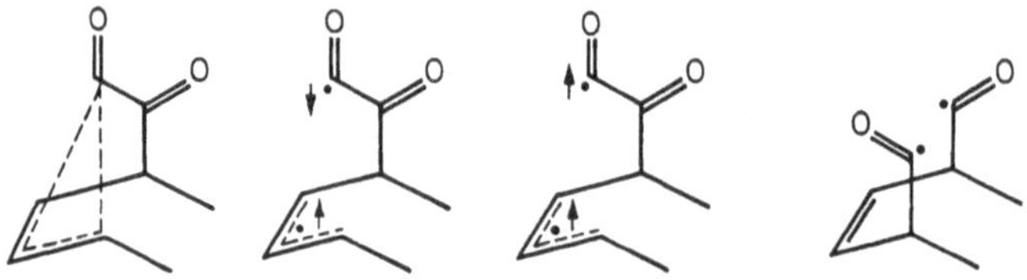

E F G H

The common experience in chemistry that exceptions crop up as soon as a reaction appears to be general has been found to apply. The two exceptions in the present instance are 7-oxabicyclo[2.2.1]heptenedione [23] (*53*) and bicyclo[2.2.2]octenedione [55] (*55*). In the case of *53* (λ_{max} 486 nm) this may be due to the abnormal spectral properties of oxadiketones in general. Irradiation of *53* at 436 nm did not result in the appearance of new long wavelength absorption characteristic of the expected cyclobutanedione *54*, although a significant increase was observed in the region of 350 nm. The final product, formed in quantitative yield, was furan. Analysis of spectroscopic changes using Mauser (extinction coefficient ratios) diagrams [70] indicated that reaction of *53* proceeds via an intermediate. Nmr experiments did not provide additional information. A possible explanation is that *54* is formed but is thermally unstable and that its absorption is hidden under that of *53*.

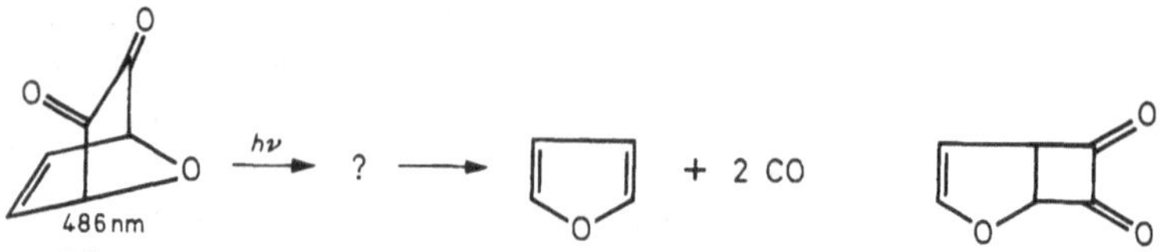

53 *54*

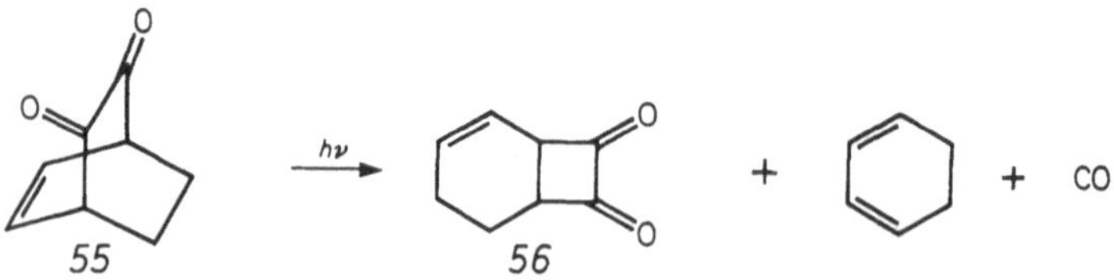

55 *56*

The second exception, *55*, does undergo rearrangement to *56* but, for reasons which are unclear, the major process is bisdecarbonylation to cyclohexadiene.

The results summarized in Table 1 and discussed above were all obtained upon excitation with visible light into the S_1 band of diones. Excitation into the S_2 band using ultraviolet light results in bisdecarbonylation but interpretation is complicated by the fact that possible intermediates and product dienes also absorb in this region of the spectrum. Mechanistic details have not been investigated yet.

B Benzo-Derivatives of Mono-enes

If the isomerization of enediones to unsaturated cyclobutanediones described in the preceding were to occur with their benzo-derivatives, migration of a double bond out of the aromatic ring would be required, as illustrated below. Not surprisingly, this has not been observed. Instead, bisdecarbonylation and α-cleavage have been described in a number of preliminary reports.

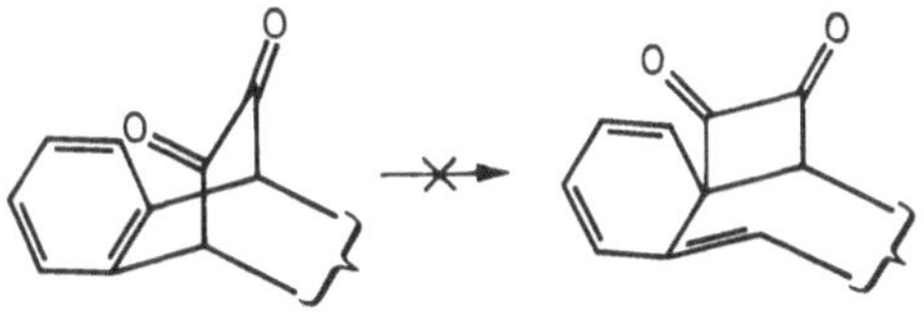

In the first case reported [71], benzonorbornedione (*57*), wavelength dependent behaviour was observed. Ketoaldehyde *58* and minor amounts of indene (*59*) were obtained above 300 nm while *59* was the major product at 254 nm. Indene was shown not to be formed from *58* under these conditions although the presumed precursor

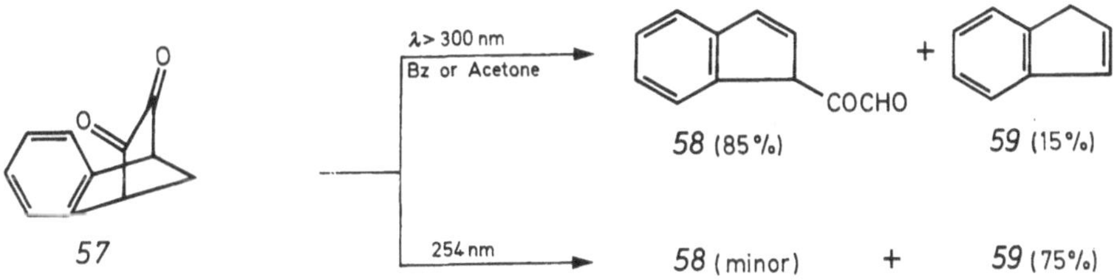

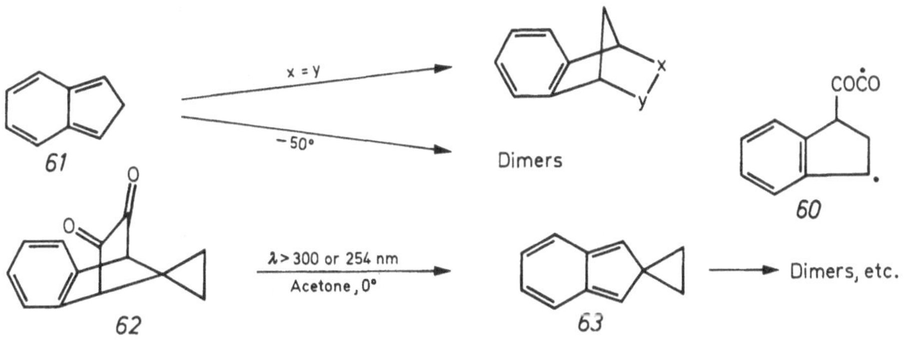

of *58*, biradical *60* may not be excluded as a common intermediate. Intermediacy of isoindene (*61*) in this and related reactions was suggested by trapping with dienophiles and by isolation of dimers of *61* from irradiation at —50 °C (where isomerization *61* → *59* is assumed to be slow). A number of 7,7-disubstituted benzonorbornenediones have also been investigated [72]. Formation of compounds analogous to *58* is not possible in these cases and the observed products were dimers, rearrangement products, or adducts derived from disubstituted isoindene derivatives (*63*). The chemistry of spirocyclopropylindene *63* was particularly interesting. Benzophenone-sensitized reactions of *62* provided significant amounts of monoketonic products, suggested to result from monodecarbonylation of a triplet biradical. The special case of compound *68a* shown below is noteworthy.

Analogous mono- and bisdecarbonylations were observed with benzobicyclo-[2.2.2]octenedione *65* where monoketone *66* was obtained in low yield and a number of products derived from presumed o-quinodimethane *67* were isolated; the material balance was incomplete. In contrast to the behaviour observed with *63*, sensitization by benzophenone gave the same product mixtures as direct irradiation.

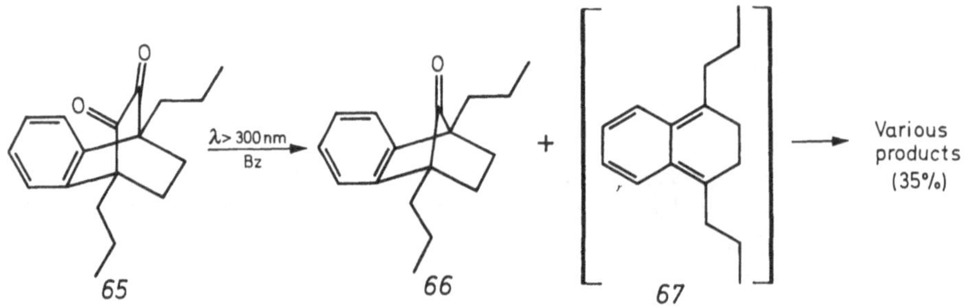

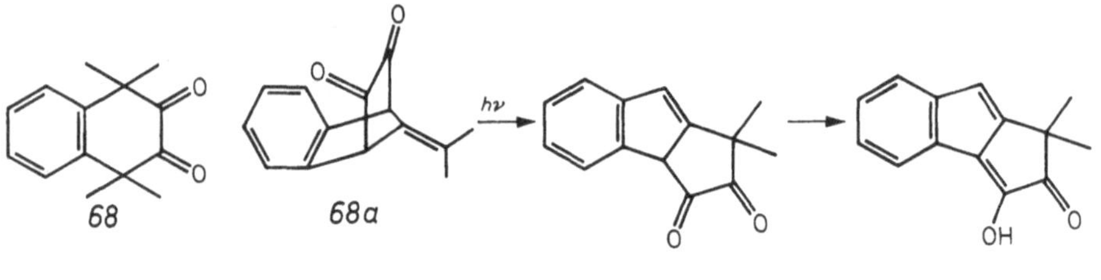

Mechanisms of these reactions have not been established. It should be noted that all reported experiments were performed in the notoriously poor hydrogen-donating solvents benzene or acetone with excitation into S_2 of the dione. Reactions at long wavelengths in good H-donating solvents or in the presence of olefins might exhibit "normal" α-diketone photochemistry as has been observed [73] with *68*.

The two isomeric epoxides *69* and *70* (structures determined by X-ray crystallography) provide an interesting contrast [74]. The product compositions shown were independent of the wavelength of the exciting light and were also obtained using benzophenone as sensitizer. Again, α-cleavage has been suggested to be the primary process in both cases but mechanisms have not been established by any means other than product identification.

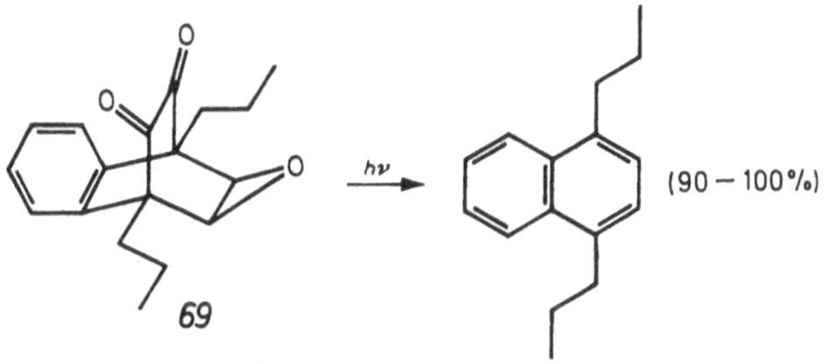

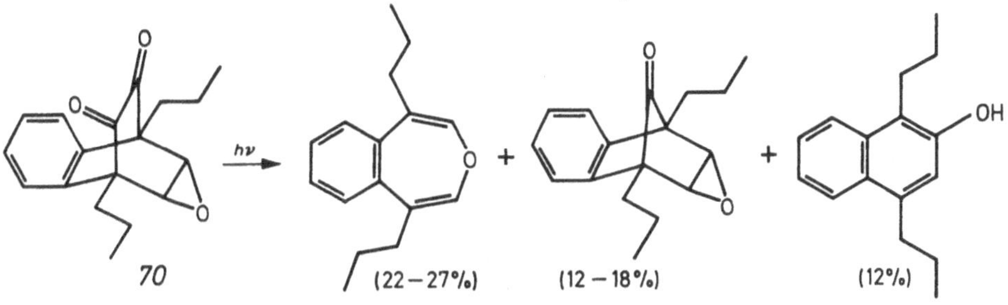

C Dienediones and their Benzo-Derivatives

The dienediones (*71*) whose photochemistry has been investigated are summarized in Table 3. In all cases bisdecarbonylation occurred to give aromatic products. Reactions were performed at room temperature using Pyrex-filtered or longer wavelength light without evidence for any intermediates. However, if 1,3-acyl migration did occur in such systems, the cyclobutanedione (*72*) formed would be expected to undergo rapid thermal bisdecarbonylation at room temperature and therefore not be observed. This argument is based on kinetic studies [55] of thermal bisdecarbonylations of cyclobutanediones. Key results, as summarized in Table 4, show that such reactions involve a transition state which reflects the stabilization of the incipient product.

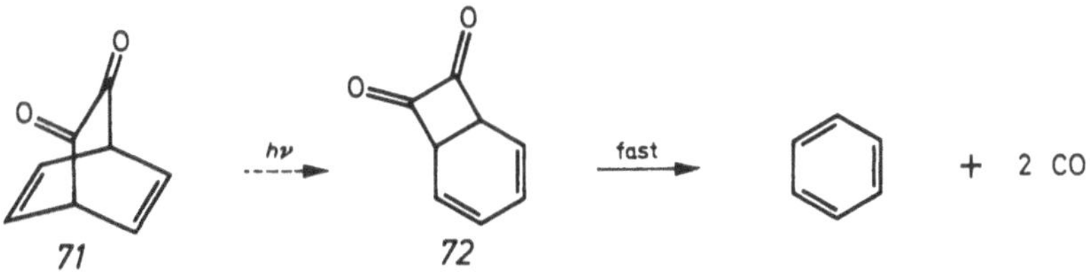

Thus, it might be possible to observe intermediate *72* by irradiating at low temperature. This result has been obtained in preliminary studies [55] although results have not been nearly so clear-cut as might be desired. Compound *73*, which is converted quantitatively into 1,2,3,4-tetrachloro-5,6-diphenylbenzene (*76*) upon irradiation in the visible at room temperature, was irradiated in a glass at 436 nm and 77 K. A new

Table 3. Photolyses of Bicyclo[2.2.2]octadienediones

Entry	R_1	R_2	R_3	R_4	R_5	R_6	Ref.
1	H	H	H	H	H	H	77)
2	Cl	Cl	Cl	Cl	Ph	Ph	78)
3	n-Pr	H	H	n-Pr	$COOCH_3$	$COOCH_3$	61) a
4	n-Pr	H	H	n-Pr	$COOCH_3$	H	61) a
5	n-Pr	H	H	n-Pr	Ph	H	61) a
6	Cl	Cl	Cl	Cl	Ph	H	39)
7	Cl	Cl	Cl	Cl	$4\text{-}ClC_6H_4$	H	81)
8	Cl	Cl	Cl	Cl	$2,4\text{-}Cl_2C_6H_3$	H	81)
9	Cl	Cl	Cl	Cl	$2,5\text{-}Cl_2C_6H_3$	H	81)
10	Cl	Cl	Cl	Cl	$3\text{-}ClC_6H_4$	H	81)
11	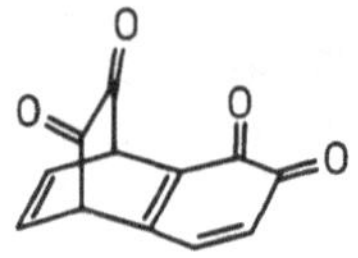						39)
12	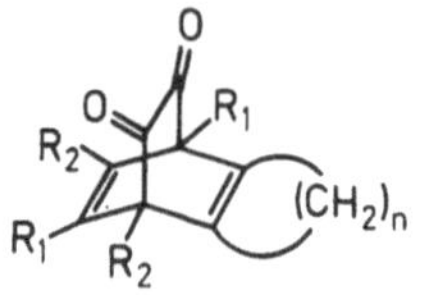			$R_1 = R_2 = Cl$, n = 5,6			79) b
				R_1 = t-bu, R_2 = H, n = 6			80)

[a] Identical results in benzophenone-sensitized reactions.
[b] Similar results were obtained with a monoimine.

long wavelength absorption ($\lambda_{max} \sim 500$ nm), suggestive of cyclobutanediones *74* and/or *75*, was observed. This new absorption was stable at 77 K and disappeared upon warming. However, the maximum intensity of the absorption attributed to *74* and/or *75* was on the order of 5–10% of what was expected for such compounds. It should be

Table 4. Activation Parameters for Thermolysis of Cyclobutanediones in Dodecane

Entry	Compound	Product	ΔG^{*}(kcal/mol)	ΔS^{*}(e.u.)	$t_{1/2}^{25°}$(h)
1	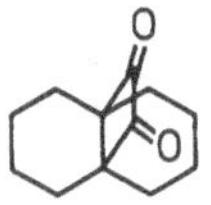	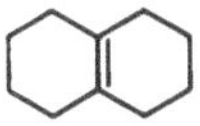	36	+13	7×10^{9}
2	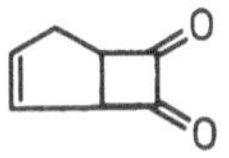		32	+4	4×10^{6}
3	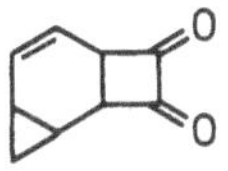		31	+10	1×10^{6}
4	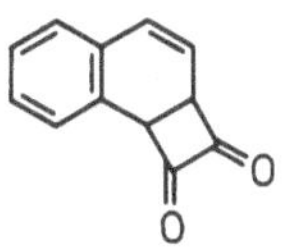	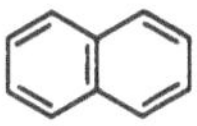	24	-1	7

noted that *74* and *75* have extended conjugated systems. It is not clear whether isomerization of *73* is a minor path which competes to a small extent with direct conversion *73* → *76*. The parent member of the series, bicyclo[2.2.2]octadienedione has not yet been available for low temperature studies.

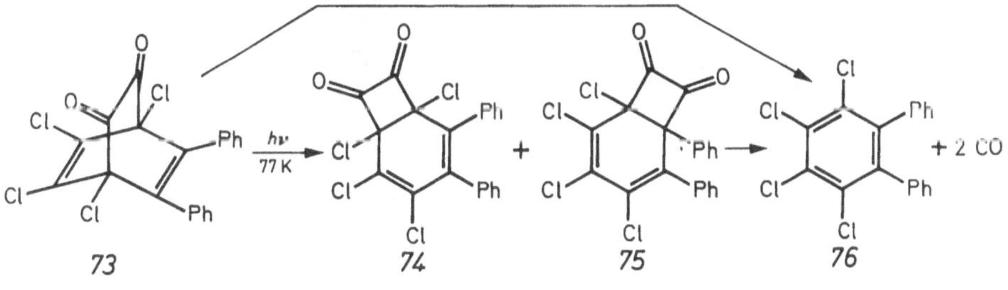

73 *74* *75* *76*

The same question arises in photolysis of benzo-bicyclo[2.2.2]octadienediones (*77*) which are summarized in Table 5. In all cases where long wavelength light was used at room temperature, a new maximum at about 500 nm was observed. As with *73*, intensity of this maximum built up to a low level at the beginning of irradiation, remained constant, and finally decreased. The simplest explanation for the results is that isomerization of *77* to cyclobutanedione *78* competes with direct bisdecarbonylation to naphthalene (*79*). This is supported by an nmr-study [61] in which it was shown that, within the limits of sensitivity of the method, naphthalene (R = n-propyl) was

Table 5. Photolyses of Benzobicyclo[2.2.2]octadienediones

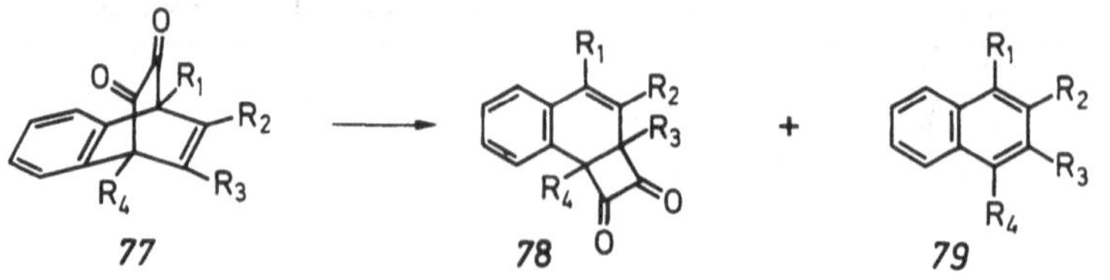

Entry	R_1	R_2	R_3	R_4	λ_{irr} (nm)	Products	Ref.
1	H	H	H	H	436	*78* + *79*	[23]
2	t-bu	H	t-bu	H	254	*79*	[147] a
3	n-Pr	H	H	n-Pr	451	*78* + *79*	[61]
4	H	H	n-Pr	n-Pr	451	*78*[b] + *79*	[61]
5	Cl	Cl	Cl	Cl	436	*78* + *79*	[55]

[a] Similar results were obtained with a monoimine.
[b] Only one isomer, *78*, $R_1 = R_2 =$ n-Pr, was obtained.

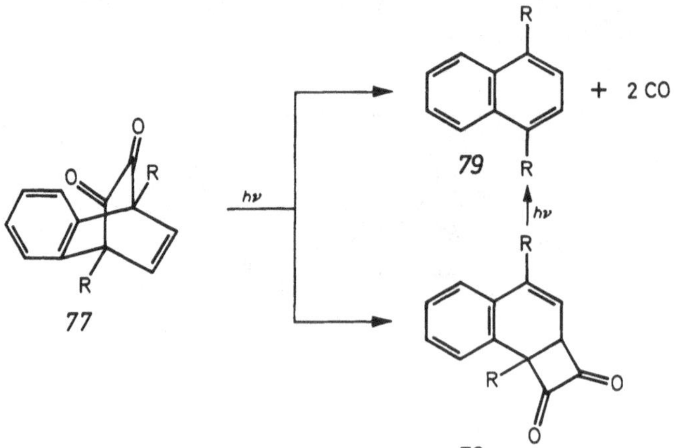

formed from the beginning of the irradiation. Analysis of Mauser diagrams for *77*, R = H also suggests [70] the occurrence of two parallel reactions.

Finally, dibenzobicyclo[2.2.2]octadienediones (*80*) under all conditions of irradiation [39, 82], even at 77 K and 436 nm [23], gave only anthracene without evidence for any intermediate.

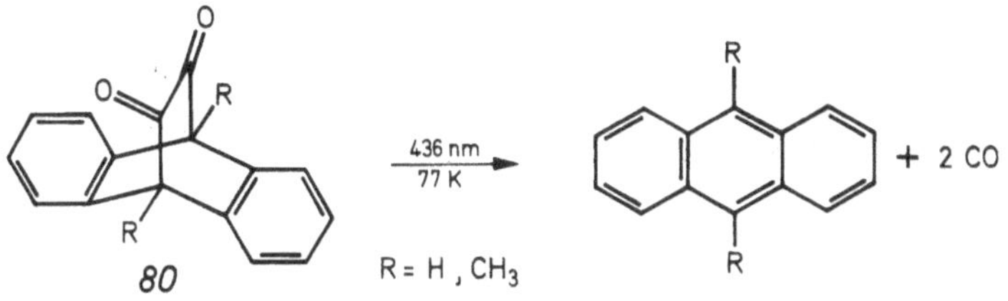

D Synthetic Aspects of Photobisdecarbonylation

Irradiation of bridged cyclohexenediones with a broad spectrum of visible light results in a two-step sequence in which the starting material is converted via a cyclobutanedione into a diene. Three variants of the Diels-Alder reaction are available for synthesis of the starting diones. As illustrated below, these are (1) Reaction of o-benzoquinones with olefinic and acetylenic dienophiles [85]. (2) Use of masked quinones (prepared from the corresponding catechols) whose Diels-Alder adducts can be hydrolyzed to α-diketones [84,85]. (3) Addition of dichlorovinylene carbonate (DCVC) to cyclic dienes followed by hydrolysis to diketones [53,86]. Appropriate choice of quinones, dienophiles, etc. allows considerable flexibility in synthesis of desired starting materials. Stereochemistry has generally been assigned on the basis of shifts observed in proton nmr-spectra of derived quinoxalines and confirmed by X-ray crystallographic analysis in a few cases [55]. Proper choice of wavelength and temperature allows irradiations of these diketones to be performed under conditions such that clean conversion to cyclic dienes can be achieved without problems arising from overirradiation or thermal instability of products. Rearrangement to cyclobutanediones has been

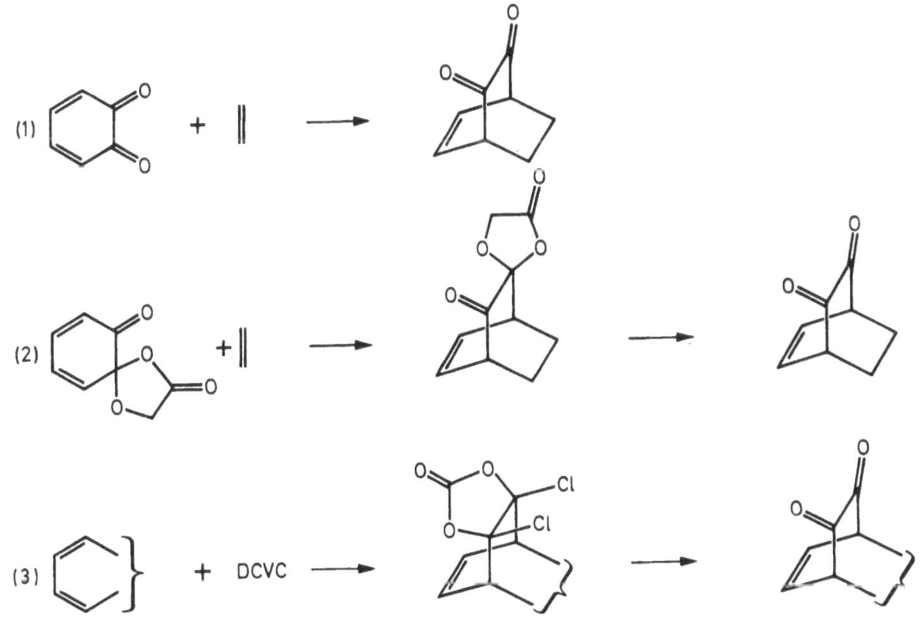

shown to precede decarbonylation in all cases where this point has been checked. The overall process is equivalent to annelation of an olefin (or acetylene) with a four-carbon diene fragment.

(1) This method was first applied [87] in 1972 for the synthesis of disubstituted barrelene *83* from dione *81*, which was prepared as illustrated. Photolysis of *81* in dilute solution afforded 63% of *83* in addition to the other products shown. Neither a cyclobutanedione nor the purported diene intermediate *82* were observed although this might be possible under appropriate conditions.

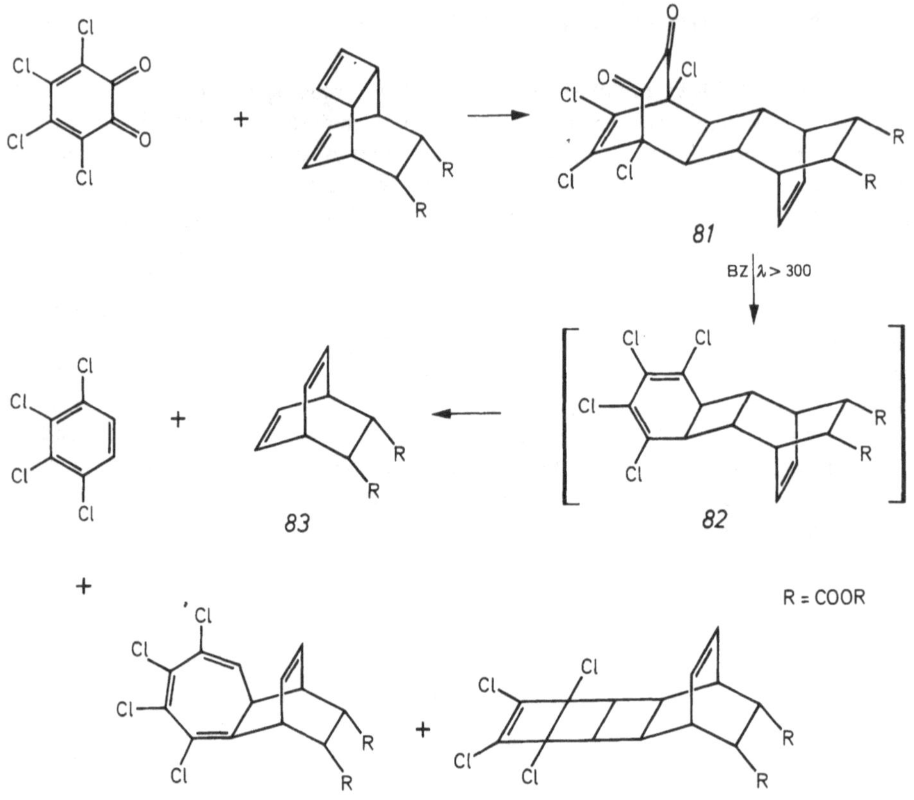

(2) Use of cyclopropabenzene as a dienophile allowed synthesis [88] of dione *84* which afforded the methanodibromoannulene *85* upon photolysis. This in turn served as precursor for a 10π analogue of tropolone. A number of interesting intermediates may be involved (vide infra). When the irradiation was performed at 436 nm at room temperature [23], direct conversion (isosbestic behaviour) of *84* to *85* was observed.

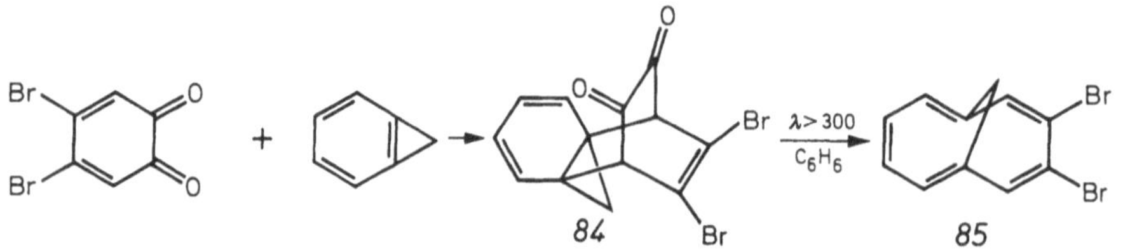

(3) Three applications relating to cyclooctatetraene chemistry have been reported. In the first of these[89], tetrachlorodiketone *86* was synthesized by a multistep sequence and photolyzed to give stable 2,3,4,5-tetrachlorobicyclo[4.2.0]octatriene (*87*) whose photochemical and thermal isomerization could then be studied. In a second approach[76], cyclobutadiene was generated in the presence of 3,5- or 3,6-di-t-butyl-1,2-benzoquinone to yield precursor diones *88*. Irradiation of *88* afforded the corresponding di-t-butylbicyclo[4.2.0]octatrienes (*89*) which isomerized thermally or photochemically to di-t-butylcyclooctatetraenes (*90*). The formation of cyclobutane-diones in long wavelength irradiations of diones *88* has been established[55]. Structures for one of these di-t-butyl series are illustrated below.

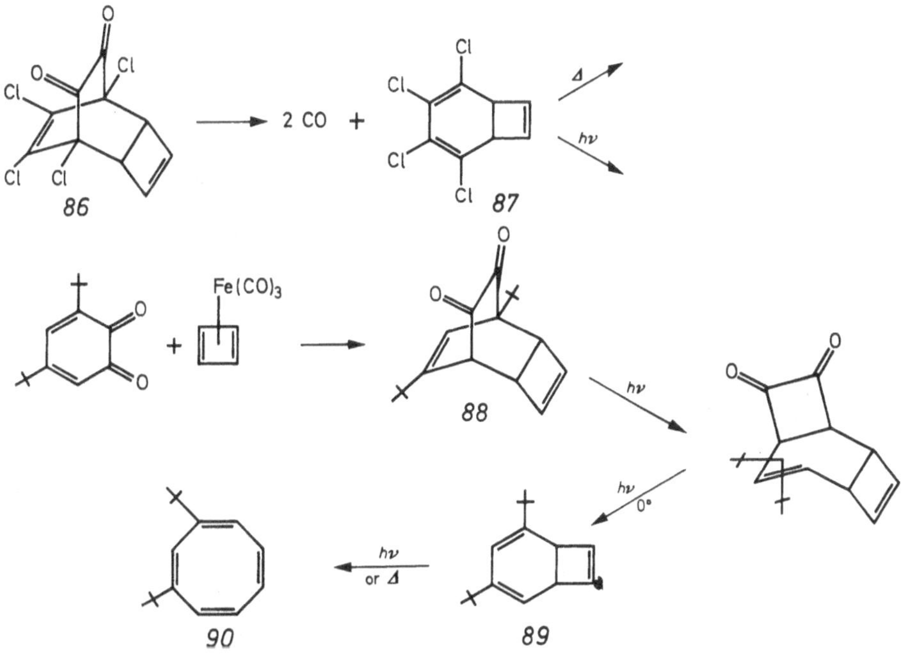

(4) The mode of reaction of fulvenes with o-quinones varies with structure. As illustrated below, 7,7-disubstituted fulvenes (*91*) react with o-benzoquinone and alkyl substituted o-quinones to give bridged diones[90,91] (*92*). The regiochemistry of these reactions was determined by photobisdecarbonylation to *93*. Subsequent oxidation with chloranil provided a synthesis of benzofulvenes (*94*).

(5) Benzvalene (*95*) could be used as dienophile with o-quinones (*96*) leading to diones *97* which were photolyzed to dienes *98*. The unsubstituted compound *98a* is yet another member of the $(CH)_{10}$ family. Irradiation of *97b* at 436 nm led to the cyclobutanedione[55].

(6) Another example is the synthesis[73] of tetrachloro Dewar anthracene *100* via diketone *99*. Photolysis of *99* produced the desired skeleton which was transformed into *100* by conventional methods.

(7) All of the preceding syntheses exploited the reactions of o-benzoquinones with reactive (strained) olefins. A different approach led to the successful synthesis[62]

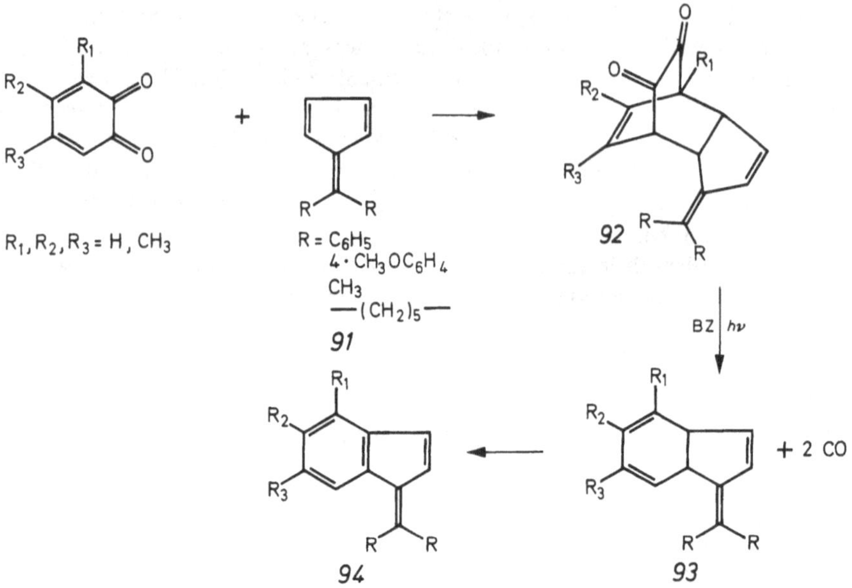
R1
R2
R3
O
O
R1,R2,R3 = H, CH3
+
R
R
R = C6H5
4-CH3OC6H4
CH3
—(CH2)5—
91
92
BZ
hν
93
+ 2 CO
94

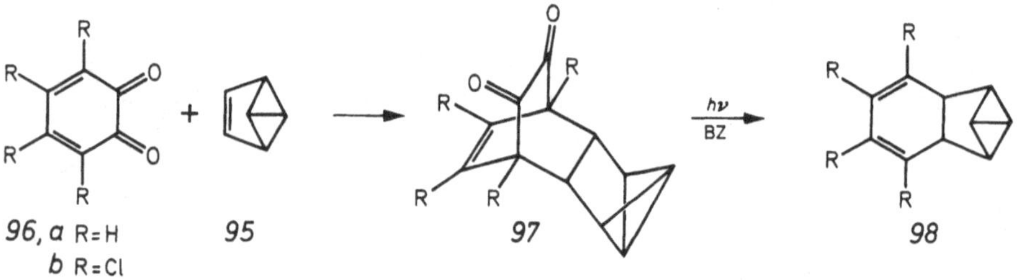
96, a R=H
b R=Cl
+
95
97
hν
BZ
98

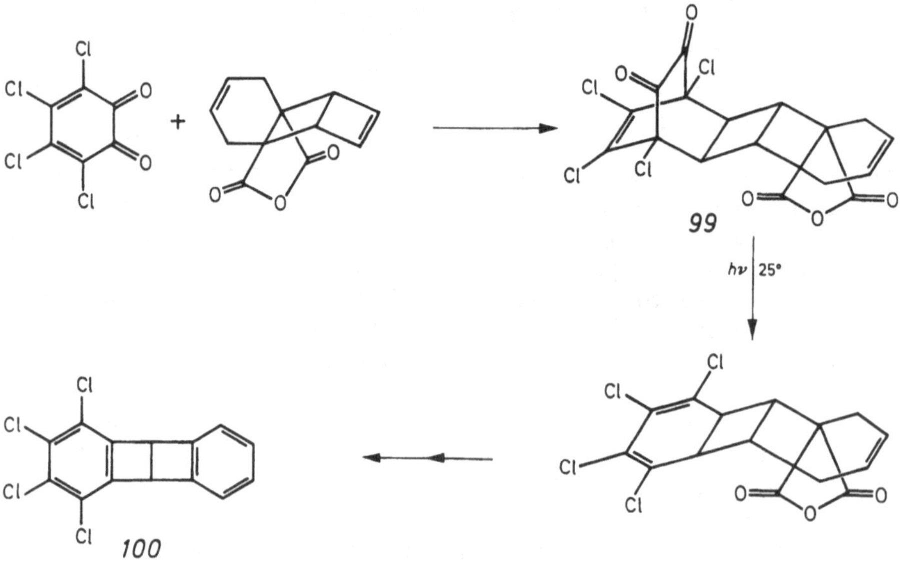
Cl
O
+
99
hν
25°
100

of norcaradiene (*103*). In this case, the diketone precursor (*102*) was obtained via the reaction of DCVC with cycloheptatriene [86] (*101*). Photolysis of *102* at short wavelength and 77 K produced *103* whose kinetics of isomerization to *101* could then be studied. Spectroscopic evidence suggests that the conversion *102* → *103* proceeds via an intermediate. Interestingly, the cyclobutanedione *104* obtained from irradiation

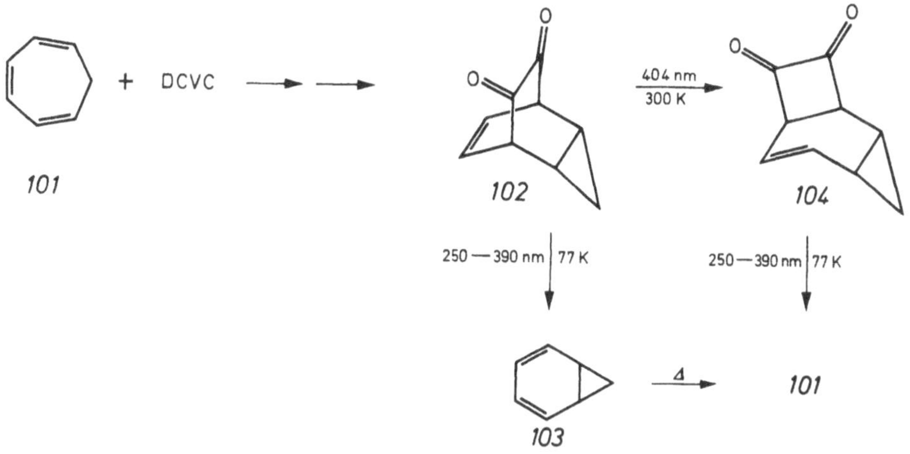

of *102* at 404 nm and room temperature was converted directly to *101* upon irradiation. Synthesis of substituted norcaradienes has been achieved [55] using the reaction of o-quinones with cyclopropene to obtain the dione precursors *105a* and *105b*. In both these cases, the norcaradienes were not photochemically stable at 77 K but were obtained by photolysis of cyclobutanediones at long wavelength. As mentioned earlier, quantum yields for reaction of tetrachlorosubstituted diones show only a small temperature dependence so that conversion of *105a* could be performed without prior isomerization to cyclobutanedione.

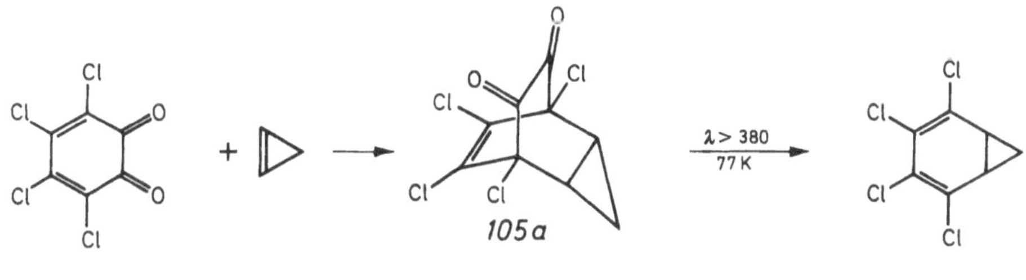

Recent results [55] suggest that "bisnorcaradiene", the valence tautomer of 1,6-methano[10]annulene, may also be accessible by such a route.

(8) Acetylenic compounds are generally good dienophiles with o-benzoquinones and the resulting dienediones are readily bisdecarbonylated to aromatic compounds. This sequence has been exploited for synthesis of a number of benzocycloalkenes [79, 80] and specifically chlorinated biphenyls [81]. This approach could be of general utility.

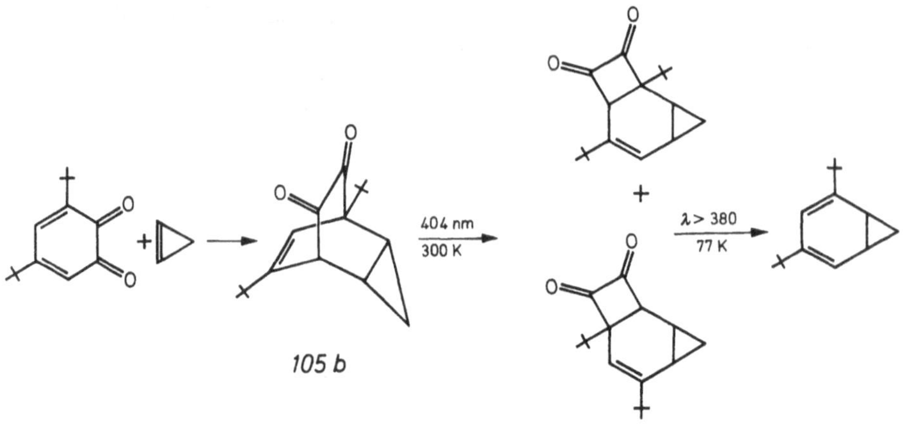

(9) An interesting application of dione bisdecarbonylation for degradative purposes has been reported recently[94] in which C_2 and C_3 of a 1,3-butadiene moiety are excised leaving C_1 and C_4 joined by a double bond. Starting from tricyclic triene *106*, photolysis produced the cyclobutene *107* which was converted via dione *108* into bisdecarbonylated product *109*. Photolysis or thermolysis of *109* produced tetrachlorobenzene and [4+4]antidicyclopentadiene (*110*), the missing member of the family of dimers of cyclopentadiene.

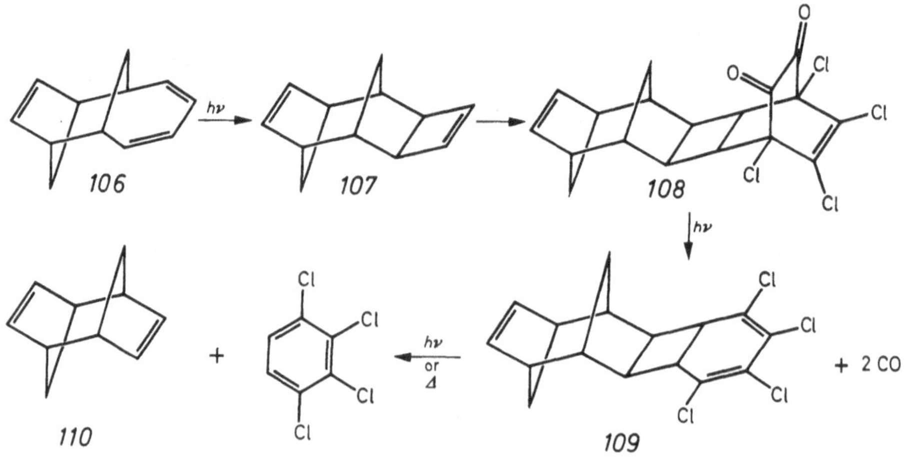

VII Reactions of Diones with Oxygen

The photochemistry summarized in Scheme I at the beginning of this review is that observed in the absence of oxygen. It has been known for many years [1] that preparative reactions of α-diketones should be performed in an inert atmosphere and that reproducible results in mechanistic studies can only be obtained if careful degassing procedures are employed. The products formed in the presence of oxygen are anhydrides (or the derived carboxylic acids) and, in special cases, lactones derived from decarbonylated intermediates. An obvious explanation for this behaviour was that triplet states of diketones act as sensitizers for formation of singlet oxygen which then reacts with ground state diketone.

This explanation was shown to be incorrect by investigation [95] of reactions of a number of α-diketones (benzil, biacetyl, 1-phenyl-1,2-propanedione) in the presence of olefins in oxygen saturated solutions. Slow consumption of diketone was observed with relatively rapid consumption of olefin and concomitant formation of epoxides, often in high yield. Many of the olefins which underwent this reaction do not form epoxides at all with singlet oxygen. For example, tetraphenylporphin-sensitized photooxygenation of tetramethylethylene afforded hydroperoxide *111* quantitatively while a biacetyl-sensitized reaction yielded the epoxide *112*. Further, it was shown that the

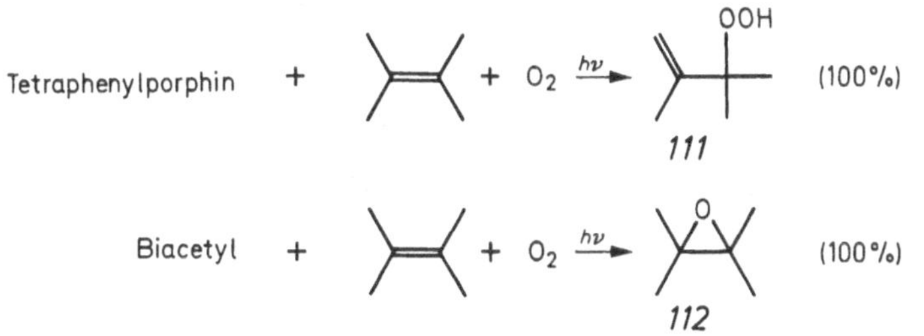

stereochemical integrity of the olefin was not necessarily preserved in the product epoxides, both cis- and trans-alkenes yielding epoxides of predominantly trans-configuration. This "Bartlett Photoepoxidation" has been investigated with a number of diketones in the presence of cycloheptatrienes [96] and a detailed study has been reported [97] of the reaction of acenaphthenequinone with olefins in the presence of oxygen. In contrast to other epoxidation procedures, the reaction is performed in neutral medium using visible light. These mild conditions have been exploited [98] for epoxidation of aflatoxin B where conventional epoxidation was impractical because of the extreme sensitivity of the substrate. The conversion of propylene to propylene oxide in dichlorobenzene using biacetyl, 1-phenyl-1,2-propanedione, or benzil has been reported recently [155]. Increasing propylene and oxygen concentrations by pressurization resulted in significant rate enhancement, diketone destruction was negligible, and quantum yields were approximately *unity*. An oxidative cleavage reaction was observed at very high (> 3.2 M) propylene concentrations.

The present state of mechanistic understanding of these reactions is ideal for the speculator since the encumbrance of facts is still quite limited. Any mechanistic proposal must, however, account for the lack of stereospecificity, for the survival of diketone even though it plays an essential role, for the absence of any correlation between diketone reduction potentials and reaction rates [156], and for the results of the following experiments using $^{36}O_2$, unlabelled biacetyl (or benzil) and tetramethyl ethylene. At 94% conversion to epoxide, 98% of unlabelled biacetyl was recovered and labelled oxygen was incorporated into *112*.

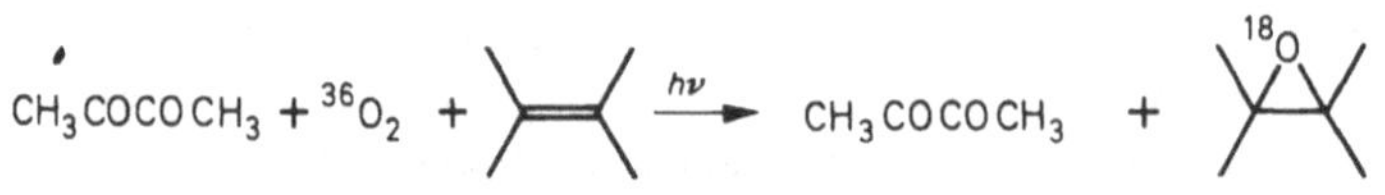

A mechanism which accounts for all the above and seems eminently reasonable is initial addition of triplet dione to olefin (vide infra) to form biradical *113* which then reacts with ground state oxygen to yield the new biradical *114*. A variety of routes can be envisaged from *114* to products. Alternative possibilities suggested include addition of oxygen to triplet dione to give *115* or *116* which could serve as precursors of peranhydrides (*117*) (or peracids) suggested to be the actual epoxidation reagents, or *115* or *116* could add to olefin giving biradicals *118* or *119*. The latter would ultimately lead to products.

Further investigations of these intriguing reactions will undoubtedly be forthcoming. We note that those reactions of diones originating from singlet states proceed (cf. Sections IV, V, VI) equally well in the presence or absence of oxygen.

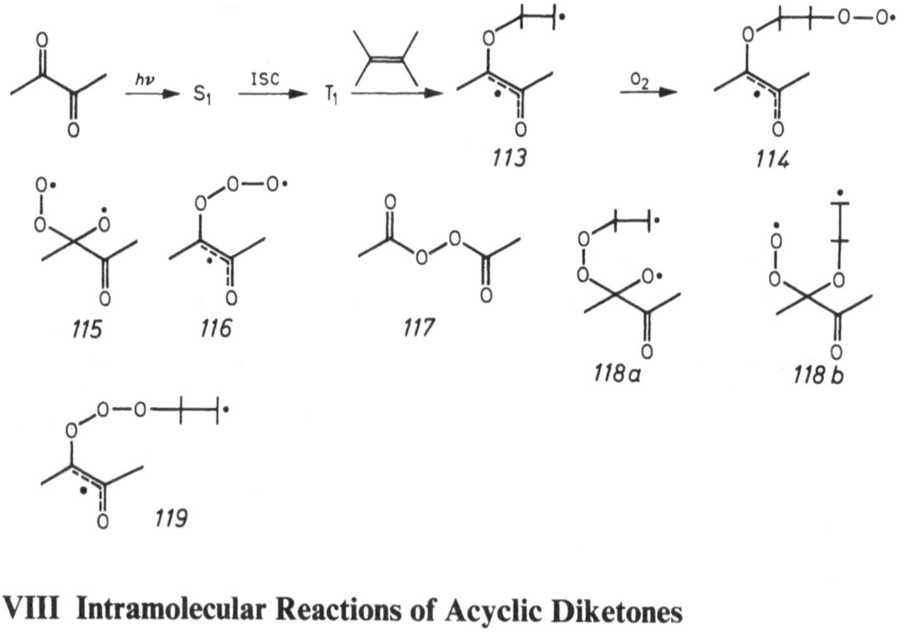

VIII Intramolecular Reactions of Acyclic Diketones

Acyclic diketones *120* (and sufficiently flexible cyclic compounds) possessing a hydrogen atom at C-4 of the 1,2-diketo system undergo intramolecular reactions via their lowest triplet states to form cyclobutanolones *122* in high chemical and quantum yields [1]. The proposal that excited states of diones assume, whenever possible, a planar s-trans conformation provides a rationale for the regiospecificity observed in these reactions. Abstraction of a hydrogen atom from C-4 via a six-membered transition state leads to 1,4-biradicals *121* which then cyclize to *122* (as a mixture of stereoisomers). Disproportionation of *121* to regenerate starting material, an important reaction of analogous monoketones, is generally of minor significance as shown by high quantum yields, small differences in results between hydrocarbon and hydroxylic media [100–102], and very little racemization of recovered starting material [102] when optically active dione was investigated.

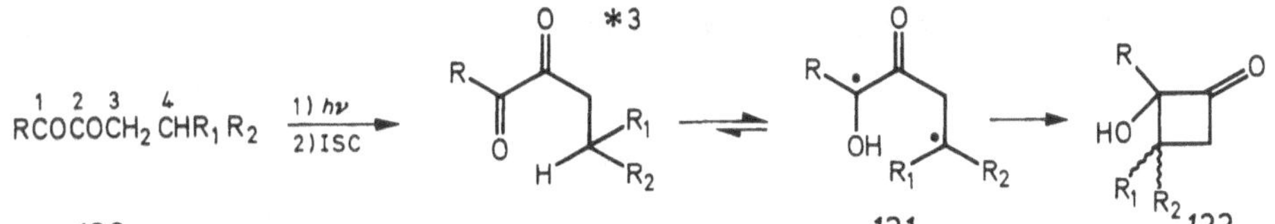

120 *121* *122*

a[100], $R = CH_3$, $R_1 = H$, $R_2 = H$. b[100], $R = CH_3$, $R_1 = H$, $R_2 = CH_2CH_3$. c[100], $R = CH_3$, $R_1 = R_2 = CH_3$. d[103], $R = CH_2Br$, $R_1 = CH_3$, $R_2 = H$. e[103], $R = CH_2Br$, $R_1 = R_2 = CH_3$. f[100,105], $R = Ph(CH_2)_3$, $R_1 = H$, $R_2 = PhCH_2$.

An alternative hydrogen abstraction from a different conformation of the excited state, again involving a six-membered cyclic transition state, is also possible, as illustrated below, and would result in the formation of acyl cyclobutanols *124* via biradical *123*. Products of type *124* have never been detected in these reactions. In the most favorable case for formation of *123*, 1,8-diphenyloctane-4,5-dione (*120f*), in which abstraction of a benzylic hydrogen would be involved, cyclobutanolone *125* was obtained [104] in 90% yield. In addition to a trace of fragmentation product [105], the cyclopentanolone *126* was the minor product. In view of the considerable tendency for rearrangement in systems of this type [106], the possibility that *126* is formed by rearrangement of acyl cyclobutanol *127* cannot be neglected.

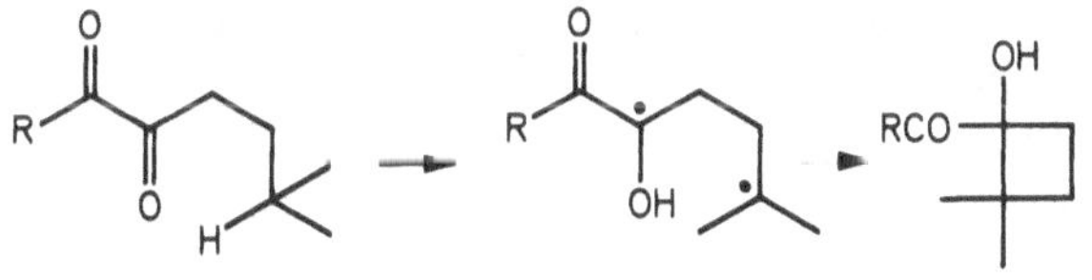

123 *124*

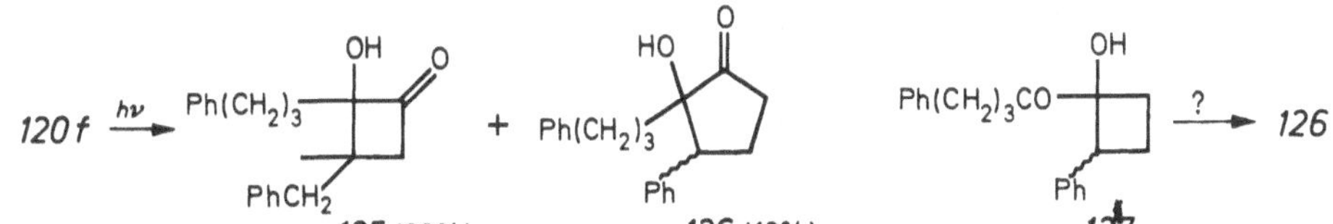

125 (90%) *126* (10%) *127*

The quantum-yields for reaction of compounds *120a*, *120b*, and *120c* in benzene solution were 0.054, 0.50, and 0.57 (0.62 in t-butyl alcohol) respectively [100]. Rates of hydrogen atom transfer were obtained from Stern-Vollmer treatment of quenching results and gave relative reactivities (per hydrogen atom) 1:80:1000 for primary, secondary, and tertiary hydrogens. This ratio reflects earlier observations [107] that cyclobutanolone formation from octane-4,5-dione was barely affected when irradiation was performed in butyraldehyde solution while irradiation of hexane-3,4-dione in propionaldehyde afforded 23% of 2-ethyl-2-hydroxycyclobutane. Nonetheless, 2,2,5,5-tetramethylhexane-3,4-dione (*128*, di-t-butyldiketone) gave cyclobutanolone *129* in nearly quantitative yield in 2-propanol, acetonitrile or benzene [34]. This has been attributed to the fact that, given a multitude of primary hydrogen atoms, one of the eighteen will always be in the right place at the right time. Favorable orientation of a methyl group in 3,3,8,8-tetramethylcyclooctane-1,2-dione has been suggested [57] to

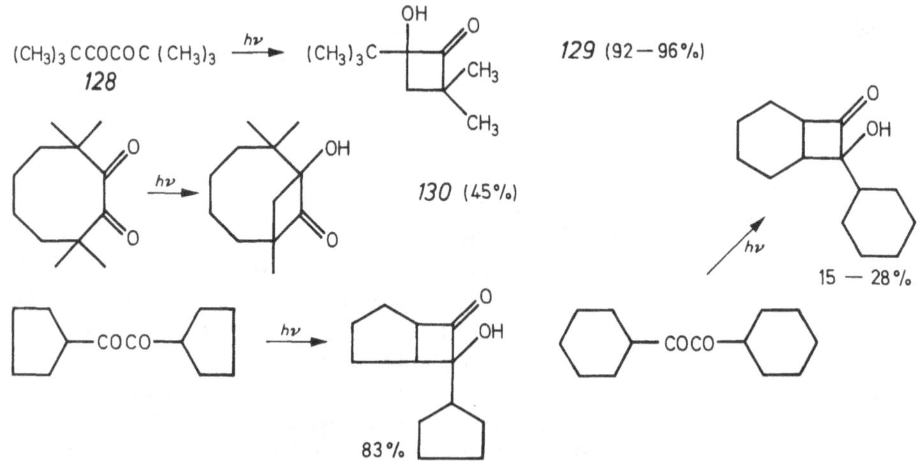

account for cyclization to *130*. Conformational factors have also been suggested to account for the difference in behaviour between dicyclopentyl diketone (83% cyclization) and dicyclohexyl diketone (15–28% cyclization).

As part of the synthesis of 1,6-dimethyldodecahedrane [108], an interesting case has been reported [109] of cyclization of a polycyclic diketone in which two new cyclopentanol rings were formed, as illustrated by the partial structures shown below. X-ray crystallographic analysis showed that conformational factors were particularly favorable for this unique result.

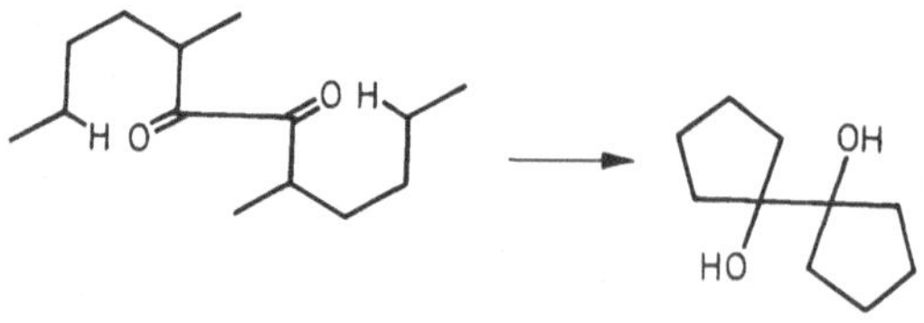

The key role played by conformations in these reactions prompted investigation of results in liquid crystals [110]. However, results obtained with octane-4,5-dione and 1,6-diphenylhexane-3,4-dione in several types of liquid crystals did not differ appreciably from those obtained in hexane solution.

The absence of intramolecular H-abstraction in cyclopropyl diketones reflects a combination of conformational factors, difficulty in formation of cyclopropyl radicals, and strain in the cyclization product.

1-Phenylalkane-1,2-diones possessing the requisite hydrogen atom at C-4 may also form cyclobutanolones upon irradiation [101,102,112]. However, chemical yields ranged from close to 100% for 1-phenyl-3,3-dimethylbutane-1,2-dione (*131*, nine primary C-4 H atoms, cf *128*) to 3% for 1-phenyl-3-ethylpentane-1,2-dione (*132*, four secondary C-4 H, one tertiary C-3 H) [112]. Further, concentration-dependent quantum yields for cyclobutanolone formation (ϕ_{CB}) were generally lower than quantum yields for the disappearance of starting diketone (ϕ_{-DK}) in these reactions. For example, values of ϕ extrapolated to zero diketone concentration in benzene for *131* were $\phi_{CB} = 0.015$ and $\phi_{-DK} = 0.53$, for 1-phenylbutane-1,2-dione 0.04 and 0.50, respectively, and for 1-phenylpentane-1,2-dione 0.52 and 0.73. These results require that an additional reaction of diketones be involved. This was shown by spectroscopic data and chemical reactions [112] to be formation of the enol (*133*) of starting diketone. Rate constants, obtained from quenching studies, gave relative ratios primary:secondary:

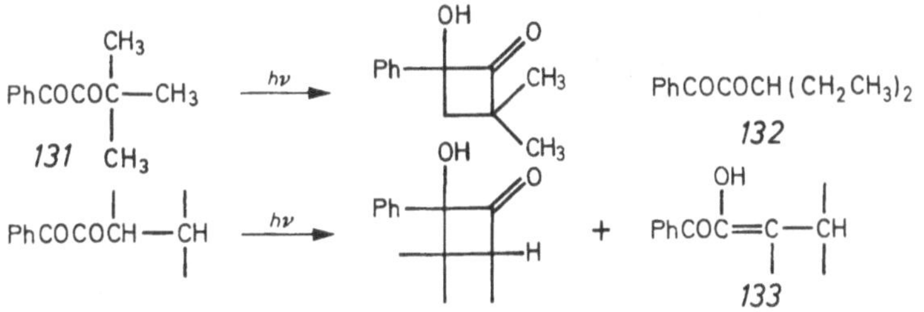

tertiary of 1:60:390 for cyclobutanolone formation (abstraction of C-4 hydrogen) and $1:1\times10^3:3\times10^5$ for enol formation (abstraction of C-3 hydrogen?). Both types of products as well as phosphorescence of starting material were shown to arise from the lowest triplet state of diketone.

While the mechanism for cyclobutanolone formation appears to be similar to that of aliphatic diketones, "the mechanism of enolization is not altogether obvious" [101]. It has been shown by a kinetic deuterium isotope effect that the cleavage of the C_3—H bond is involved in the rate determining step. The simplest explanation, abstraction of

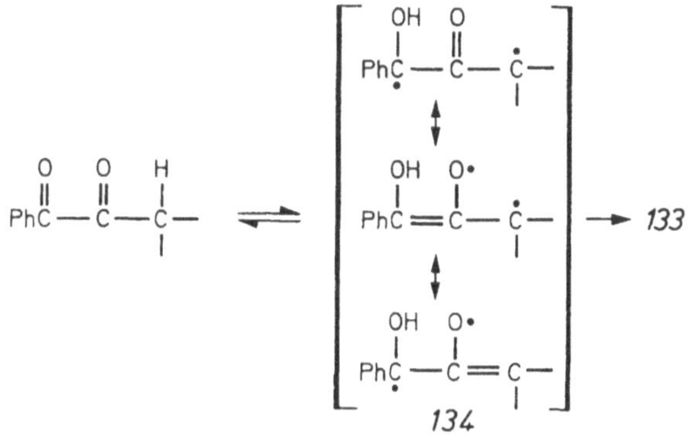

C-3 hydrogen to form the biradical *134* which partitions between disproportion to enol and reversal to starting material, has been questioned on a number of grounds [101]. An alternative which has been suggested is that excited triplet states of these diones have zwitterionic character involving some positive charge on C-2 which facilitates loss of a proton from C-3 followed by reprotonation on oxygen. A bimolecular mechanism is not consistent with the observed results.

Enol formation has also been observed [113] ($\phi \sim 0.05$) with 5,5-dimethylheptane-2,3-dione (*135*) which has no abstractable C_4 hydrogen. Interestingly, 5,5-dimethyl-6-phenylhexane-2,3-dione (*136*) gave fragmentation products (biacetyl and olefin) whose formation is best explained by H-abstraction to the biradical shown below.

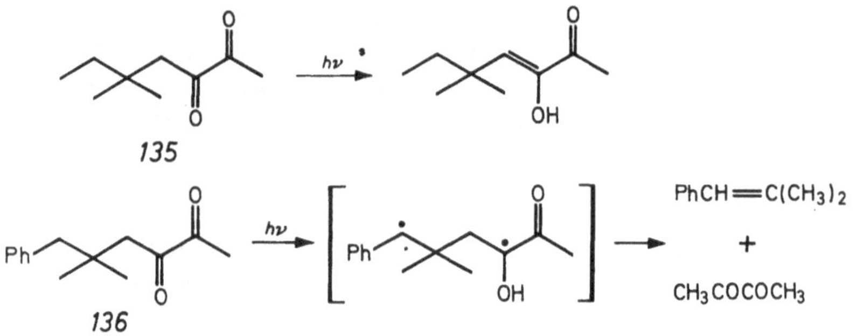

135

136

Arylpropanediones (except o-tolyl) were photochemically inert in benzene solution but underwent photopinacolization and photoaddition in 2-propanol [114]. Reactions occurred at the carbonyl group adjacent to the aromatic ring as has also been observed with 1,2-indanediones [115].

The presence of unsaturation in acylic diones may enhance hydrogen abstraction reactions when allylic or benzylic positions are involved. Thus 6-methylheptene-2,3-diones (*137a, b*) gave nearly quantitative yields of cyclopentenolones upon irradiation with visible light [103, 116]. The most intensively investigated systems have been ortho-substituted phenylpropanediones (*138*) which gave hydroxyindanones (*139*), again in very high yields. Mechanisms proposed for this transformation involve hydrogen ab-

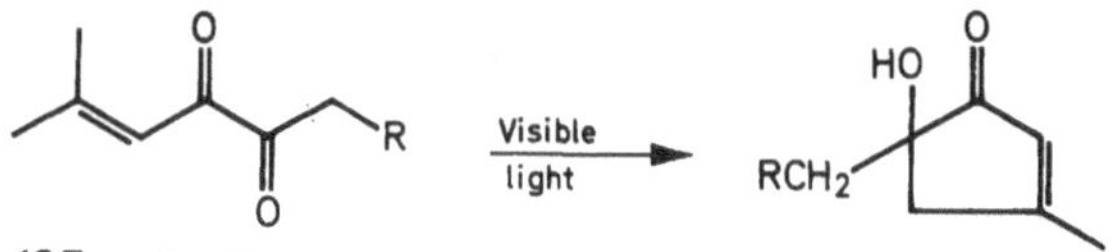

137 a R = H
b R = Br

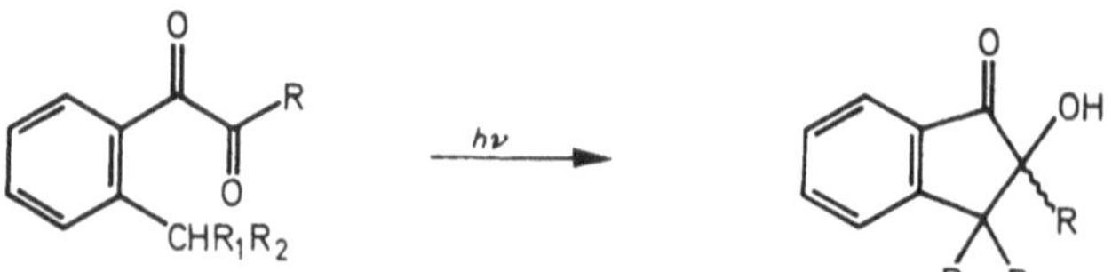

138 *139*

straction by triplet dione to give either biradical *140* or *141* or the enol *142* (after intersystem crossing).

The principal argument for intermediacy of *142* is the report [114, 117] of isolation of a Diels-Alder adduct in low yield when 1-(o-tolyl)propane-1,2-dione was irradiated in the presence of dimethyl acetylenedicarboxylate. A considerable accumulation of evidence against this proposal and in favor of mechanisms proceeding via *141* and *142* has accumulated. These include failure to incorporate deuterium when irradiations were performed in CH_3OD [104, 118], stereochemistry of reactions [119], failure to obtain Diels-Alder adducts from irradiations performed in the presence of maleic anhydride, and trapping of biradicals with sulfur dioxide [120]. The results with sulfur dioxide are quite convincing since SO_2 is known to be a poorer dienophile than maleic anhydride but has been used successfully for trapping biradicals in other systems.

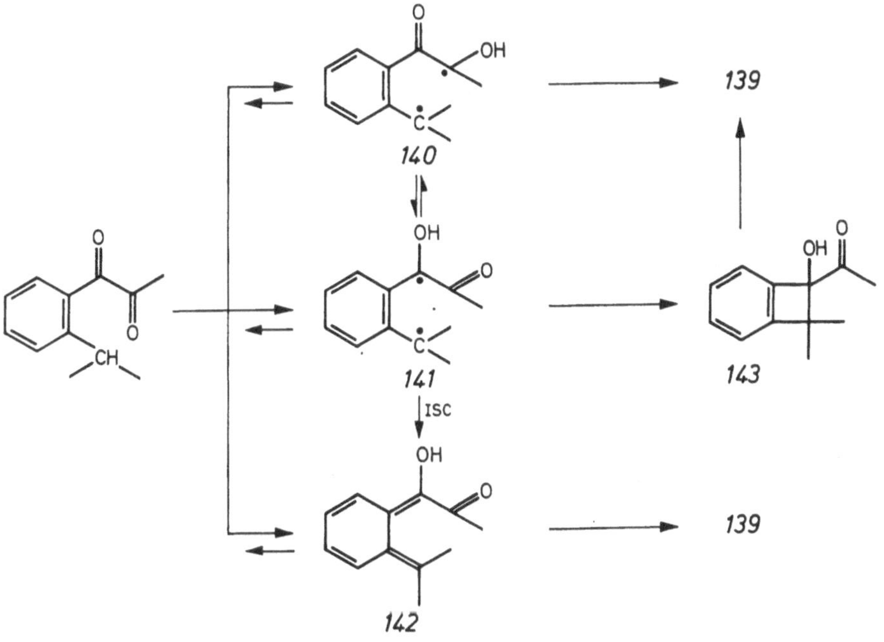

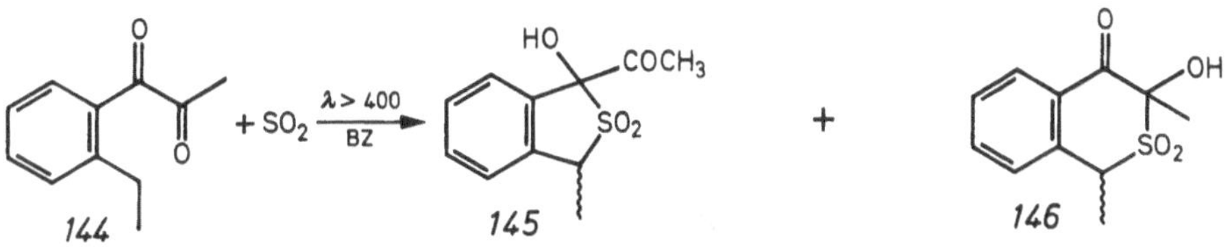

Irradiation of 1-(o-ethylphenyl)propane-1,2-dione (*144*) in the presence of sulfur dioxide gave two interconverting sulfones *145* and *146* in 75% yield. This indicates the occurrence of rapid tautomerism in semidione radicals such as *140* and *141* and emphasizes the point that product structures do not necessarily provide information on the initial hydrogen abstraction step (cf. also the photo reduction of unsymmetrical benzils [121]). Thus, granting the assumption of facile isomerization

143 → *139* (this point might be checked by low-temperature photolysis), the distinction between *140* and *141* as key intermediates is not relevant to product structures.

An additional type of product, the chromanone *148* was obtained from irradiation of disubstituted *147*. The result is explicable in terms of biradical *141*. Quantum yields were much lower than with less substituted compounds.

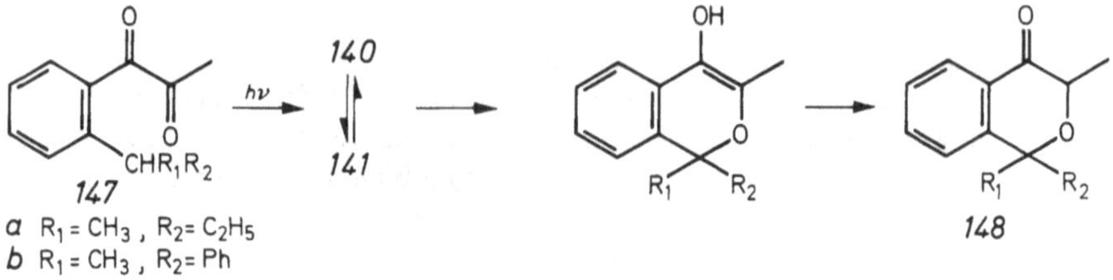

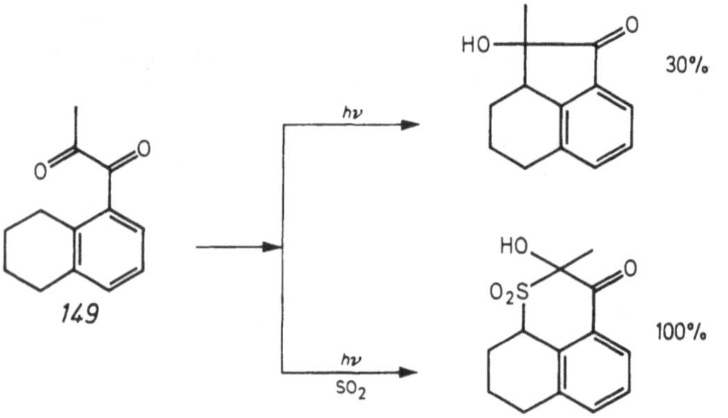

This type of cyclization is not limited to propanediones but has also been observed with ortho substituted benzils [104] and with tetralin derivative *149*. In the latter case, a cyclic sulfone was also obtained in quantitative yield.

The exceptional case with conjugated dienones is 1-(1′-cyclohexenyl) propane-1,2-dione (*150*) where hydrogen abstraction would not lead to a reasonable product. Instead, the furanone *152* was formed [104] probably by way of the biradical *151*.

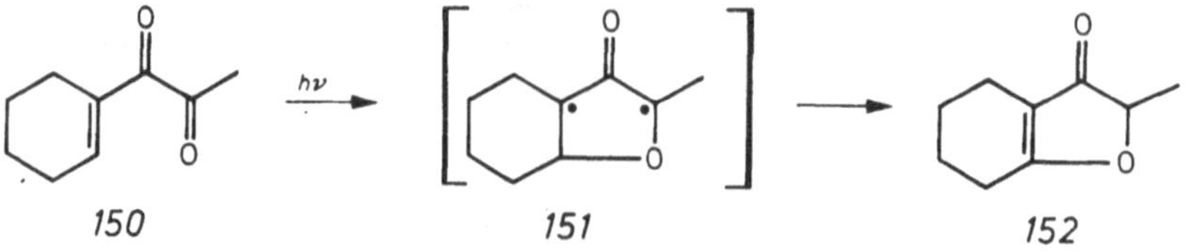

Hydrogen abstraction via a six-membered transition state is no longer feasible when the double bond is moved one carbon out of conjugation with the dione as in 4,4,5-trimethyl-5-hexene-2,3-dione (*153*). The product, obtained in over 90% yield [113], was the bicyclic ketooxetane *154*. This appears to be an intramolecular example of the well-known photoaddition of diones to alkenes (cf. Section IX A). The structure of *154* appears to be well-established and to exclude any possibility of a fast isomerization, involving 1,3-acyl migration, to *155*.

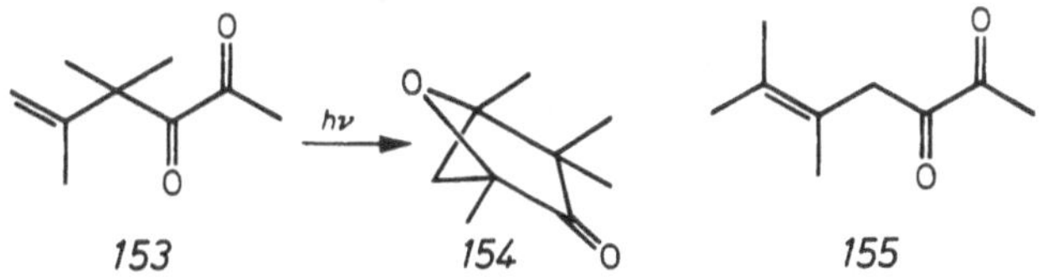

153 154 155

Removing the double bond one carbon further from the dione results in the presence of an allylic hydrogen atom which is suitably disposed for abstraction. This was indeed observed with *156a, b* when photolysis was performed at temperatures below 5°. The vinyl cyclobutanolones (*157*, mixtures of stereo-isomers) formed were thermally unstable and isomerized to diketones *158*. When irradiations were performed without cooling, the products observed were bicyclic ketooxetanes *159* as expected by analogy with *153*. When abstractable hydrogens were not available as in *156c*, inefficient formation of bicyclic ketooxetane *160* was observed.

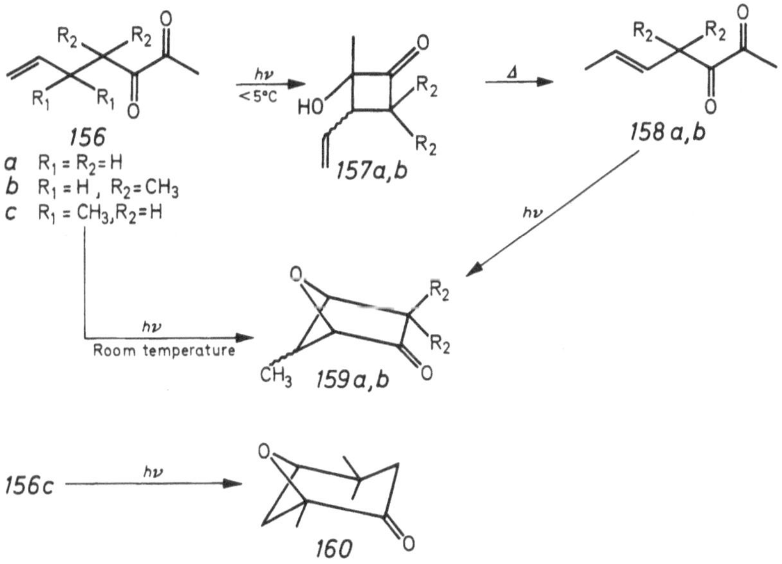

In summary, intramolecular reactions of a considerable variety of diones provide interesting mechanistic questions as well as a route for construction of four- and five-membered rings in high photochemical and chemical yields. Except for a synthesis of cis- and trans-bicyclo[7.1.0]decan-2-one from cyclodecane-1,2-dione [122], this potential has been ignored.

IX Additional Reactions of Diones

A With Olefins

Reactions of α-diketones (generally via triplet states) with olefins may involve two competing processes [1]: (1) addition to give a biradical (*161*) or (2) abstraction of an allylic hydrogen (when available) to give semidione and allylic radicals. The latter pair may disproportionate back to starting materials (or isomerized olefin) or couple to give adducts (*164* and/or *165*). Biradical *161* can, after intersystem crossing, fragment to starting materials or collapse to ketooxetanes (*162*) or dioxenes (*163*) (dioxoles have also been observed).

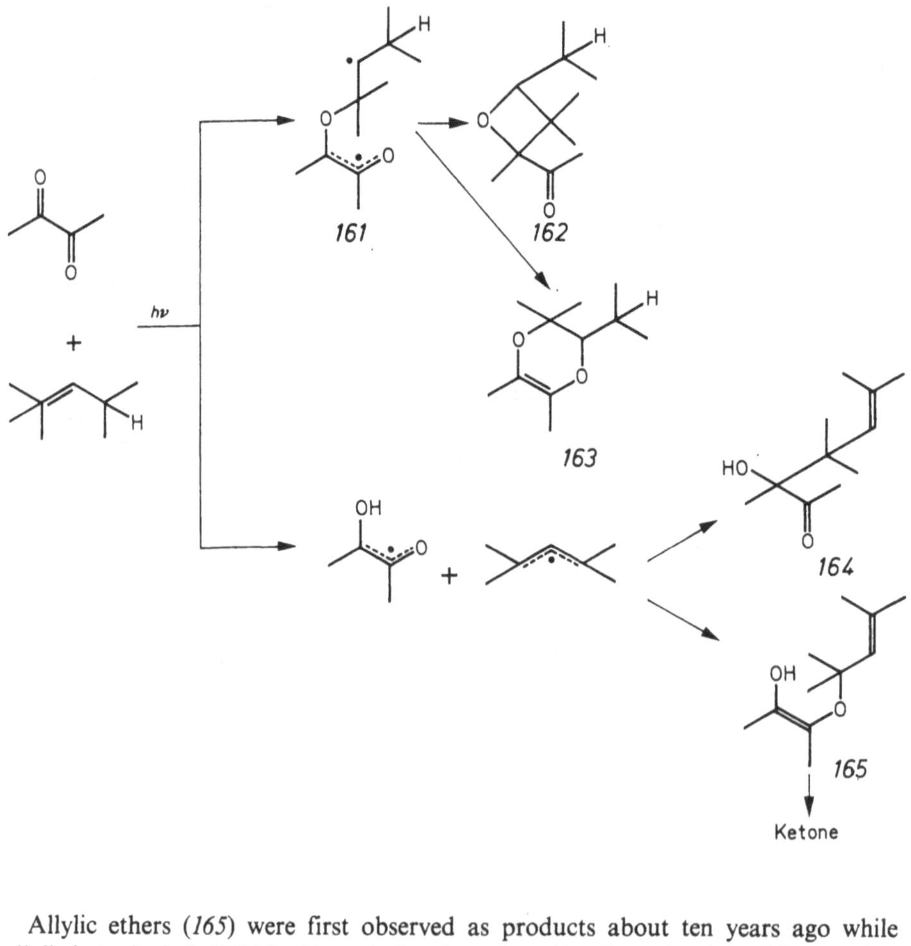

Allylic ethers (*165*) were first observed as products about ten years ago while allylic ketoalcohols (*164*) had been isolated much earlier. Investigation of the reaction of biacetyl with α-trideuteriomethylstyrene showed that the simple radical coupling mechanism illustrated above is not correct [124] since the deuterium scrambling expected from reaction of a symmetrical 2-phenylallyl radical was not observed in the product (*166*). The intermediate involved must then be the biradical illustrated (analogous to *161*) which proceeds to product by a 1,5-hydrogen migration.

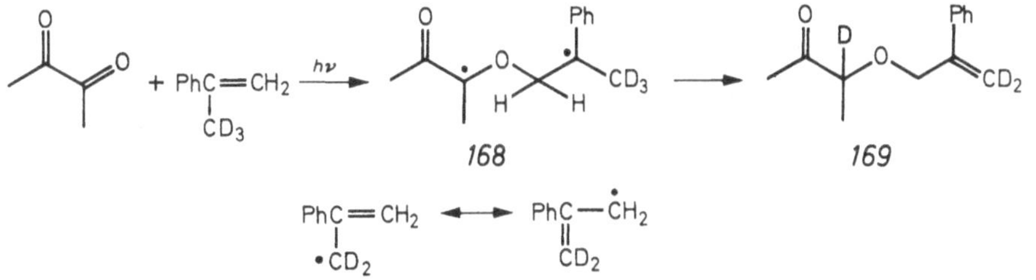

Thus, four types of products (often involving stereoisomers) may be formed in dione-olefin reactions. The situation is further complicated by the fact that product compositions show solvent and temperature-dependence [125] (effects of other variables have not been investigated) so that results from different laboratories have sometimes been at variance. Prediction of results is hazardous.

The most extensively studied diketone has been biacetyl [124–128]; a detailed mechanistic study of its reactions with a number of olefins has appeared [128]. It was suggested that a triplet exciplex is the precursor for formation of *161*. The occurrence of electron transfer, presumably subsequent to exciplex formation, has been demonstrated [157,158] in the reaction of biacetyl with tetramethyl-1,3-dioxole (*167*). Esr-spectroscopy of irradiated mixtures indicated the presence of biacetyl radical anion and dioxole radical cation. This reaction, which produced a complex mixture of products, was suggested to involve the excited singlet state of biacetyl.

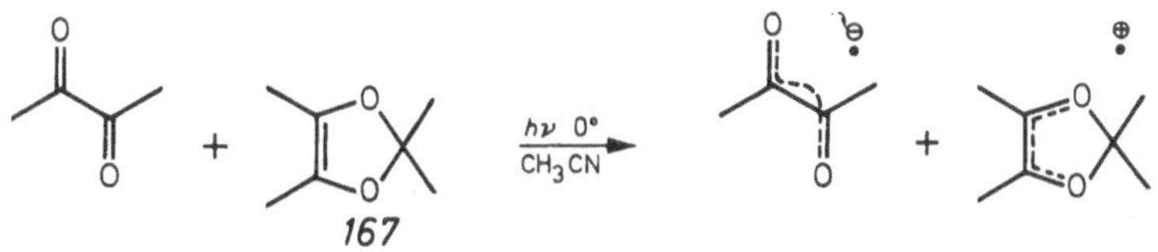

The quantum yields for dione-olefin reactions have generally been low (<0.1) and rate constants for excited state reactions were of the order of $1\text{–}10 \times 10^4\ M^{-1}\ sec^{-1}$. An interesting exception [127b] was compound *168* where intramolecular addition to the product shown proceeded with a rate constant estimated to be greater than $1 \times 10^{10}\ M^{-1}\ sec^{-1}$ suggesting a singlet state reaction. Stereochemistry of additions of biacetyl to 7-substituted norbornenes showed the expected dependence on nature of the 7-substitutent [127].

Other diones whose reactions with olefins have been reported include 1-phenylpropane-1,2-dione [129] (nearly exclusive oxetane formation), 2,2,5,5-tetramethyltetrahydrofuran-3,4-dione [130], 1,1,4,4-tetramethyltetralin-2,3-dione [131], and others, as well as the series of tetramethylcycloalkanediones [8,131] mentioned earlier. A number of intramolecular additions to give oxetanes were mentioned in the previous section.

Additions of diketones (biacetyl, benzil, phenylpropanedione, phenylglyoxal) to thioacetylenes also occur [132]. As in additions of monoketones to acetylenes, the

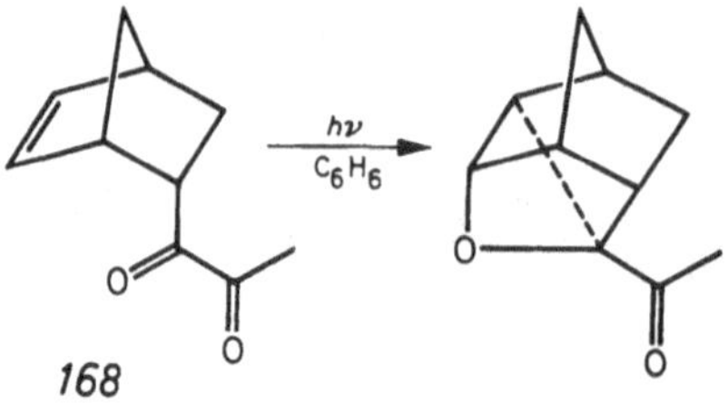

168

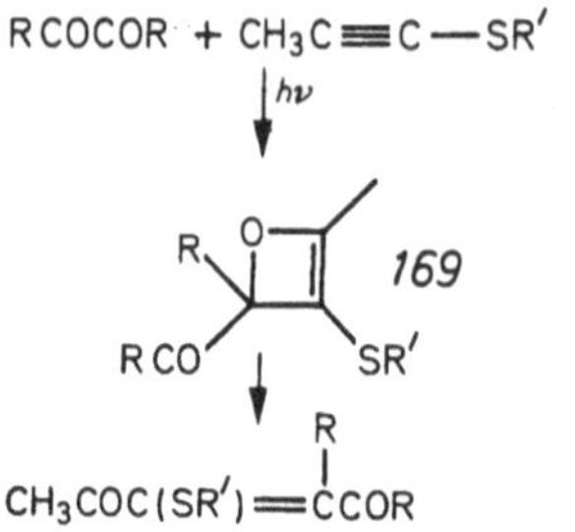

products were presumed to arise from initially formed oxetes *169* which underwent ring opening under the conditions of reaction.

B Hydrogen Atom Abstraction Reactions

Reactions of this type have been referred to in connection with intramolecular cyclizations of acyclic diones. They are exceedingly common with all types of diketones (excepting cyclobutane derivatives and bridged cyclohexenediones) since almost any C—H bond is susceptible to attack by triplet diketone. The products are semidione radical (*164*) and substrate derived radical.

$$\text{RCOCOR} + -\overset{|}{\underset{|}{\text{C}}}-\text{H} \xrightarrow{h\nu} \mathit{164} + -\overset{\cdot}{\underset{|}{\text{C}}}-$$

Subsequent events can include radical coupling at carbon or at oxygen of *164* to give addition products, dimerization, and disproportionation. Amines have been added to the list of substrates which can participate in such abstraction reactions [133–137, 150] and benzene also appears to be able to participate, albeit very inefficiently, as shown by formation of biphenyl in certain irradiations of diones in benzene solution [138,139].

An interesting effect of light intensity has been observed in the reaction of bicyclo[2.2.2]octanedione (*170*) with aromatic aldehydes [140]. These reactions produced esters of α-ketols (*171*) and labile α-aroyl-α-hydroxyketones (*172*). At first

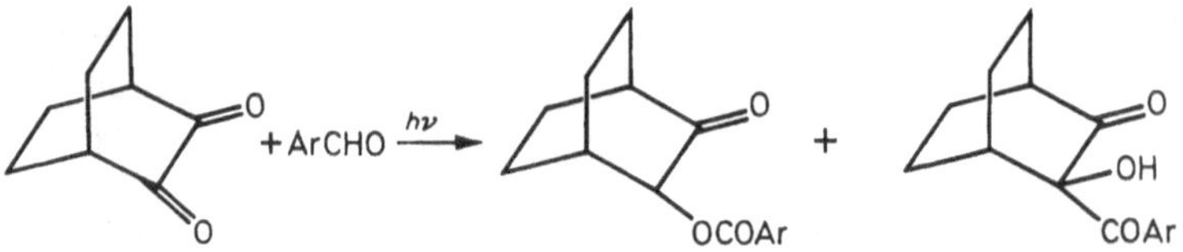

170 *171* *172*

glance, the two products would seem to arise from the two possible modes of coupling of semidione with aroyl radicals. However, the ratio *171*:*172* in the reaction of *170* with p-chlorobenzylaldehyde showed a marked dependence on light intensity, ranging from 14.8:1 at the lowest intensity used to 0.4:1 at the highest.

The following explanation was proposed. At high light intensity, with concomitantly high radical concentrations, the predominant reaction is radical coupling which leads mainly (or exclusively) to *172*. At low intensity, addition of aroyl radicals to *170* results in the formation of a new radical *173* which terminates by transfer of a hydrogen atom followed by ketonization

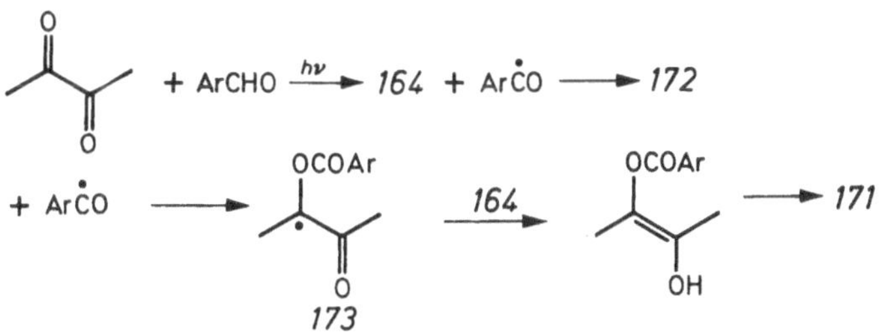

Rapid hydrogen transfer occurs between semidione radicals and ground state diketones. This was illustrated by irradiation of a mixture of camphorquinone (*174*, λ_{max} 470 nm, $E_T = 51$ kcal/mole) and tetralindione *68* (λ_{max} 388 nm, $E_T = 54$ kcal/mole) in 2-propanol at various wavelengths [141]. The exclusive product, no matter what the wavelength, was reduced *68* indicating that the equilibrium below lies to the right. Ketyl radicals derived from monoketones, such as benzophenone, transfer

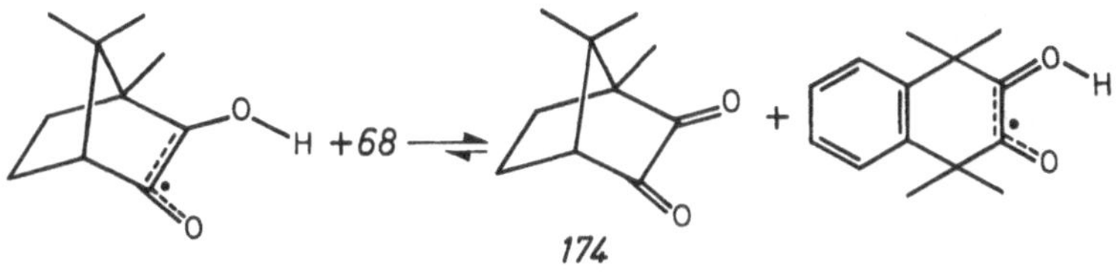

hydrogen rapidly to α-diketones. It is thus possible to generate the same radicals as obtained by direct irradiation of a diketone in the presence of a hydrogen donor by adding benzophenone and irradiating at a wavelength where the monoketone absorbs most or all of the incident light. Problems of energy transfer can be minimized by use of low dione concentrations. Since quantum yields for hydrogen abstraction by benzophenone are much higher than for α-diketones, the overall process can be much more efficient. This effect, sometimes referred to as chemical

$$Ph_2CO + SH \xrightarrow{h\nu} Ph_2\dot{C}OH + S\cdot$$

$$Ph_2\dot{C}OH + RCOCOR \rightarrow Ph_2CO + R\dot{C}(OH){-}COR$$

$$R{-}\dot{C}(OH){-}COR + S\cdot \rightarrow \text{product(s)}$$

sensitization [142], has been exploited to show that the photolysis of benzophenone in p-xylene involves free radicals while a considerable part of the reaction of camphorquinone with p-xylene involves a caged pair of radicals [143]. Another application was in the study of dione-amine reactions [135].

Detailed studies of H-atom abstraction reactions of cyclopropyl diketones [111], 2,2,5,5-tetramethyltetrahydrofuran-3-4-dione [130] cycloalkanediones [57] and substituted benzils [121], among others, have appeared. The cyclopropyl diketones reacted normally via semidione radicals without ring enlargement.

Reactions of *68* or *170* with α-benzyl-α-hydroxyketones resulted in decarbonylated cleavage products [144]. Based on deuterium labelling studies, it appears that the initial step involves abstraction of a hydroxylic hydrogen atom.

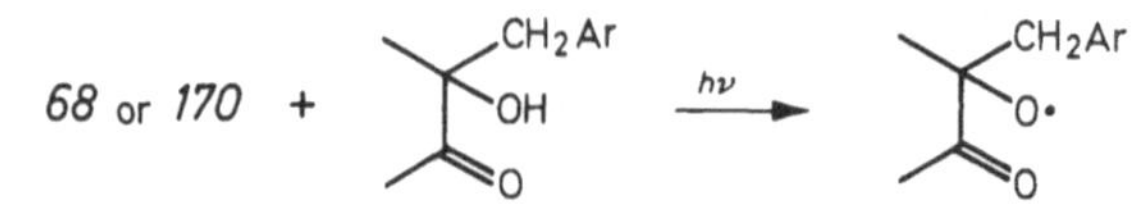

C Reactions in Inert Medium

Irradiations in "inert" solvents, generally benzene, of diketones which normally participate in hydrogen abstraction have been examined by a number of workers. Such reactions involve a number of difficulties including very low quantum yields for disappearance of starting material (necessitating extremely long irradiation times), formation of complex mixtures, and unsatisfactory material balances. Careful degassing to reduce oxygen concentration to the absolute minimum possible is essential in the study of reactions of this type. Under these circumstances mechanistic studies become very difficult and it is not surprising that details of reaction mechanisms are obscure at best.

The situation with 3,3,7,7-tetramethylcycloheptane-1,2-dione (*175*) is illustrative. Irradiation of *175* in 2-propanol produced the α-hydroxyketone *176* in quantitative yield [57]. In one study of its reaction in benzene [139], irradiation through quartz produced *176* in 60% yield plus biphenyl in unstated yield. In a second study [57], irradiation at 366 nm to 20% conversion gave *176* in 36% yield, no biphenyl, and the additional products shown below.

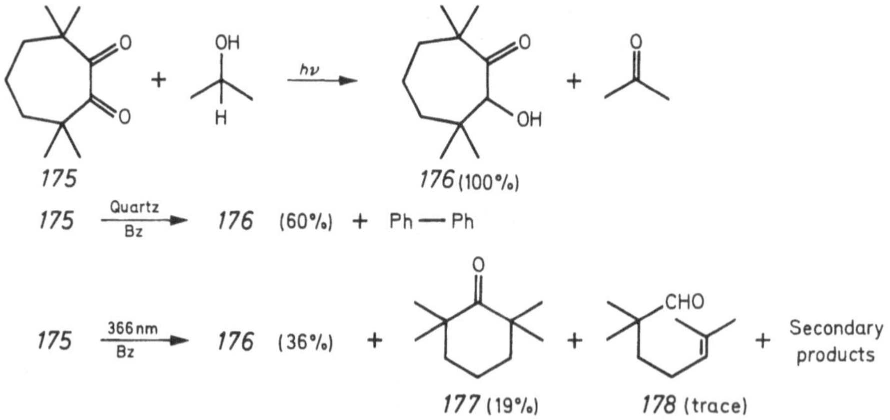

In the first case, formation of biphenyl was taken to indicate that benzene serves as a source of hydrogen atoms for the reduction of *175* to *176*. In the second case, where no biphenyl was formed, ground state *175* was assumed to serve as the hydrogen donor for reduction to *176*. The other products were suggested to arise from cleavage of the intercarbonyl bond of *175* followed by loss of carbon monoxide to form the biradical *179* which proceeds by conventional reactions to *177*, *178* and the secondary products. However, we note that, if excited *175* abstracts hydrogen from ground state *175*, one of the radicals produced can undergo β-scission followed by decarbonylation to produce radical *180* which is also a reasonable precursor for the products observed. Such a mechanism appears to account for the efficient regio-selective acetylation of adamantane when mixtures of biacetyl and adamantane are photolyzed [149].

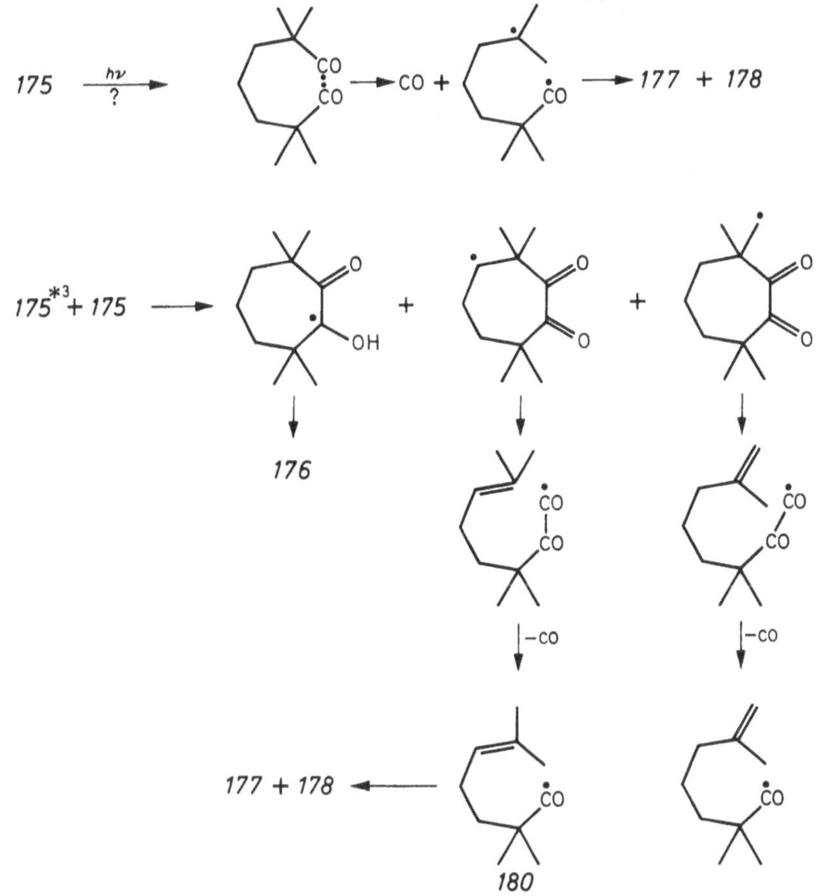

In fact, as has been noted earlier [2], no clear-cut proof exists as yet for the occurrence of α-cleavage in photolysis of diketones with the exception of the special case of four-membered ring diones.

The sulfur analog, *181*, of *175* has been investigated by two groups [139,145] with similar results although the interpretations differ. In a variety of solvents (benzene, t-butyl alcohol, cyclohexane, methanol) and wavelengths (250, 300 and 350 nm), the major product was β-thiolactone *184* (15–50%), isobutylene, products derived from dimethylketene, and decarbonylation products analogous to those described above for *175*. One group proposed [139] an α-cleavage mechanism similar to that shown above while the second group [145] suggested that *184* was formed by way of an electron transfer mechanism to give bicyclic zwitterion *182* which fragmented to isobutylene and dione *183*, the precursor of *180*. The sulfoxide and sulfone of *179* were also investigated [139].

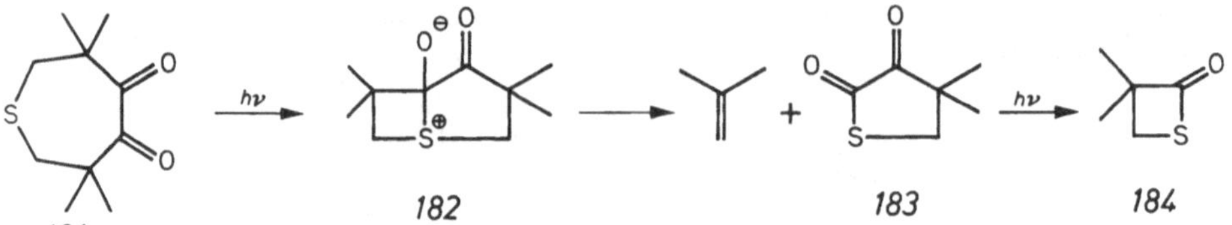

181 *182* *183* *184*

Alternative mechanisms involving β-scission of radicals formed by hydrogen abstraction are also possible.

Irradiation of *185* in benzene produced acetone (80–100% yield) and a complex mixture of products which did not include the dimer of dimethylketene [144].

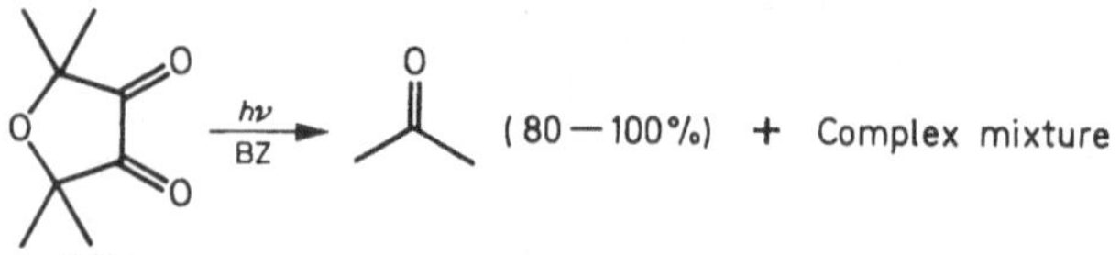

185

The use of carbon tetrachloride as a solvent in diketone irradiations is dangerous since intermediate radicals can abstract chlorine giving trichloromethyl radicals which can result in formation of complex mixtures. In a study [146] of the irradiation of *68* in CCl_4, products included (inter alia) reduced *68* and the trichloromethyl compound *186*.

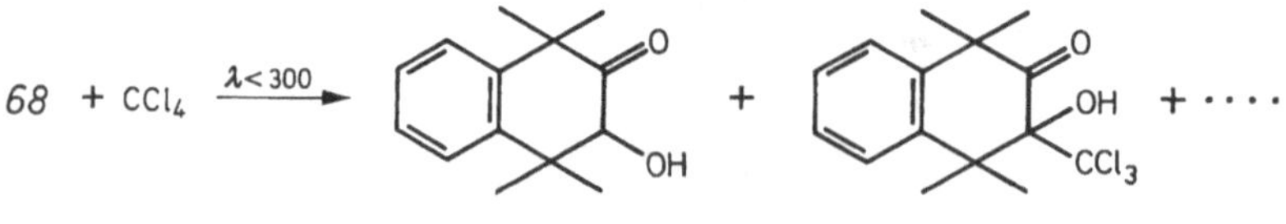

21% *186* (14%)

Photochemical reaction of a benzil moiety attached to a crown ether was inhibited in the presence of sodium ions [147].

X Addendum

The following addendum, added in proof, includes literature up to mid-1984 arranged in the order of presentation of the various topics in the text.

II Spectroscopy

The study of intramolecular energy transfer between the aromatic and a-diketo moieties of cyclic compounds *14b* has been extended to ortho (n = 3, 4) and meta (n = 3, 4) isomers and the full paper describing these results is in press [159] as is the description of synthesis and determination of conformations of these compounds [160]. A good correlation, based on the Dexter equation, was found between efficiency of singlet energy transfer and interchromophore distance allowing evaluation of the parameters of the Dexter equation. Preliminary results [161] on triplet energy transfer show the expected complementarity.

An ENDOR investigation [162] of the lowest triplet state of benzil in the crystal has led to the conclusion that the torsion angle between carbonyl groups in the triplet is 157°. While this angle is appreciably larger than the ground state value of 111°, it is significantly smaller than the generally accepted value of 180° for triplet benzil in solution. In view of the relatively small magnitude of the rotational barrier, this "discrepancy" may well be due to to special factors operating in the crystal.

A theoretical treatment [163] of absorption spectra of open-chain diketones has led to the conclusion that the barrier to internal rotation "must be larger in the $^1n^*$ state than it is in the ground state possibly by as much as a factor of two".

VI A Bridged Cyclohexenediones

The exceptional case of bicyclo[2.2.2]octene-2,3-dione (*55*) which exhibits competing bisdecarbonylation to cyclohexadiene and isomerization to cyclobutanedione *56* has been shown [164] to involve an interesting concentration effect. In the range of 1×10^{-5} M *55* the only reaction was isomerization to *56* but this reaction competed less and less effectively with bisdecarbonylation as the concentration of *55* increased. Triplet sensitized reactions resulted exclusively in bisdecarbonylation; correspondingly, this reaction, but not rearrangement was quenched by added anthracene. It was suggested that intersystem crossing involves the intermediacy of an excimer and that, for reasons which are not clear, rearrangement *55*→*56* is significantly slower than in the numerous other examples studied.

VII Reactions of Diones with Oxygen

Further work on the mechanism of dione reactions with oxygen includes a detailed study [165] of reactions of biacetyl and benzil with emphasis on the latter. The earlier conclusion[2] that photolysis of benzil does not produce benzoyl radicals directly received further support. Products of photolysis of benzil in benzene solution in the presence of oxygen were biphenyl, phenyl benzoate, and perbenzoic acid. Addition of an olefin led to reduction or elimination of these products and to formation of epoxides and allylic hydroperoxides. Earlier results using labelled oxygen were reproduced with the additional observation that scrambling of oxygen atoms (from $O^{16}O^{18}$) was not observed. The mechanism proposed for benzil involves initial addition of ground state oxygen to triplet benzil followed by cleavage of the resulting biradical to benzoyl and benzoylperoxy (PhC(O)OO) radicals. The latter then react with olefin to form an epoxide and a benzoyloxy radical in a stepwise process. The benzoyl radicals may be oxidized to benzoylperoxy radicals or undergo a variety of reactions with solvent. While this mechanism appears to account well for the reactions of benzil, it is not satisfactory for biacetyl where ratios of epoxide formed to dione consumed were much larger than the value 1–3 observed with benzil. At this point it seems as if every dione requires a separate mechanistic scheme.

This point illustrated by results of an investigation [166] of reactions of phenanthrenequinone and acenaphthenequinone in the presence of oxygen. Major products in the presence of olefins were the dione-olefin 1:1 adducts (cf section IXA) with litle if any epoxide formed. Biacetyl remains the dione of choice for photoepoxidation.

IX A Reactions of Diones with Olefins

Continuing work [158] on photoreactions of electron-rich olefins with biacetyl shows that the complexity of product mixtures obtains in these reactions also. Effects of solvent polarity provide further support for the importance of ionic intermediates in these reactions. The reactions of biacetyl with 1,1-diethoxyethylene are proposed to proceed via the triplet state (in contrast to reactions with dioxoles). The reversal of regiospecifity between thermal and photochemical cycloaddition of this olefin with biacetyl is nicely explained by the assumption of excited state electron transfer from olefin to dione to give the corresponding radical ions.

XI References

1. a) Rubin, M. B.: Top. Curr. Chem. *13*, 251 (1969)
   b) Monroe, B. M.: Adv. in Photochem. *8*, 77 (1971)
2. Rubin, M. B.: in: Excited States in Organic Chemistry and Biochemistry, ed. Pullman, B., Goldblum, N., Reidel, D.: Dordrecht, Holland 1977, p. 381
3. Swenson, J. R., Hoffmann, R.: Helv. Chim. Acta *53*, 2331 (1970)
4. Cowan, D. O. et al.: Angew. Chem. Int. Ed. Engl. *10*. 401 (1971)
5. Dougherty, D., Brint, P., McGlynn, S. P.: J. Am. Chem. Soc. *100*, 5597 (1978)
6. Arnett, J. F. et al.: ibid. *96*, 4385 (1974)
7. Schang, P. S., Gleiter, R., Rieker, A.: Ber. Bunsenges. Phys. Chem. *82*, 629 (1978)
8. a) Verheijdt, P. L.: Ph. D. Thesis, University of Amsterdam 1981. A copy of this thesis was kindly provided by Prof. H. Cerfontain
   b) Verheijdt, P. L., Cerfontain, H.: J. Chem. Soc. Perkin II, 1343 (1983)
9. Sarphatic, L. A., Verheijdt, P. L., Cerfontain, H.: Rec. Trav. Chim. Pays-Bas *102*, 9 (1983)
10. Gleiter, R., Schang, P., Seitz, G.: Chem. Phys. Lett. *55*, 144 (1978)
11. Gleiter, R. et al.: Angew. Chem. Int. Ed. Engl. *16*, 400 (1977)

12. Dougherty, D., Blankespoor, R. L., McGlynn, S. P.: J. Electron. Spectrosc. Relat. Phenom. *16*, 245 (1979)
13. Martin, H.-D. et al.: Chem. Ber. *111*, 2557 (1978)
14. Leonard, N. J., Mader, P. M.: J. Am. Chem. Soc. *72*, 5388 (1950)
15. Mori, T., Nakahari, T., Nozaki, H.: Can. J. Chem. *47*, 3266 (1979)
16. a) Rubin, M. B., Welner, S.: J. Org. Chem. *45*, 1847 (1980)
    b) Rubin, M. B., Migdal, S.: unpublished results
17. Bloomfield, J. J., Moser, R. E.: J. Am. Chem. Soc. *90*, 5625 (1968)
18. Neely, S. C. et al.: ibid. *93*, 4903 (1971)
19. Bartetzko, R. et al.: ibid. *100*, 5589 (1978)
20. Dougherty, D. et al.: J. Phys. Chem. *80*, 2212 (1976)
21. Gleiter, R., Leismann, H.: University of Heidelberg, to be published
22. Martin, H.-D., Albert, B., Schiwek, H.-J.: Tetrahedron Lett., 2347 (1979); Israel J. Chem, in press
23. Weiner, M.: D. Sc. Thesis, Technion 1977
24. Legon, A. C.: J.C.S. Chem. Comm., 612 (1973); J.C.S. Faraday Trans. II, 651 (1979)
25. Hagen, K., Hedberg, K.: J. Am. Chem. Soc. *103*, 5360 (1981)
26. a) Getz, D. et al.: J. Phys. Chem. *84*, 768 (1980); Speiser, S. et al.: Chem. Phys. Lett. *61*, 199 (1979)
    b) Hassoon, S. et al.: ibid. *98*, 345 (1983);
    c) Speiser, S., Katriel, J.: ibid. *102*, 88 (1983)
27. Weinreb, A., Werner, A., Lebovitz, Z.: Photochem. and Photobiol. *29*, 755 (1979)
28. Bera, S. C., Mukherjee, R., Chowdhury, M.: J. Chem. Phys. *51*, 754 (1969)
29. Kaftory, M., Rubin, M. B.' J.C.S. Perkin Trans. II, 149 (1983)
30. a) Morantz, D. J., Wright, A. J. C.: J. Chem. Phys. *54*, 692 (1971)
    b) Sandross, K.: Acta Chem. Scand. *27*, 3021 (1973)
31. Bera, S. C., Karmakar, B., Chowdhury, M.: Chem. Phys. Lett. *27*, 397 (1974)
32. Arnett, J. F., McGlynn, S. P.: J. Phys. Chem. *79*, 626 (1975)
33. a) Fang, T.-S., Brown, R. E., Singer, L. A.: J.C.S. Chem. Comm. *116* (1978)
    b) Inoue, H., Sakurai, T.: J.C.S. Chem. Comm., 314 (1983)
    c) Flamigni, L. et al.: J. Photochem. *21*, 237 (1983)
34. Sarphatie, L. A., Verheijdt, P. L., Cerfontain, H.: Recl. Trav. Chim. Pays-Bas *102*, 9 (1983)
35. Charney, E., Tsai, L.: J. Am. Chem. Soc. *93*, 7123 (1971)
36. For leading references see Russell, G. A. et al.: J. Org. Chem. *44*, 2780 (1979); Russell, G. A., Schmitt, K. D., Mattox, J.: J. Am. Chem. Soc. *97*, 1882 (1975)
37. Staudinger, H., Anthes, E.: Ber. *46*, 1426 (1913)
38. a) Haddon, R. C., Poppinger, D., Radom, L.: J. Am. Chem. Soc. *97*, 1645 (1975) and references therein
    b) Raine, G. P., Schaefer III, H. F., Haddon, R. C.: J. Am. Chem. Soc. *105*, 194 (1983)
39. Strating, J. et al.: Tetrahedron Lett., 125 (1969)
40. Chapman, O. L., McIntosh, C. L., Barber, L. L.: J.C.S. Chem. Comm., 1162 (1971)
41. Obata, N., Takizawa, T.: ibid. 587 (1971)
42. Toda, F., Todo, E.: Bull. Chem. Soc. Japan *48*, 583 (1975)
43. Dehmlow, E. V.: Tetrahedron Lett., 1271 (1972)
44. Jung, M. E., Lowe, J. A.: J. Org. Chem. *42*, 2371 (1977)
45. Staab, H. A., Ipaktschi, J.: Chem. Ber. *101*, 1457 (1968)
46. Kolc, J.: Tetrahedron Lett., 5321 (1972)
47. Chapman, O. L.: Pure and Applied Chem. *40*, 511 (1974)
48. Eggerding, D., West, R.: J. Am. Chem. Soc. *97*, 207 (1975); *98*, 3641 (1976)
49. Schmidt, A. H.: Synthesis 961 (1980)
50. Maier, G., Jung, W. A.: Tetrahedron Lett. 3875 (1980)
51. Toda, F. et al.: Bull. Chem. Soc. Japan *46*, 1737 (1973)
52. Bloomfield, J. J., Owsley, D. C., Nelke, J. M.: in: Organic Reactions (ed. Dauben, W. G.), Wiley, N.Y. 1976, Vol. 23, Chap. 2
53. Scharf, H.-D.: Angew. Chem. Int. Ed. Engl. *13*, 520 (1974)
54. Conia, J. M., Robson, M. J.: ibid. *14*, 473 (1975)
55. Rubin, M. B.: unpublished results
56. Ron, A.: Technion, unpublished results

57. Verheijdt, P. L., Cerfontain, H.: Recl. Trav. Chim. Pays-Bas *102*, 173 (1983)
58. Bloomfield, J. J., Irelan, J. R. S., Marchand, A. P.: Tetrahedron Lett. 5647 (1968)
59. Rubin, M. B.: Chimia *35*, 406 (1981)
60. Rubin, M. B., Harel, Y.: unpublished
61. Liao, C. C. et al.: National Tsing Hua University, unpublished results. We thank Professor Liao for communicating these results prior to publication
62. Rubin, M. B.: J. Am. Chem. Soc. *103*, 7791 (1981)
63. Heine, H.-G.: Chem. Ber. *104*, 2869 (1971)
64. Denis, J. M., Conia, J. M.: Tetrahedron Lett. 461 (1973)
65. de Groot, Ae., Oudman, D., Wynberg, H.: ibid. 1529 (1969)
66. Rubin, M. B., Weiner, M., Scharf, H.-D.: J. Am. Chem. Soc. *98*, 5699 (1976)
67. For reviews of the photochemistry of β,γ-unsataurated ketones see Hixon, S., Mariano, P. S., Zimmerman, H. E.: Chem. Rev. *73*, 531 (1973); Dauben, W. G., Lodder, G., Ipaktschi, J.: Top. Curr. Chem, *54*, 73 (1975); Houk, K. N.: Chem. Rev. *76*, 1 (1976); Schaffner, K.: Tetrahedron *32*, 641 (1976)
68. Rubin, M. B. et al.: J. Photochem. *11*, 287 (1979)
69. Speiser, S., Rubin, M. B., Hassoon, S.: J. Photochem., *26*, 297 (1984)
70. Scharf, H. D., Fleischhauer, J.: Houben Weyl "Methoden der Organischen Chemie" (ed. Muller, E.), Thieme, Stuttgart 1975, Vol. IV 5a, p. 24ff.
71. Warrener, R. N., Russell, R. A., Lee, T. S.: Tetrahedron Lett., 49 (1977); Warrener, R. N., Pitt, I. G., Russell, R. A.: J.C.S. Chem. Comm., 1136 (1982)
72. Warrener, R. N., Harrison, P. A., Russell, R. A.: Tetrahedron Lett., 2031 (1977); idem. J.C.S. Chem. Comm. 1134 (1982); Warrener, R. N. et al.: ibid. 546 (1984)
73. Gream, G. E., Paice, J. C., Ramsay, C. C. R.: Aust. J. Chem. *20*, 1671 (1967)
74. Liao, C. C. et al.: J. Am. Chem. Soc. *104*, 292 (1982)
75. Kunze, M.: Ph. D. Thesis, Univ. Würzburg 1980. We thank Prof. H.-D. Martin for making a part of this thesis available
76. a) Paquette, L. A. et al.: J. Am. Chem. Soc. *103*, 2262 (1981)
b) Paquette, L. A. et al.: J. Org. Chem. *48*, 1262 (1983)
77. Scharf, H.-D., Klar, R.: Chem. Ber. *105*, 575 (1972)
78. Bryce-Smith, D., Gilbert, A.: J.C.S. Chem. Comm. 1702 (1968)
79. Sakanishi, K., Shigeshima, T., Hachiya, H.: Suzuka Kogyo Koto Semmon Gakko Kivo *8*, 269 (1975); Chem. Abstr. *88*, 74103z (1978)
80. Verboom, W., Bos, H. J. T.: Recl. Trav. Chim. Pays-Bas *100*, 207 (1981)
81. Pyle, J. L., Schaffer, A. A., Cantrell, J. S.: J. Org. Chem. *46*, 115 (1981)
82. Hart, H., Dean, D. L., Buchanan, D. N.: J. Am. Chem. Soc. *95*, 6294 (1973); Vaughan, W. R., Yoshimine, M.: J. Org. Chem. *22*, 528 (1957)
83. Horner, L., Merz, H.: Justus Liebigs Ann. Chem. *570*, 7 (1950)
84. Berney, D., Deslongchamps, P.: Canad. J. Chem. *47*, 515 (1969); Pure and Appl. Chem. *49*, 1329 (1977)
85. Chang, T. H. et al.: J. Chinese Chem. Soc. *27*, 97 (1980)
86. Scharf, H.-D., Friedrich, P., Linckens, A.: Synthesis, 256 (1976)
87. Nunn, E. E., Wilson, W. S., Warrener, R. N.: Tetrahedron Lett., 175 (1972)
88. Vogel, E., Ippen, J., Buch, V.: Angew. Chem. Int. Ed. Engl. *14*, 566 (1975)
89. Warrener, R. N., Nunn, E. E., Paddon-Row, M. N.: Tetrahedron Lett., 2355 (1976)
90. Friedrichsen, W., Schroer, W.-D., Schmidt, R.: Ann., 793 (1976); Tetrahedron Lett., 2469 (1974); Allmann, R. et al.: Tetrahedron *32*, 147 (1976)
91. Friedrichsen, W. et al.: Justus Liebigs Ann. Chem., 440 (1978)
92. Christl, M. et al.: Chem. Ber. *110*, 3745 (1977)
93. Pritschins, W., Grimme, W.: Tetrahedron Lett., 4545 (1979)
94. Grimme, W. et al.: Angew. Chem. Int. Ed. Engl., *20*, 113 (1981); cf. also Chem. Ber. *114*, 3197 (1981)
95. Shimizu, N., Bartlett, P. D.: J. Am. Chem. Soc. *98*, 4193 (1976);
Bartlett, P. O., Roof, A. A. M., Shimizu, N.: ibid. *104*, 3130 (1982)
96. Mori, A., Takeshita, H.: Chem. Lett., 395 (1978)
97. Koo, J.-Y., Schuster, G. B.: J. Org. Chem. *44*, 847 (1979)
98. Buchi, G., Fowler, K. W., Nadzan, A. M.: J. Am. Chem. Soc. *104*, 544 (1982)

99. Bartlett, P. D., Becherer, J.: Tetrahedron Lett., 2983 (1978)
100. Turro, N. J., Lee, J.-J.: J. Am. Chem. Soc. *91*, 5651 (1969)
101. Zepp, R. G., Wagner, P. J.: ibid. *92*, 7466 (1970)
102. Wagner, P. J. et al.: ibid. *98*, 8125 (1976)
103. Hamer, N. K.: J.C.S. Perkin Trans. I, 61 (1983)
104. Burkoth, T. L., Ullman, E. F.: Tetrahedron Lett., 145 (1970)
105. Bishop, R.: J.C.S. Chem. Comm., 1288 (1972)
106. Urry, W. H., Duggan, J. C., Pai, M. H.: J. Am. Chem. Soc. *92*, 5785 (1970)
107. Urry, W. H., Trecker, D.: ibid. *84*, 118 (1962)
108. Paquette, L. A., Balogh, D. W.: ibid. *104*, 74 (1982)
109. Balogh, D. W. et al.: ibid. *103*, 226 (1981)
110. Nerbonne, J. M., Weiss, R. C.: Israel J. Chem. *18*, 266 (1979)
111. Kelder, K., Cerfontain, H., van der Wielen, F. W. M.: J.C.S. Perkin Trans. II, 710 (1970)
112. Turro, N. J., Lee, T. J.: J. Am. Chem. Soc. *92*, 7467 (1970)
113. Bishop, R., Hamer, N. K.: J. Chem. Soc. (C), 1197 (1970); idem, J.C.S. Chem. Comm., 804 (1969)
114. Ogata, Y., Tagaki, K.: Bull. Chem. Soc. Japan *47*, 2255 (1974)
115. Rigaudy, J., Paillous, N.: Bull. Soc. Chim. France, 576 (1971)
116. Bishop, R., Hamer, N. K.: J. Chem. Soc. (C), 1193 (1970)
117. Ogata, Y., Tagaki, K.: J. Org. Chem. *39*, 1385 (1974)
118. Bishop, R., Hamer, N. K.: JCS Chem. Comm., 804 (1969)
119. Hamer, N. K., Samuel, C. J.: J.C.S. Perkin Trans. II, 1316 (1973)
120. Hamer, N. K.: J.C.S. Perkin Trans. I, 508 (1979)
121. Ogata, Y., Tagaki, K., Fuji, Y.: J. Org. Chem. *37*, 4026 (1972)
122. Paukstelis, J. V., Kao, J.-L.: Tetrahedron Lett., 3691 (1970)
123. Maruyama, K., Ono, K., Osugi, J.: Bull. Chem. Soc. Japan *45*, 847 (1972)
124. Ryang, H.-S., Shima, K., Sakurai, H.: J. Am. Chem. Soc. *93*, 5270 (1971); J. Org. Chem. *38*, 2860 (1973)
125. Turro, N. J. et al.: Tetrahedron Lett. *21*, 2775 (1980)
126. Chow, Y. L. et al.: Can. J. Chem. *48*, 3045 (1970)
127. a) Sauers, R. R., Valenti, P. C., Tavss, E.: Tetrahedron Lett., 3129 (1975);
b) Sauers, R. R., Valenti, P. C., Crichlow, C. A.: J. Am. Chem. Soc. *104*, 6378 (1982)
128. Jones, G., II, Santhanam, M., Chiang, S.-H.: ibid. *102*, 6088 (1980)
129. Shima, K. et al.: Bull. Chem. Soc. Japan *50*, 761 (1977)
130. Rubin, M. B., Ben-Bassat, J. M., Weiner, M.: Israel J. Chem. *16*, 326 (1977)
131. Gream, G. E., Mular, M., Paice, J. C.: Tetrahedron Lett., 3479 (1970)
132. Mosterd, A., Matser, H., Bos, H. J. T.: ibid. 4179 (1974)
133. Turro, N. J., Engel, R.: J. Am. Chem. Soc. *91*, 7113 (1969)
134. Greame, G. E., Paice, J. C., Uszynski, B. S. J.: J.C.S. Chem. Comm. 895 (1970)
135. McLauchlan, K. A., Sealy, R. C., Wittmann, J. M.: J.C.S. Perkin Trans. II, 926 (1977); Mol. Physics *36*, 1397 (1978)
136. Mehrotra, K. N., Pandey, G. P.: Bull. Chem Soc. Japan *53*, 1081 (1980)
137. Scaiano, J. C.: J. Phys. Chem. *85*, 2851 (1981)
138. Rubin, M. B., Neuwirth-Weiss, Z.: J. Am. Chem. Soc. *94*, 6048 (1972)
139. Kooi, J., Wynberg, H., Kellogg, R. M.: Tetrahedron *29*, 2135 (1973)
140. Rubin, M. B., Inbar, S.: J. Am. Chem. Soc. *100*, 2266 (1978)
141. Rubin, M. B., Hershtik, Z.: J.C.S. Chem. Comm., 1267 (1970)
142. Engel, P. S., Monroe, B. M.: Adv. Photochem. *8*, 245 (1971)
143. Rubin, M. B.: Tetrahedron Lett., 3931 (1969)
144. Gutmann, A. L.: M.Sc. Thesis, Technion 1978
145. Johnson, P. Y., Zitsman, J., Hatch, C.: J. Org. Chem. *38*, 4087 (1973); Tetrahedron Lett., 1991 (1972)
146. Greame, G. E., Paice, J. C.: Aust. J. Chem. *22*, 1249 (1969)
147. Hirano, H., Kurumaya, K., Tada, M.: Bull. Chem. Soc. Japan *54*, 2708 (1981)
148. Boyer, J. H., Srinivasan, K. G.: J.C.S. Chem. Comm., 699 (1973)
149. Tabushi, I., Kojo, S., Fukunishi, K.: J. Org. Chem. *43*, 2370 (1978)
150. Bunbury, D. L., Chan, T. M.: Can. J. Chem. *50*, 2499 (1972)

151. Steinberg, N., Gafni, A., Steinberg, I. Z.: J. Am. Chem. Soc. *103*, 1636 (1981)
152. Broomhead, E. J., McLauchlan, K. A., Roe, J. C.: J.C.S. Perkin Trans II, 796 (1980) and references therein
153. Monroe, B. M., Lee, C., Turro, N. J.: Mol. Photochem. *6*, 271 (1974)
154. Schuh, M. D.: J. Phys. Chem. *82*, 1861 (1978)
155. Shepherd, J. P.: J. Org. Chem. *48*, 337 (1983)
156. Clennan, E. L., Speth, D. R., Bartlett, P. D.: ibid. *48*, 1246 (1983)
157. Mattay, J. et al.: J. Photochem., *23*, 319 (1983)
158. Mattay, J. et al.: Angew. Chem. Int. Ed. Engl., *23*, 249 (1984); Tetrahedron Lett., *25*, 817 (1984); Proceedings Xth IUPAC Symposium on Photochemistry, Interlaken, 1984, p. 227
159. Hassoon, S. et al: J. Phys. Chem., in press
160. Rubin, M. B. et al.: Isr. J. Chem., in press
161. Speiser, S., Hassoon, S., Rubin, M. B., Proc. Xth IUPAC Symposium on Photochemistry, Interlaken, 1984, p. 227
162. Chan, L. Y., Heath, B. A.: J. Chem. Phys., *171*, 1070 (1979)
163. Pawlikowski, M., Zgierski, M. Z., Orlando, G.: Chem. Phys. Lett., *105*, 612 (1984)
164. Rubin, M. B., Proc. Xth IUPAC Symposium on Photochemistry, Interlaken, 1984, p. 173
165. Sawaki, Y., Foote, C. S.: J. Org. Chem., *48*, 4934 (1983)
166. Sawaki, Y.: Bull. Chem. Soc. Japan, *56*, 3464 (1983)

Photochemistry in Micelles

Nicholas J. Turro, G. Sidney Cox, and Mark A. Paczkowski

Department of Chemistry, Columbia University, New York, New York 10027, USA

Table of Contents

Organic photochemical reactions conducted in micellar solutions are reviewed from the standpoint of systematizing and correlating published results. Five common effects are found to distinguish and characterize micellar photochemistry relative to conventional solution photochemistry: "super" cage effects, local concentration effects. viscosity effects, polarity effects, and electrostatic effects. These effects can contribute to the occurence of enhanced selectivity and efficiency of photoreactions relative to those in conventional homogeneous solution.

Abbreviations

SDS	sodium dodecyl sulfate
HDTCl	hexadecyltrimethylammonium chloride
HDTBr	hexadecyltrimethylammonium bromide
Brij 35	polyoxyethylene (23) lauryl ether
DTAC	dodecyltrimethylammonium chloride
TTACl	tetradecyltrimethylammonium chloride
HDPB	hexadecylpyridinium bromide
KDC	potassium decanoate
SPFO	sodium perfluorooctanoate
KTC	potassium tetradecanoate

1 Introduction

A surfactant (*sur*face *act*ive material) or detergent is a molecule whose structure possesses both polar (or ionic) and nonpolar moieties. A typical detergent structure is RX where R is a straight chain hydrocarbon of 8–18 carbon atoms or some other hydrophobic residue, and X is a hydrophilic group. Depending on the nature of X, detergents may be classified as nonionic, cationic, or anionic. In aqueous solution the polar portion of the detergent is hydrophilic and the nonpolar portion is hydrophobic. The result of these antagonistic chemical features is a tendency for cooperative self-association of detergent monomers to form aggregates. The term "micelle" refers to such aggregates of colloidal dimensions, and is generally accepted to mean a situation in which these aggregates are in a mobile equilibrium with the molecules or ions from which they are formed [1].

It is commonly observed that there is a relatively small range of concentrations below which micelles are absent or present in very low concentrations and above which virtually all the detergent molecules exist as micelles. The relatively small range of concentrations defines a critical micelle concentration (CMC), which may be extracted from an experimental plot of some observable property *versus* detergent concentration. Figure 1a shows the conventional representation of the micelle formed by ionic detergents. Although represented as a spherical aggregate, the instantaneous or even the average shape may be ellipsoidal or irregular. The topological shape (b) is preserved, however. The inside or core of the micelle consists of the hydrocarbon

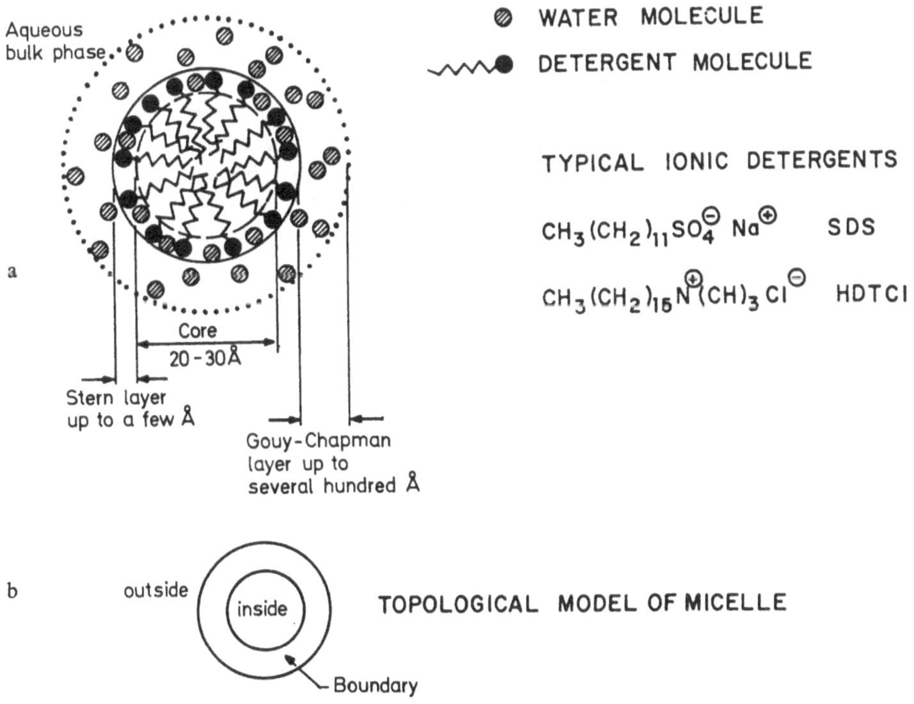

Fig. 1a and b. Schematic representation of a typical ionic micelle

chains of the detergent molecules. The boundary or Stern layer consists of the detergent head group and bound gegenions. The immediate outside or Gouy-Chapman region is a diffuse double layer containing unbound gegenions. Depending on the temperature, concentration, and other experimental variables, the micelle may be roughly spherical, ellipsoidal, disk-like, or rod shaped. At higher concentrations and especially in the presence of additives, liquid crystalline structures are formed.

During the past decade aqueous micellar solutions have been extensively used as media for photochemical reactions and have been found in many situations to dramatically change the nature and/or rate of a reaction compared to homogeneous media [2-7]. Often, the most relevant property of micelles in moderating reactivity is their ability to solubilize hydrophobic molecules in a bulk aqueous solution. Because of the small size of conventional micelles (20–30 Å), this solubilization process allows the organization of solutes on a molecular level. The change in products and/or rates of reactions in micellar systems compared to homogeneous solutions can be used to help identify the mechanism of a reaction. Micelles can also be effective in the catalysis of reactions, thereby making their use attractive in synthetic chemistry. Lastly, because of structural similarities, micelles provide convenient models for both enzymes and bilayer membranes [1f].

This review will focus on a variety of photochemical reactions which have been studied in micellar solutions, and will show how the micelle affects the outcome of these reactions. Photochemical reactions can be altered by solubilization into micellar solutions, such that the products or the relative yield of products can change relative to homogeneous solution. Furthermore, an increase or decrease in the dynamics or the efficiencies of photochemical reactions may also occur upon solubilization in micelles.

In general, there are five common types of effects that micelles can produce on reactions: cage, local concentration, viscosity, polarity, and electrostatic effects. *Cage effects* involve the ability of micelles to hold two reactive intermediates together long enough for reaction to occur. For example, photochemical fragmentation reactions often produce two geminate radicals which can either recombine or diffuse apart. In micelles, large cage effects are observed relative to homogeneous solvents whose magnitude cannot be explained by the microviscosity inside the micelle. The main reason for this observation is that the hydrophobicity of the solutes inhibit diffusion into the aqueous phase, thereby increasing the time spent by the intermediates in the restricted space of the micelle.

Local concentration effects result from the high concentration of solutes which may exist in the small volume of the micelle. This effect is commonly observed for bimolecular reactions between hydrophobic solutes. In some cases, bimolecular photoreactions can occur in micellar solutions where the same total macroscopic concentration in organic solvents would yield no reactions; such phenomena may result simply from a higher local concentration in the micelle.

Microviscosity effects inside a micelle result from an unusually higher viscosity in a micelle than the overall viscosity of the solution. This phenomenon leads to modification of photochemical reactions which are sensitive to viscosity. The "microviscosity" in common micelles composed of ionic detergent typically vary from 15 cP to 100 cP. In some cases, the viscosity effect on a reaction can be used to calibrate the microviscosity of micelles.

Polarity effects exist because the interior of a micelle is less polar the aqueous phase, but more polar (because of water penetration or exposure) than hydrocarbon solvents. Therefore, reactions affected by polarity will have different reactivities in micelles compared to aqueous solution or organic solvents. The actual polarity or viscosity observed by a solute will be a time average value of the various locations which each solute experiences. Therefore, the average location of a solute can be quite different for solutes with different hydrophobicities.

Electrostatic effects result from the presence of charged species associated with the Stern-Layer of micelles. The chemical reactivity can be altered by changing the charge on the surfactant. The phenomenon is also observed for biomolecular reactions where one or more solutes are charged, and is observed for photochemical reactions with charged intermediates.

2 Photophysics

A wealth of information about the structure and dynamics of micellar systems has been obtained with fluorescence and phosphorescence molecules. The use of these

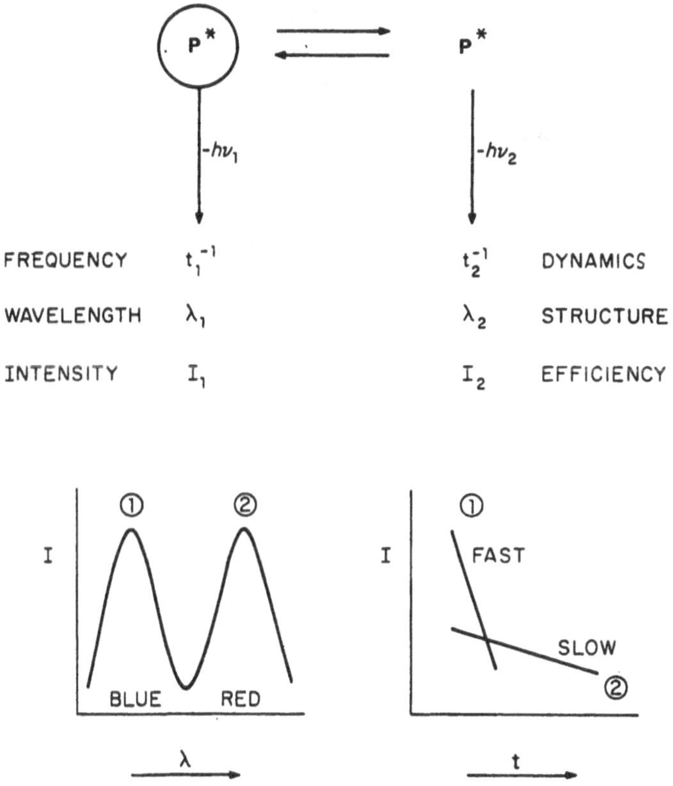

Fig. 2. Schematic representation of a luminescent probe (P*) in micellar ① and aqueous ② environments

photophysical probes as reporters of micellar properties has been previously reviewed [3,8]; therefore, this paper will only briefly discuss the use of these systems. In general, any luminescent molecule whose emission properties are influenced by its environment can be used as a probe of micelles. For example, a luminescent probe may have a different emission spectra, emission lifetime, emission efficiency, and/or emission polarization parameters in the micellar phase compared to the aqueous phase (Fig. 2). Measurable differences in emission properties can be related to molecular properties of the various environment which are compared.

The micropolarity that a probe experiences in a micelle has been determined by changes in emission spectra [7a)], fluorescent lifetime [7b)], and quantum yield [7c)]. By comparing the polarity of several probes, the relative site of solubilization can be deduced. The microviscosity in a micelle can be determined by excimer formation [7d)], and by fluorescence depolarization [7e)]. The environments reported by a probe will be an average of the environments experienced by the probe during its excited state lifetime. If diffusion of solutes during the lifetime of a probe is small, one obtains a measure of the instantaneous environments experienced by the probe, and not an average of the various regions occupied by the probe.

The rates of diffusion of solutes and surfactants in and out of micelles have been measured using photophysical techniques. The most commonly used method is to measure the deactivation of excited states of the probe by added quenchers, which are only soluble in the aqueous phase. The measurement of either the decrease in emission intensity or a shortening of the emission lifetime of the probe can be employed to determine exit and entrance rates out of and into micelles [7d)]. The ability of an added quencher to deactivate an excited state is determined by the relative locations and rates of diffusion of the quenchers and excited states. Incorporation of either the quencher or excited state into a surfactant allows one to determine the rates of diffusion of surfactants. Because of the large dynamic range available with fluorescent and phosphorescent probes (Fig. 3), rates as fast as

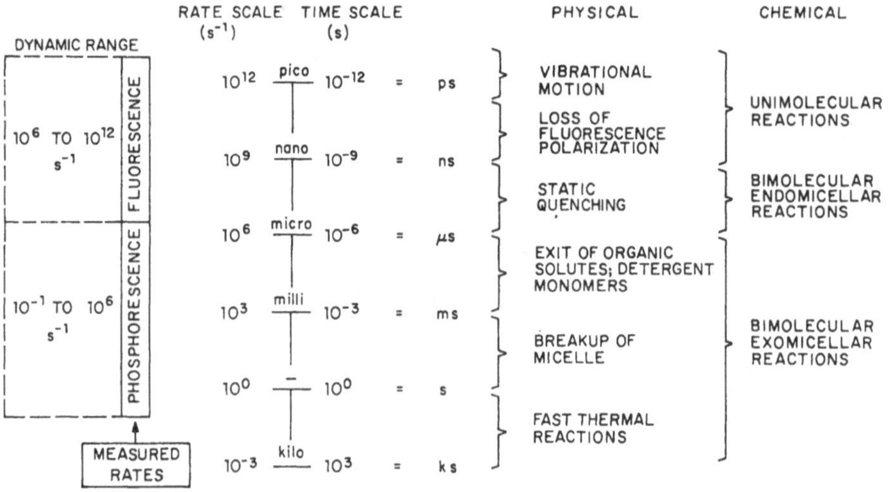

Fig. 3. The dynamic range for fluorescent, phosphorescent, and photochemical probes

diffusion of a probe from one part of the micelle to another (ps to ns) and as slow as the rate of complete break-up of the micelle (ms to s) can be measured. In some cases, the quenching process leads to either excimer or exciplex emission [8]. Studying the kinetics of appearance and disappearance of excimer emission after pulsed excitation of a monomer can extend the type of rates that are measurable with a single system [8].

Photochemical reactions in micelles may also be employed as probes of the structure and dynamics of the micelle. In general, any change in a reaction in micelles compared to organic solvents or aqueous solutions, indicates a difference in the nature of the media. From these studies, various microscopic properties of the micelle can be inferred. For example, the site of solubilization of the solute, its molecular motions within a micelle, and its rate of diffusion in and out of the micelle can be deduced by studying reaction dynamics. Structural information, such as the framework of the micelle in terms of the location of the charged head groups to the hydrophobic chains, and the amount and depth of penetration of water into the hydrophobic regions can be inferred by monitoring the reaction efficiency, the product ratios, or the formation and decay of reaction intermediates.

The use of chemical reactions to deduce the location of reagents in the micelle aggregate is a complicated business and is fraught with difficulties. It must be remembered that probes in micelle aggregates are dynamic species, from the standpoint of detergent and probe exchange, from the standpoint of detergent conformation changes and probe positional changes within the aggregate and from the standpoint of complete disintegration of a probe bound aggregate. However, since the timescales of each of these various processes are different and span the range of nanoseconds to seconds, the *lifetime* of an electronically excited probe may be a critical factor in determining the response of the probe to the environment. For example, probes that have lifetimes of the order of several nanoseconds do not stray far from the sites that they occupy just before excitation, i.e., the positions of the probe in the ground state. On the other hand, probes that persist for milliseconds may have experienced motion throughout the entire micellar volume and even may have exchanged many times between various micelles.

The interpretation of rate (or product) data in terms of probe location in a micelle is similar conceptually to the interpretation of rate (or product) data in terms of conformational shapes. In effect, the various positions of a probe in a micelle can be viewed as "conformations" of the probe micelle associate. In this regard, the Curtin-Hammett principle [9a,b] states that if two or more conformations are in rapid equilibrium relative to reaction from either shape, and in which each form gives a characteristic product (or if only one form yields a product), the ratio of products (or the rate of formation of a single product) is independent of the energy difference between the conformers and depends only on the relative free energy levels of the transition statutes.

Applied to the use of reactions to locate the position of a probe in a middle, the Curtin-Hammett principle suggests that not only the site of the probe but its reactivity at various sites must be considered. For example, if the reactivity of a probe at various sites along a detergent chain were independent of site, reactivity would reflect probe position, i.e., the extent of product formation would related to the "concentration" of probe at each site. If on the other hand, if the reactivity of the probe were much greater near the head of the detergent chain or near the tail of the detergent chain, the extent of product formation would no longer be related to the "concentration" of probe along the chain. For example, if 1% of the probe is located on the average near the detergent head, but the reactivity is 10^4 times greater near the detergent head than it is at other portions of the detergent, nearly all of the probe products will be generated near the detergent head, in spite of the low steady state average concentration of probe near the head.

3 Photochemical Reactions

3.1 Norrish Type I Reactions

Homolytic α-cleavages (Type I reactions to produce a radical pair) involving the n, π* states of ketones, have been studied extensively in homogeneous solutions [9c)], but only recently have ketones been employed to investigate this important class of photoreactions in micellar environments. For ease of discussion, we can classify reactions of this type in terms of the resulting products:

1) photodecarbonylation, which involves the loss of carbon monoxide from the acyl fragment to yield radicals (Scheme I) which form, in turn, coupling products;
2) photofragmentation, which involves internal hydrogen transfer from a radical pair (disproportionation) (Scheme II).

$$ACH_2COCH_2B \rightarrow ACH_2\dot{C}O\ \dot{C}H_2B \rightarrow ACH_2\cdot\ \cdot CH_2B \rightarrow ACH_2CH_2B$$

(Scheme I)

$$AC(CH_3)_2COB \rightarrow AC(CH_3)_2\quad COB \rightarrow AC\begin{matrix}\diagup CH_3 \\ \diagdown\!\!\diagdown CH_2\end{matrix} + HCOB$$

(Scheme II)

We shall now show how the reaction dynamics and reaction products can be affected by micellar environments, how the knowledge of the reaction mechanism in homogeneous solution can assist in determining the reaction mechanism in micelles and how understanding of the mechanism in a micelle can be employed to enhance our understanding of the structure and dynamics of micelles.

3.1.1 Photodecarbonylations

In homogeneous solutions, the photochemistry of dibenzyl ketone (DBK, *1*) and substituted DBK's proceeds through the mechanism illustrated in Scheme III [10)]. The quantum yield for the formation of diphenyl ethylene (DPE, *2*) is quite high

$$\underset{1}{PhCH_2COCH_2Ph} \xrightarrow{h\nu} PhCH_2\dot{C}O\dot{C}H_2Ph \xrightarrow{-CO} Ph\dot{C}H_2\dot{C}H_2Ph \downarrow \underset{2}{PhCH_2CH_2Ph}$$

(Scheme III)

(*ca.* 0.7) in homogeneous solution, so that cage recombination of the original radical pair cannot be a highly efficient process. In homogeneous solutions unsymmetrical DBK's have been found to form products (via a free radical mechanism) of all possible combinations in the ratios one would expect for statistical coupling of spin non-correlated radicals (Scheme IV). Therefore, the solvent cage effect in homogeneous solutions is *ca.* 0%. The cage effect remains *ca.* 0 and is not sensitive to the viscosity of the solvent over the range of viscosities of 0.6 (benzene) to 41 (cyclohexanol)[10c].

ACOB →	AA	+AB	+BB
homogeneous	25%	50%	25%
HDTCl	12%	76%	12%

(Scheme IV)

However, in aqueous solution containing hexadecyltrimethylammonium chloride (HDTCl) micelles, the ratio of products favors the coupling of the initially formed radical pair, resulting in a cage effect of ~50% (cage = AB − [AA + BB]/[AA + AB + BB])[11]. As illustrated in Fig. 4, the cage effect is highly dependent on the concentration of surfactant, with a sharp increase in the cage occurring at the critical micellar concentration (CMC) of the surfactant. The reason for the dramatic increase in the cage effect with the increase in the surfactant concentration is attributed to an important property of the micelle, i.e., the ability to sequester small organic molecules in its hydrophobic core for a period of time from microseconds to milliseconds[12]. In homogeneous solvents, the analogous cage of solvent

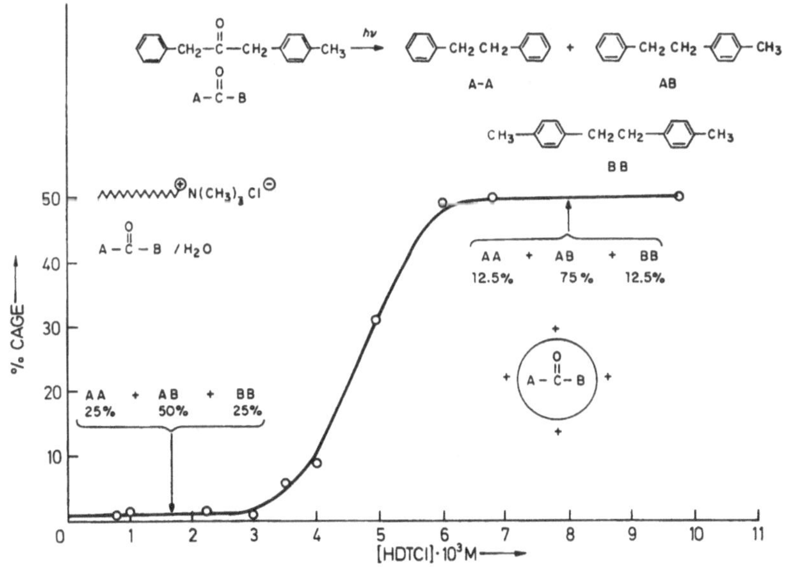

Fig. 4. The cage effect for benzyl-1-methylbenzyl ketone a function of concentration of the surfactant HDTCl shows a typical CMC profile

Table 1. Quantum Yields for Photolysis of Dibenzyl Ketones in HDTCl [19, 26]

Ketone	Φ_{-K} [a]	Cage effect [b]	
		0 Gauss	kGauss
DBK	0.3	33%	16%
DBK-1-^{13}C	0.22	33%	16%
DBK-2,2'-^{13}C	0.25	46%	22%
DBK-2,2'-$^{2}H_4$	0.32	29%	14%
DBK-$^{2}H_{10}$	0.32	28%	N/A
DBK-$^{2}H_{14}$	0.33	27%	N/A
4-Me-DBK	0.16	59%	31%
4-Cl-DBK	0.21	52%	3%
4-Br-DBK	0.11	70%	70%
4,4'-diMe-DBK	0.16	59%	31%
4,4'-di-tbutyl-DBK	0.13	95%	76%
1,2-diphenyl-2-methylpentanone	—	71%	52% [c]

[a] Quantum yield for disappearance of ketones in the absence or presence of $CuCl_2$ as scavenger.
[b] Cage effects refer to the efficiency of combination of geminate benzyl radical pairs.
[c] In the presence of a 3 kGauss magnetic field.

molecules last for *ca.* 10^{-10}–10^{-11} sec. Thus, for radical reactions, the probability of spin-correlated events occurring is increased because diffusional separation of the radical fragments and the formation of spin uncorrelated free radicals is inhibited.

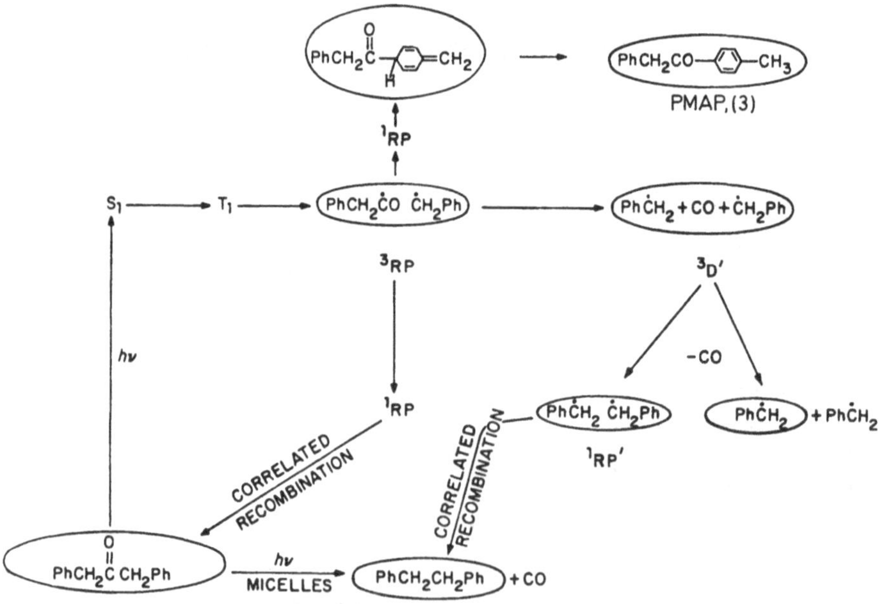

Fig. 5. The mechanism for the photodecomposition of dibenzyl ketone (DBK) in micellar solution

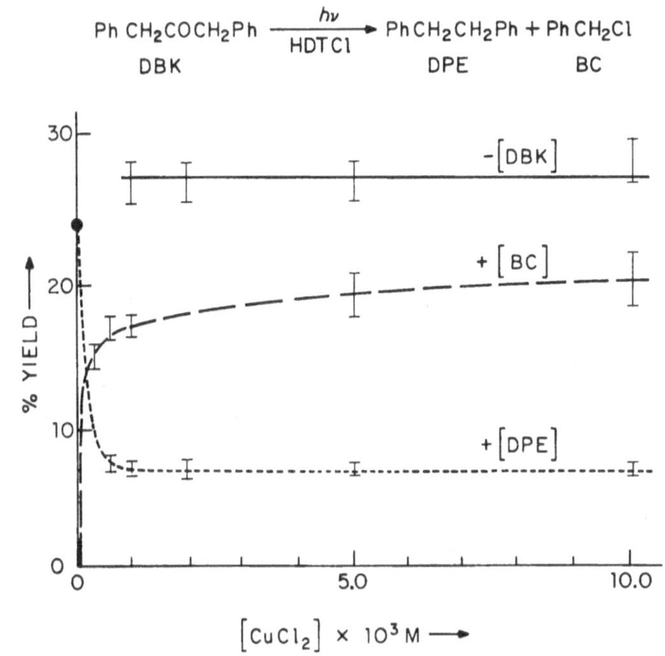

Fig. 6. Plots of the disappearance of dibenzyl ketone (DBK) and the appearance of benzyl chloride (BC) and diphenyl ethane (DPE) as a function of $CuCl_2$ concentration

Above the CMC the quantum yield for the formation of DPE is approximately $^1/_3$ the value in benzene [13,14]. (Table 1).

A minor product observed in the photolysis of DBK in aqueous HDTCl (Fig. 5) is the isomeric ketone 4-methylphenylacetophenone (PMAP, *3*) [15-18]. The question arises as to whether the formation of *3* is due to some special orienting feature of the micelle which operates on the ratio of coupling products *1* and *3*. Although *3* has not been detected by the conventional analytical methods for product analyses, (e.g., vpc), some *3* is detectable via chemically induced dynamic nuclear polarization (CIDNP) measurements [28]. Thus, it may be that the *ratio* of formation of *1* to *3* is similar in both micellar and homogeneous solutions, but that the absolute yield of both combination products is so much smaller in homogeneous solution that *3* is not detectable by conventional analyses.

Since the cage effect is less than 100% in most cases (Table 1), some of the radical pairs exit the micelle and become water solubilized free radicals. At some later time, the free radicals recombine to form DPE. This has been verified by Cu^{2+} quenching experiments (Fig. 6) [13,19]. The disappearance of DBK is not dependent upon the concentration of Cu^{2+}. However, the yield of DPE drops very rapidly with increasing copper concentration and then levels off. The leveling off region is directly related to the amount of cage reaction. The products formed by reaction with copper are benzyl chloride (*4*), and benzyl alcohol (*5*) (Scheme V) [20].

$$\underset{1}{PhCH_2COCH_2Ph} \xrightarrow[h\nu]{CuCl_2,\ HDTCl} \underset{4}{PhCH_2Cl} + \underset{5}{PhCH_2OH} + \underset{2}{PhCH_2CH_2Ph}$$

(Scheme V)

Since the quantum yield for disappearance of ketone (Φ_{-K}, Table 1) is independent of copper, it is concluded that neither the triplet ketone nor the primary radical pair $PhCH_2\dot{C}O\ \dot{C}H_2Ph$ is scavenged. As stated above, the proposed mechanism for the photolysis of DBK in micellar solution is illustrated in Fig. 5 [13,21]. The micellar environment inhibits the diffusion of radicals to the bulk aqueous phase: the radical pair's distance maximum separation is maintained to a few tens of angstroms or less. The amount of escape being reduced, the radicals can then undergo more efficient·intersystem crossing and recombination.

As seen in Fig. 5, the mechanism for α-cleavage proceeds through the triplet (n, π*) state of the ketone, and in accordance with Wigner's spin conservation rule [22], the initial radical pair formed is expected to be a triplet. This occurs in both homogeneous and in micellar solutions. In homogeneous solutions, the rate of intersystem crossing, k_{TS}, is much less than the rate of diffusion, k_{diff}, to form free radicals, and therefore the cage reaction is minimal. But due to the "restricted space" of the micellar environment, k_{diff} is reduced considerably (it actually equals the exit rate from the micelle, i.e., 10^6–10^7 s^{-1}) so that the probability for the formation of singlet intramicellar radical pairs and recombination is higher.
formation of geminate singlet intramicellar radical pairs and recombination is higher.

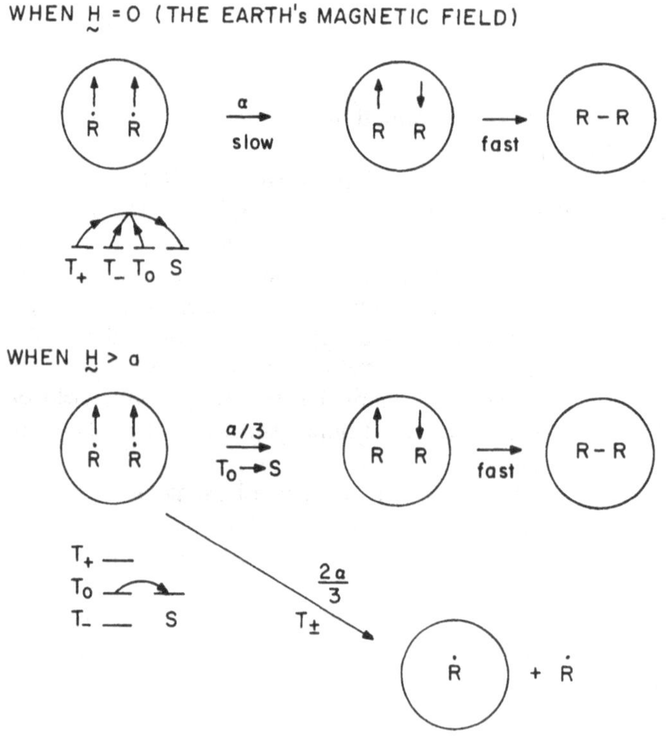

Fig. 7. A schematic description of the influence of an applied magnetic field on the cage effect of geminate triplet radical pairs. In the absence of the field, all three triplet levels can undergo intersystem crossing to yield singlet radical pairs. At high fields ($H >$ a, where a is the hyperfine coupling constant) only T_0 can intersystem cross to singlet

The "restricted space" concept of micelles provides a key towards understanding the mechanism of intersystem crossing in radical pairs [19,20]. Figure 7 shows a schematic description of the intersystem crossing mechanism of radical pairs in a micelle. If the secondary recombinations are cut-off by the addition of a free radical scavenger (e. g., Cu^{2+}), then the DPE observed is a result of the geminate recombination of radical pairs in the micelle. It has been demonstrated that the intersystem crossing mechanism is due to hyperfine (electron-nuclear) interactions [20,23]. In the vector model of spin correlated radical pairs, there are three possible triplet pairs: T_+, T_-, and T_0. In the absence of an external magnetic field, the hyperfine mechanism operates most effectively; the triplet pairs are degenerate in energy with the singlet, and therefore crossing occurs from all three states. In the presence of a magnetic field, the T_- and T_+ states are split in energy from T_0 which remains degenerate with the singlet state S (the magnitude of the splitting is determined by the strength of the field). As a result, intersystem crossing occurs from only T_0. Since intersystem crossing is cut-off from T_+ and T_-, these two states remain as radical pairs whose fragments are able to escape into the aqueous phase and then react with scavenger. The net result is a reduction in the yield of DPE. Table 1 summarizes the values of cage effects obtained for various substituted DBK's. Note that in all cases the value of the cage is decreased in the presence of a magnetic field.

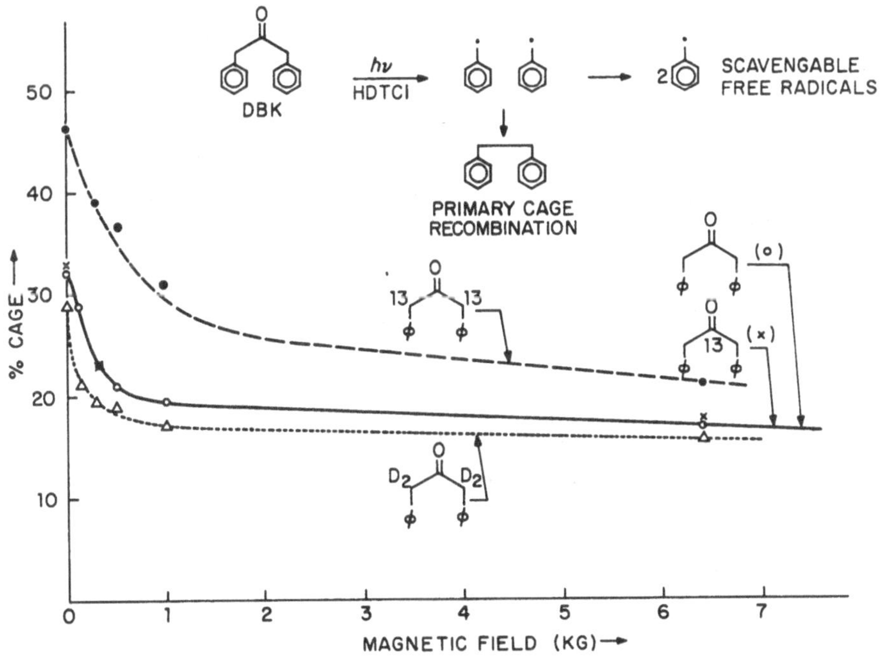

Fig. 8. The cage effect (measured in the presence of $CuCl_2$) for the formation of DPE as a function of magnetic field for DBK and various isotopically substituted isomers

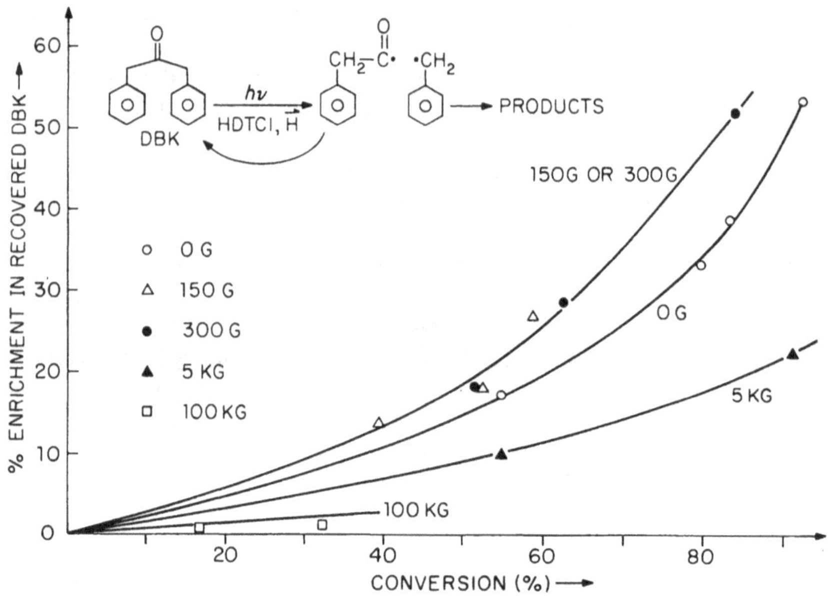

Fig. 9. The percent enrichment of ^{13}C at a single carbon of DBK as a function of conversion and magnetic field strength

An interesting and potentially useful application of micelles and magnetic field effects is isotope enrichment [11–18, 24]. As discussed above, in the case of reactions proceeding through radical pairs in micelles, the hyperfine mechanism is most effective. Since ^{13}C nuclei possess magnetic moments and ^{12}C nuclei do not possess magnetic moments, radical pairs possessing ^{13}C undergo intersystem crossing at a more rapid rate than radical pairs containing only ^{12}C nuclei. Therefore, the recovered starting material should be enriched in ^{13}C after photolysis because of the more efficient combination of pairs containing ^{13}C. In practice, it has been shown that this result occurs for a variety of DBK's. The enrichment factors which measure the efficiency of ^{13}C separation are quantitatively defined by the isotope enrichment factor, α, which is related to the quantum yields for disappearance of DBK by:

$$\alpha = \frac{\text{rate of disappearance of } ^{12}\text{C ketone}}{\text{rate of disappearance of } ^{13}\text{C ketone}} = \frac{\Phi(-\text{DBK}^{12}\text{C})}{\Phi(-\text{DBK}^{13}\text{C})} \qquad (1)$$

Table 2 summarizes the α factors for various substituted DBK's. Furthermore, this table shows, as is expected, that the application of an external magnetic field reduces the amount of isotope enrichment. Figure 8 and Fig. 9 illustrate the effect of the application of an external magnetic field on the extent of the cage effect and on the efficiency of isotope enrichment. The results can be understood in terms of a change in the relative contributions in the intersystem crossing mechanism of the hyperfine and Zeeman mechanisms [14].

Table 2. The Isotope Enrichment Factor, α, for Various Dibenzyl Ketones

Ketone	Solvent	α[a]
DBK[b]	Benzene	1.04
DBK-1-^{13}C (90%)	0.5 M HDTCl (0.5 G)	1.37
DBK-1-^{13}C (90%)	0.5 M (300 G)	1.53
DBK-1-^{13}C (90%)	0.5 M HDTCl (14,000 G)	1.16
DBK-2,2'-^{13}C (90%)	0.5 M HDTCl (0.5 G)	1.20
DBK[b]	0.05 M HDTCl (0.5 G)	1.47
DBK[b]	0.105 M HDTCl (15,000 G)	1.12

[a] Determined by quantum yield measurements according to equation (1). See text.
[b] Natural abundance.

This change is also reflected in the amount of cage in the absence and presence of an external magnetic field as seen in Table 1. The presence of a magnetic nucleus such as ^{13}C affects significantly the zero field cage when this nucleus is at the α-position to the carbonyl. Again, this illustrates the effect of the hyperfine coupling in the intersystem crossing mechanism. This is also evident in the case where deuterium is substituted for hydrogens on the α-carbon. When spin-orbit coupling is large, as in the case of halogen substituted DBK's intersystem crossing proceeds through a

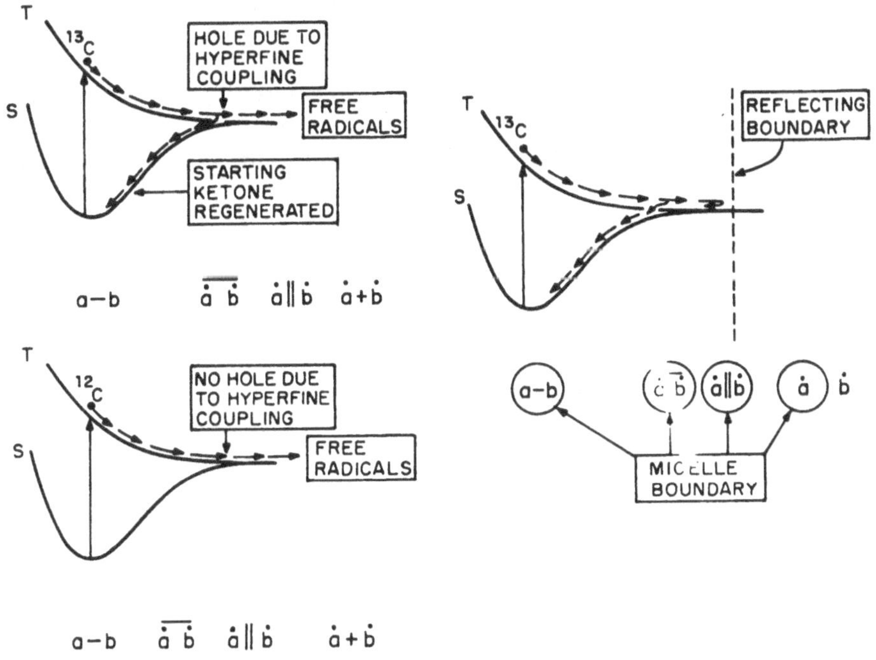

Fig. 10. A schematic representation of the basis of the ^{13}C isotope enrichment mechanism and the role micelles play in this mechanism

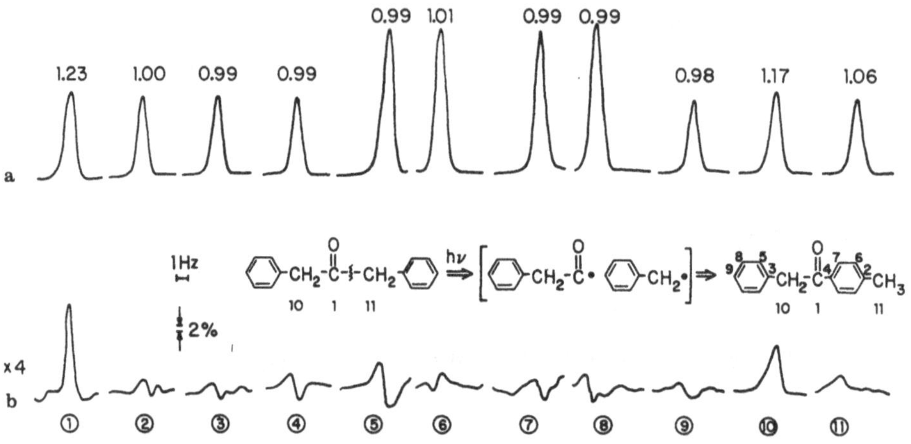

Fig. 11a and b. **a** NMR peaks of PMAP after zero-filling and Gaussian convolution. The numbers over each peak is its isotope ratio (see Table 4). **b** The difference between analytical and standard spectra. The net area indicates the enrichment at each carbon

different mechanism, i.e., one which is not sensitive to an externally applied magnetic field, and therefore, little or no reduction in the amount of cage is observed. Only the study of these processes in micelles allows this interpretation of the intersystem crossing mechanism.

The role played by the micelle in the intersystem crossing mechanism is illustrated schematically in Fig. 10. It is the "reduced dimensionality"[20] of the micellar environment which provides a reflecting boundary which gives the radical pairs multiple changes in case the "hole" which corresponds to the proper geometry for intersystem crossing is overshot[11,14,[illegible]]. Pure viscosity effects in homogeneous appear to play only a relatively minor role in the enrichment mechanism (Table 3)[15,25].

^{13}C enrichment in the minor product PMAP has also been observed. The degree of enrichment is proportional to the relative ^{13}C hyperfine coupling constants (see Table 4 and Fig. 11)[24].

Table 3. Single Stage ^{13}C Isotope Enrichment Parameters for Photolysis of Dibenzyl Ketone Under Various Conditions

Ketone	Solvent	α^b	Viscosity (cP)
DBK[a]	0.05 M HDTCl	1.68	30–40
DBK-1-^{13}C (25.4%)	Benzene	1.03	0.6
DBK-1-^{13}C (10.2%)	Benzene	1.04	0.6
DBK-1-^{13}C (10.2%)	n-Dodecane	1.05	1.35
DBK-1-^{13}C (10.2%)	Cyclohexanol	1.075	60

[a] Natural abundance.

[b] At 0 G (earth's magnetic field).

Table 4. Relative ^{13}C Isotope Ratios for Pmap by ^{13}C NMR

Carbon No.[a]	δ_c (ppm)[b]	S[c]	a_c (Gauss)	% Enrichment[d]
1	197.40	1.23 ± .01	+124	55
2	144.12	(1.00)	− 14	5 (2, 3)
3	134.93	0.99	0	
4	134.25	0.99	+ 11	5
5	129.59	0.99	0	5 (5, 6)
6	129.48	1.01	+ 12	
7	128.91	0.99	− 9	0
8	128.79	0.99	0	0
9	126.95	0.98	0	0
10	45.60	1,17	+ 51	30 (10, 11)
11	21.84	1.06	+ 24	

[a] See Fig. 11 for carbon numbering.
[b] Chemical shifts relative to $\delta_C^{CDCl_3} = 77.27$ ppm.
[c] Ratio of ^{13}C NMR intensities for analytical and standard samples divided by the ratio for C_2.
[d] See Fig. 11.

Photolysis of 2,4-diphenylpentan-3-one (meso and d, l, *6*) produces different ratios of products[26)] in homogeneous and aqueous micellar solutions (Scheme VI). While DPE is the major product in both homogeneous and micellar solutions, only in micelles is the recovered starting material isomerized (meso → d,l; d,l → meso). This indicates that while diffusion out of the micelle is hindered, rotation and diffusion inside the micelle is still facile.

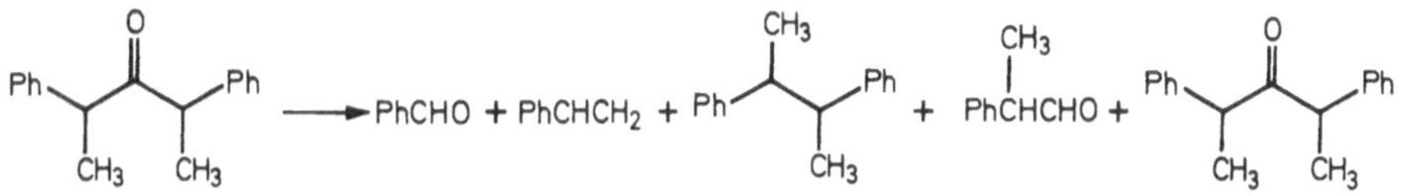

	6 (meso or d,l)	7	8	9	10	6 (d,l or meso)
Pentane		3 %	3 %	93 %	0.5 %	—
HDTCl		8 % (7 + 8)		72 %	3 %	8 %

(Scheme VI)

CIDNP studies have proven to be a valuable tool in investigating the mechanisms of decarbonylation and disproportionation reactions in micelles[27−29)]. Since the mechanisms involve the formation of triplet radical pairs, nuclear polarization of the protons near the radical centers occurs and results in the observation of emission or enhanced absorption in the NMR spectra of products of the radical pairs. For example, the photolysis of di-t-butyl ketone (*11*) in HDTCl yields both decarbonylation and disproportionation products (Scheme VII)[27,29)]. The CIDNP spectra (Fig. 12) taken at various concentrations of copper chloride (free radical scavenger) illustrates that the intramicellar product is isobutylene (*12*), while 2,2,4,4-tetramethylbutane (*13*) and 2-methyl-propane (*14*) are the extramicellar products.

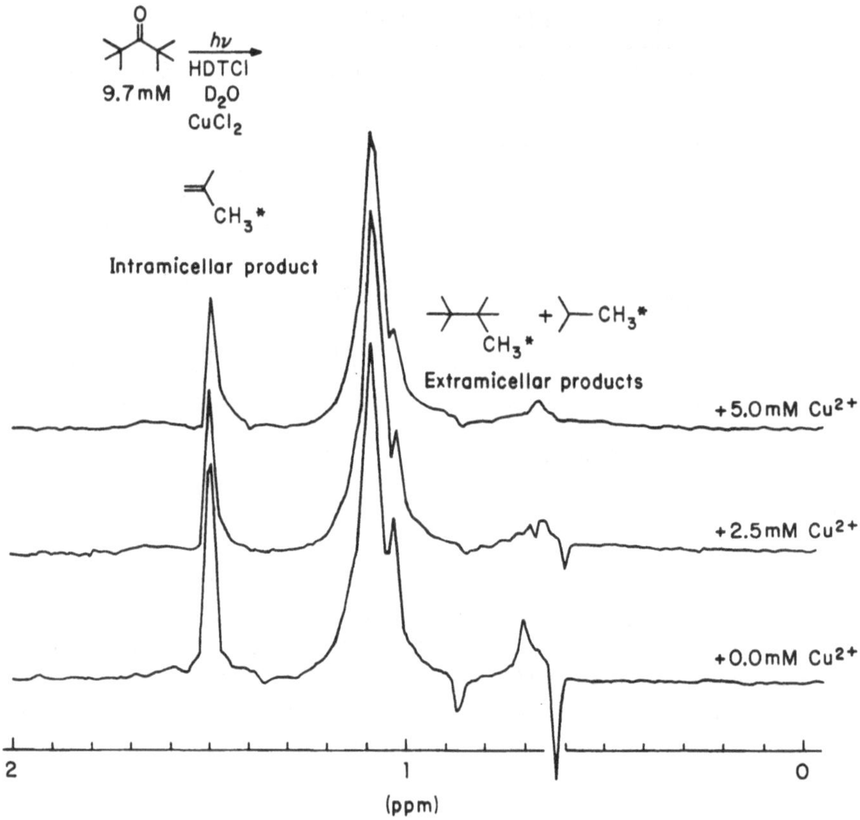

Fig. 12. The steady state CIDNP spectrum of di-t-butyl ketone in HDTCl in the presence of increasing amounts of $CuCl_2$ (free radical scavenger)

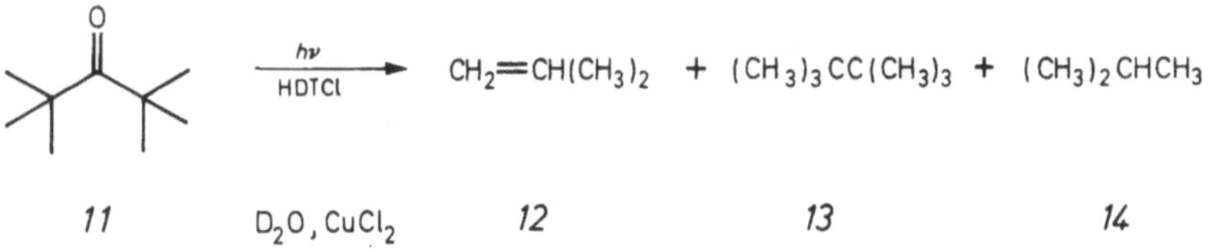

(Scheme VII)

3.1.2 Photofragmentation

When radical pairs are produced by α cleavage of ketones, and the loss of CO is prevented on energetic grounds, fragments of the reactions of the primary radical pair become important. Deoxybenzoins (*15*)[30] or adamantyl ketones (*20*)[25,30] provide examples of such systems. The photodecomposition of 1,2-diphenyl-2-methyl-propanone (*15*)[30] illustrates an example of a photofragmentation in reaction (Scheme VIII). The ratio of the major products are highly dependent on solvent environment. For example, the photochemistry of this compound in acetonitrile or benzene yields benzaldehyde (*16*) and dicumyl (*9*) as the major products with α-methyl styrene (*17*) and benzil (*18*) being minor products, while in HDTCl α-methyl styrene and benzaldehyde (disproportionation products) are the major products. This indicates that the micelle sequesters the initial radical pair so that disproportionation is strongly favored, i.e., the micelle keeps the initially formed spin correlated triplet radical pair together long enough for intersystem crossing to the singlet to occur. This fact is confirmed by the absence of a significant Cu^{2+} quenching effect on the formation of the major products in HDTCl.

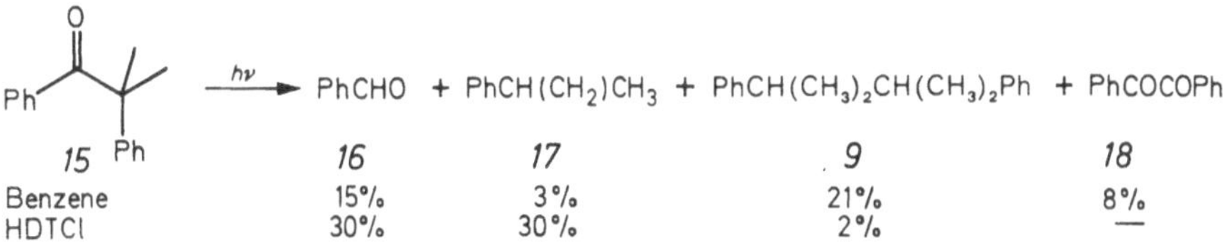

(Scheme VIII)

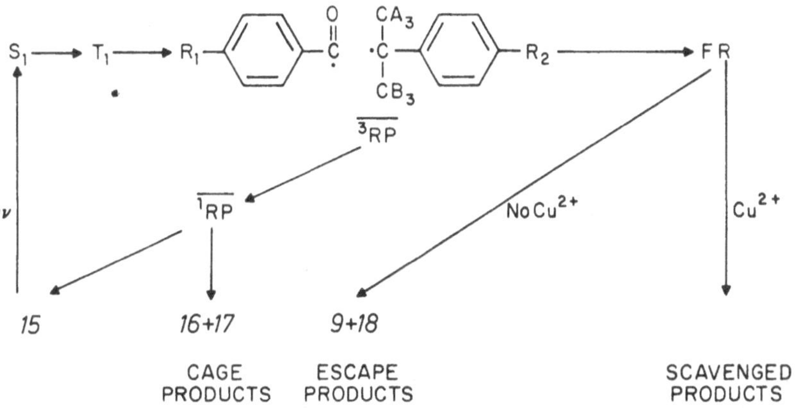

Fig. 13. The mechanism for the photodecomposition of deoxybenzoins in micellar solutions

The mechanism of photolysis of *15* follows the general lines of radical pair formation from a triplet ketone (Fig. 13). Magnetic field and cage effects in the photolysis of *15* are very similar to those found for DBK's (Fig. 14), so that the mechanistic arguments used for DBK's apply equally as well to deoxybenzoins [36], such as *15*. The cage effect for deoxybenzoins in the presence of an external magnetic field and copper quencher decreases as expected due to the decrease in the rate of intersystem crossing [30]. The triplet radicals, unable to undergo intersystem crossing, diffuse into the aqueous phase and are scavenged by copper. An interesting deuterium isotope effect is also observed. When the hydrogens on the α-methyl groups are substituted with deuterium, a decrease in the cage is observed which is proportional to the degree of deuterium substitution. This is presumably due to the less efficient hyperfine coupling in the deuterated radical pair. In contrast, the ^{13}C isotope effect is reflected by an increase in the cage whereby the magnetic isotope increases the rate

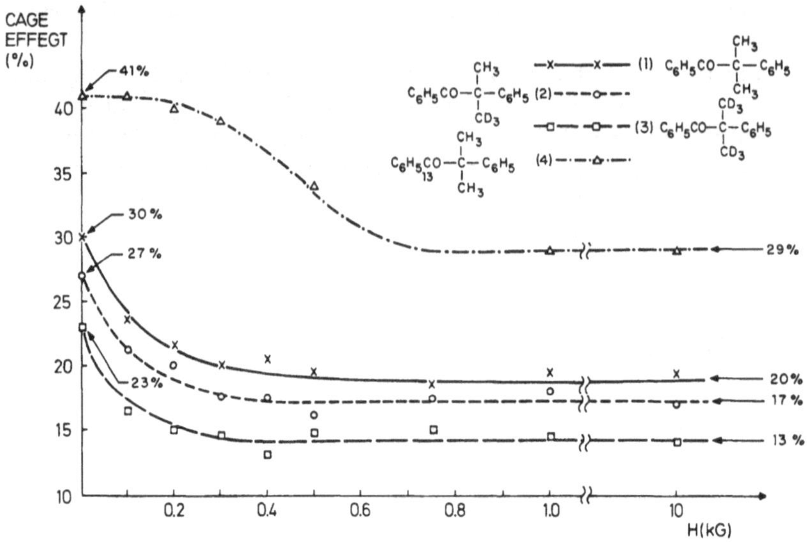

Fig. 14. The cage effect for deoxybenzoins and various isotopically substituted isomers as a function of magnetic field strengths

of intersystem crossing by electron-nuclear interactions to give singlet radical pairs. In all cases, the singlet radical pair can give recombination to starting ketone or disproportionation.

The effect of the micelle environment on the photolysis of adamantyl ketones is dramatic[31]. The major products in the photolysis of phenyladamantyl ketone (*20*) in benzene are shown in Scheme IX. However, in HDTCl only (*22*) is formed in significant amounts. Minor products are (*16*) (7%) and two new products: AdOH (*25*) (3%) and benzil (*18*) (3%). The appearance of large amounts of adamantane

PhCOAd $\xrightarrow{h\nu}$	PhCHO +	PhCOPh +	AdH +	PhAd +	AdAd +	AdOH +	PhCOCOPh
benzene	75%	6%	33%	35%	3%	—	—
HDTCl	7%	—	60%	—	—	3%	3%
20	*16*	*21*	*22*	*23*	*24*	*25*	*18*

(Scheme IX)

(*22*) (60%), but very little amounts of products incorporating the acyl radical, is attributed to the reaction of the acyl radical with the detergent backbone. This is also reflected in the poor mass balance in the reaction. Addition of copper as a water phase quencher drastically changes the products observed after photolysis (Scheme X). The change is understood if one assumes that in the absence of copper, the stable adamantyl radicals escape their original micelle, and at a somewhat later time, abstraction of a hydrogen occurs, but in the presence of copper the adamantyl radicals that escape from micelles are scavenged by copper.

$$\underset{20}{PhCOAd} \xrightarrow[h\nu]{CuCl_2,\ HDTCl} \underset{\substack{5\,\% \\ 22}}{AdH} + \underset{\substack{38\,\% \\ 25}}{AdOH} + \underset{\substack{33\,\% \\ 26}}{AdCl} + \underset{\substack{45\,\% \\ 27}}{PhCO_2H}$$

(Scheme X)

Phenyladamantyl ketone, when irradiated to 90% conversion, yields an enrichment in the ^{13}C content in both the products and recovered starting material. The recovered starting material is 200% enriched in ^{13}C content; benzaldehyde is 160% enriched, and adamantane in 180% enriched (calculated assuming selective enrichment at the benzaldehyde CO carbon and the 1-carbon of adamantane)[23]. The role of the micelles in this reaction provides "reduced dimentionality" allowing recombination of the radical pairs to occur over diffusional processes. By comparison, viscosity has only a slight effect on the isotope enrichment (Table 5).

Table 5. Effect of Solvent Viscosity on the ^{13}C Enrichment of PAK by Photolysis

	Solvent	Viscosity (cP)	α
PAK	Cyclohexane	0.9	1.02
	n-Dodecane	1.35	1.06
	Cyclohexanol	60	1.17
	HDTCl	30–40	1.75

3.2 Norrish Type II and Hydrogen Abstractions

Norrish Type II reactions have been the center of many studies in recent years, and therefore are one of the best understood photochemical reactions in homogeneous solutions. In general, the triplet n,π* state of carbonyl compounds shows a propensity to abstract a hydrogen from any number of donor molecules such as aliphatic hydrocarbons, to form ketyl radicals. It is particularly well documented that aryl alkyl ketones show a remarkable solvent dependence for the Type II process, i.e., the efficiency of the hydrogen abstraction and ratio of cyclobutanol isomers being higher in more polar solvents due to hydrogen bonding. Therefore, the study of such processes in micelles should provide much useful information on the interaction of the micelle with the solubilized ketones and reaction intermediates and the effect of the local environment on the efficiencies of this reaction.

The earliest report on the effect of micelles on Type II reactions dealt with the dependence of the ratio of disproportionation to cyclobutanol formation for the intramolecular hydrogen abstraction reaction of octanophenone (*28*: $R=C_4H_9$) and valerophenone (*28*: $R=CH_3$) (Scheme XI)[32]. The reaction in homogeneous solutions is solvent polarity dependent with cleavage and cyclization being favored over reversion to starting material in more polar solvents. For example, the quantum yield for decomposition of octanophenone increases from 0.29 in benzene to 1.00 in t-butanol. For HDTCl, the quantum yield for decomposition of the ketone is 0.71.

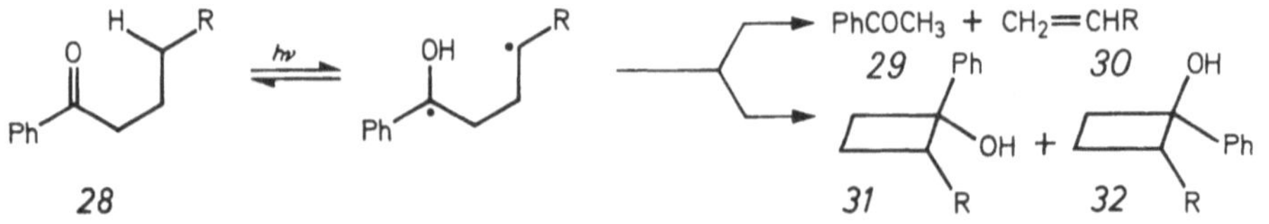

(Scheme XI)

Assuming that polarity is the major factor in determining the differences in the quantum yields, this result indicates that the ketone sees a more polar environment than benzene, but less polar than t-butanol. For valerophenone, the quantum yield for decomposition is equal to the value measured in t-butanol. This is consistent with the molecule being more polar than octanophenone, since the former has three less methylene groups. Europium quenching studies indicate that no significant amount of diradical escapes into the water layer, since no decrease in the amount of product is observed. In both cases the results indicate that the carbonyl functionality lies near the Stern Layer.

In homogeneous solution, the ratio of the cyclobutane isomers [(*31*)/(*32*)] is found to be dependent on the solvent polarity, and therefore is an indicator of the polarity which the solvated molecules sees. For example, the ratio of [(*31*)/(*32*)] is 4.7 in benzene, while this ratio decreases to 1.1 in the more polar solvent t-butanol. Therefore, the ratio found in HDTCl, i.e., 1.2, is consistent with the conclusion that the ketone is solubilized near the polar Stern layer of the micelle.

The quantum yield for the decomposition of the surfactant ketone 16-oxo-16-(p-tolyl)-hexadecanoic acid (*33*) increases when co-dissolved with sodium dodecyl sulfate (SDS) in comparison to benzene solution (0.8 *vs.* 0.2). The same effect is also

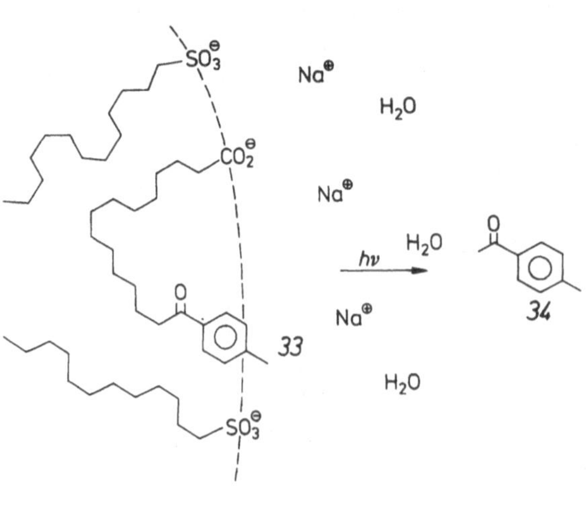

Fig. 15. The decomposition of the surfactant ketone 16-oxo-16-(p-tolyl)-hexadecanoic acid (33) in SDS micelles lead to the formation of acetophenone which escapes the micelle

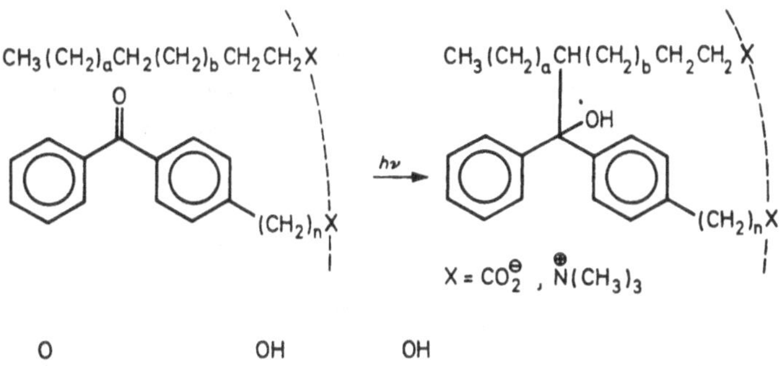

(Scheme XII)

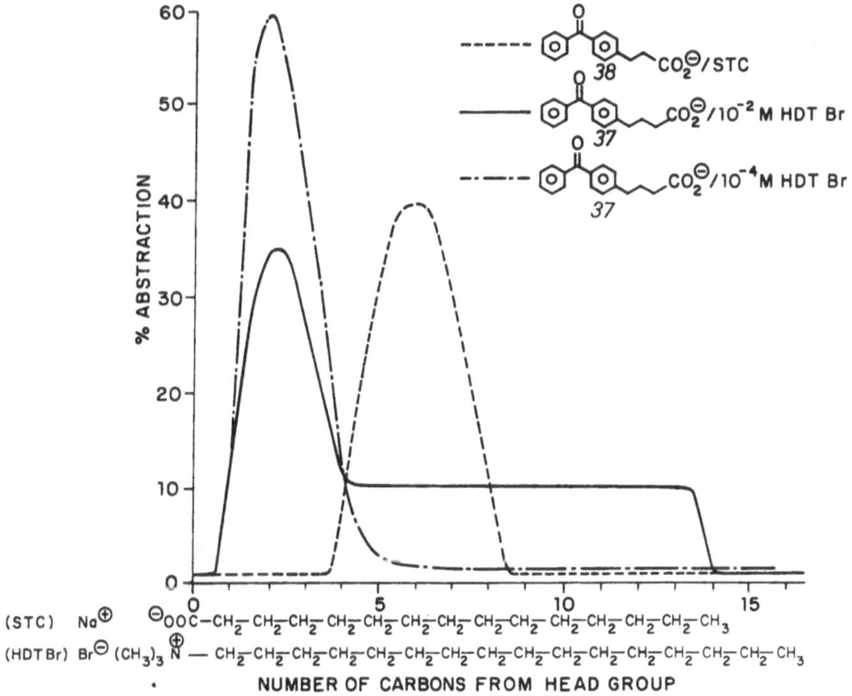

Fig. 16. The relative amounts of hydrogen abstraction as a function of the number of carbons removed from the head group for the carboxylic acid (37) in HDTBr and the carboxylic acid (38) in sodium tetradecanoate

observed for the probe surfactant with ten methylene groups. In contrast, monolayers assemblies severely restrict the Type II processes [2]. These results are interpreted as indicating that both the ketone and acid moieties lie near the Stern layer and experience a polar environment (Fig. 15). The steric requirements for the Type II reaction can be met in micelles, but not in the more highly structured monolayer assemblies.

Numerous reports on the photoreactions of benzophenones and substituted benzophenones in micelles have appeared in the literature [34-37]. Systematic studies on the regioselective functionalization of methylene groups of a detergent induced by hydrogen abstraction following benzophenone and surfactant benzophenone excitation have been performed (Scheme XII). For benzophenone and the 4-carboxylate ester derivatives, little selectivity is observed over the positions 1 through 13 of SDS surfactant, although attack at position 2 is slightly favored [34,35,37]. For 4-(4-benzoylphenyl)butyrate in HDTBr solutions below the CMC, highly selective hydrogen abstraction from the C-2 position is observed while above the CMC, the reaction becomes less specific (Fig. 16). These results indicate that the reaction center is near the head groups, and there is extensive coiling of the detergent in the micelle. However, surprisingly, for the case of 2-(4-benzoylphenyl)propionate in sodium tetradecanoate micelles [36], a large degree of selectivity was observed for the 5-8 positions on the surfactant. It is suggested that the reaction center, the carbonyl, extends further into the micelle than in the previous cases.

When benzophenone itself is dissolved in detergent solutions, the photochemical processes are affected by the structure of the micelle [38-43]. Scheme XIII represents the dynamical processes.

$$BP \rightarrow {}^1(BP)^*$$

$${}^1(BP)^* \rightarrow {}^3(BP)^*$$

$${}^3(BP)^* + RH \rightarrow {}^3\overline{R^{\cdot}K^{\cdot}}$$

$${}^3\overline{R^{\cdot}K^{\cdot}} \rightarrow {}^1\overline{R^{\cdot}K^{\cdot}} \rightarrow RK$$

$${}^3\overline{R^{\cdot}K^{\cdot}} \rightarrow \text{Escape} \rightarrow RR + RK + KK$$

(Scheme XIII)

In a fashion very similar to the observed photoprocesses of DBK's, the micelle sequesters the radical pair (ketyl radical, $K^{\cdot}$ and hydrogen donor, $R^{\cdot}$) so that efficient intersystem crossing may occur. In most cases, the hydrogen donor is the detergent itself so that when intersystem crossing occurs, the benzophenone attaches itself to the micellar backbone. The mechanism for intersystem crossing has been shown to be due to a hyperfine coupling interaction and therefore the decay of the radical pairs is sensitive to isotopic substitution and to applied magnetic fields.

The triplet state decay kinetics of benzophenone has been monitored in fluorinated surfactants (sodium perfluorooctanoate, SPFO) where the surfactant does not quench the triplet [40]. In this case, the excited benzophenone cannot react with the surfactant and the excited molecule escapes into the aqueous phase. When increasing amounts of SDS is added to solution, the observed triplet lifetime (by nanosecond transient absorption techniques) decreases indicating that hydrogen abstraction is occurring from the SDS (Fig. 17).

The magnetic isotope effect on the lifetime of the observed transients as well as the amount of cage recombination has been studied by various authors [38,41-43]. Experimentally, there are several species which can absorb light after laser excitation

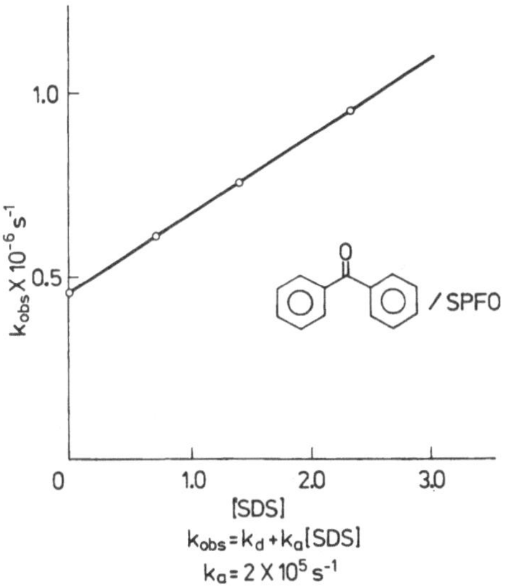

Fig. 17. The observed decay rate of the triplet state of benzophenone in SPFO as a function of SDS concentration. The slope is equal to k_a, the probability of hydrogen abstraction per unit time

of benzophenone (Scheme XIV). The influence of the micelle on the mechanism of the reaction can be monitored directly for each transient species.

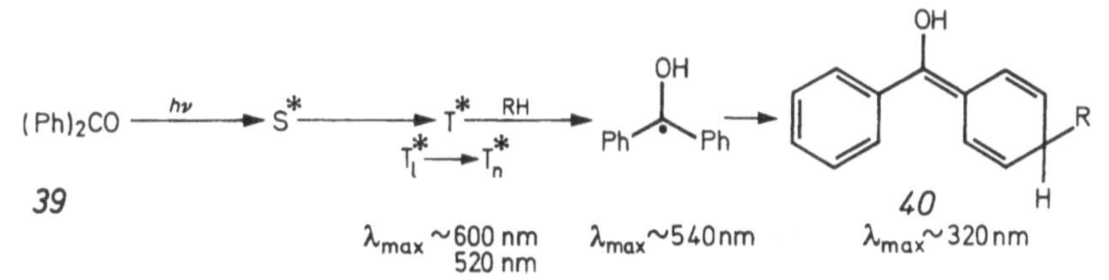

(Scheme XIV)

It was demonstrated that the formation of LAT's (Light Absorbing Transients)[43] (*40*) is due to recombination of singlet radical pairs within the micelle, and that their increased presence in micellar solution reflects on the most probable orientation of the ketyl radical where the hydroxy group is near the head groups leaving the para position more accessible to the alkyl radical. Furthermore, the amount of escape of the ketyl radical increases and the relative amount of LAT's decreases in the presence of an external magnetic field due to the Zeeman splitting of the triplet sublevels, therefore decreasing the intersystem crossing rate (Fig. 18)[42]. Similar results are obtained when an external hydrogen source, 1,4-cyclohexadiene is used as triplet quencher[38]. Substitution of the carbon at the carbonyl position with ^{13}C isotope results in a decrease in the amount of escape of ketyl radical from the micelle.

The photoreduction of quinones has also been studied in micellar environment[44-46]. The relative quantum yields for disappearance of benzoquinone is various surfactants have been found to be influenced by the nature of the micelle. For example, the rate of reaction (Scheme XV) is accelerated in anionic surfactants[44],

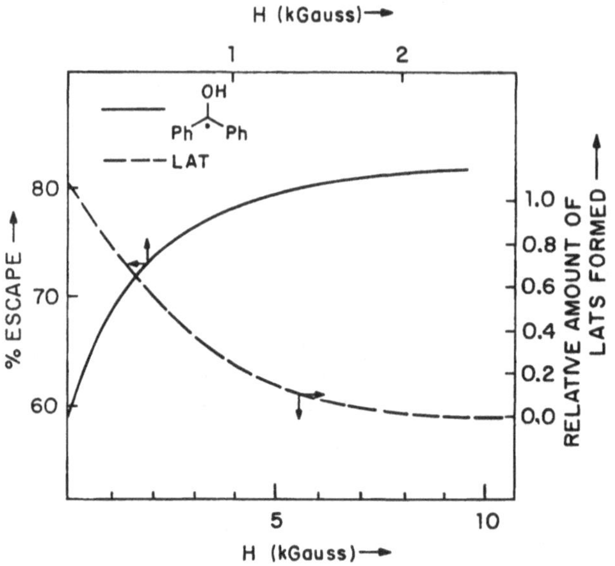

Fig. 18. A comparison of the escape of ketyl radical (——) and the formation of LAT's (——) as a function of magnetic field. The escape of the ketyl radical is much more sensitive to external magnetic field than is the function of LAT's

unaffected in non-ionic surfactants, and is inhibited in cationic surfactants. It is suggested that the photoreduction in water proceeds by an ionic mechanism, and that the cationic intermediate is stabilized by binding with the anionic micelles.

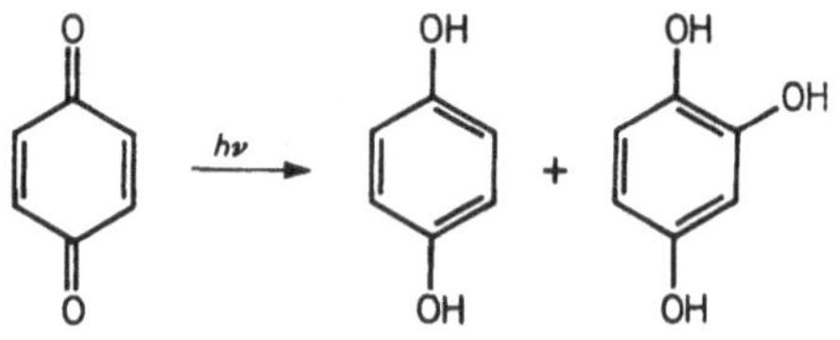

41 *42* *43*

(Scheme XV)

It is also observed that the photolysis of p-xyloquinone (*44*) is SDS results in the formation of the semiquinone radical whose decay consists of a fast and slow component (Fig. 19)[45]. The fast component is attributed to a combination of the rates of intersystem crossing (hyperfine interactions) and escape of the triplet radical pair from the micelle, and the slow decay is due to recombination following escape. In the presence of a magnetic field, the ratio of the fast/slow decays decreases (Fig. 19). This can be explained by the increase of the escape of triplet radicals from the micelle due to Zeeman inhibition in intersystem crossing. The rates for triplet-singlet intersystem crossing of anthraquinone semiquinone radicals (Scheme XVI) in SDS are found to decrease with increasing magnetic field [46]. This result is interpreted in terms of an inhibition of $T_{\pm}$—S intersystem crossing due to Zeeman slitting of the triplet sublevels while the rates for escape from the micelles showed no dependence on external magnetic field.

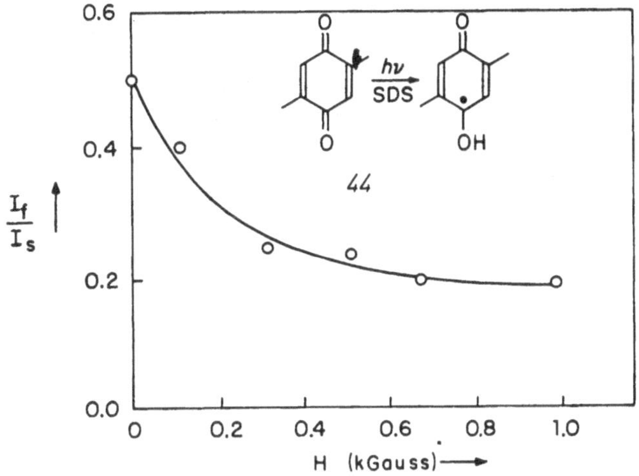

Fig. 19. The ratio of the fast to slow decay intensities as a function of applied magnetic field. The decrease is indicative of an increase in the escape of radicals from the micelle

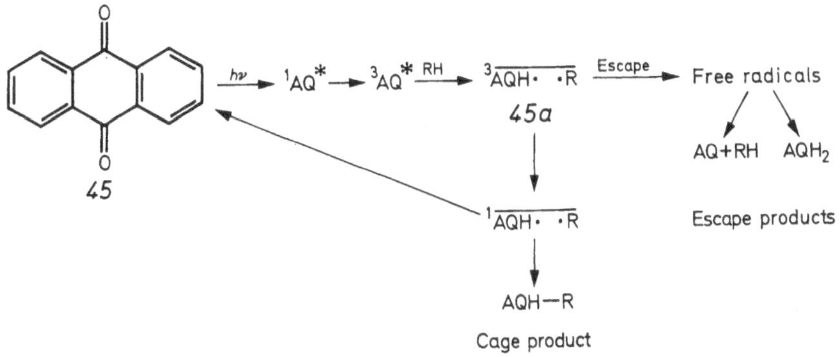

(Scheme XVI)

3.3 Photodimerization and Photocyclizations

An extremely useful property of micelles is the ability to obtain high local concentrations of bound organic substrates at low macroscopic concentrations. Therefore, it is expected that there will be an increase in the quantum efficiencies of intermolecular cycloaddition reactions in micellar solutions over homogeneous solutions. Estimated rate enhancements up to 10^3 over homogeneous rates are expected so that reactions that are relatively inefficient in homogeneous solutions may become quite efficient in the presence of micelles. In addition, orientational effects of solutes by the micelle may also lead to the observation of unique products.

Micelles are found to affect the dimerization of uracil by changing the relative amounts of the dimers (*47–50*) (Scheme XVII). For example, the distribution of dimers formed from the irradiation of uracil dissolved in water is in the order of (*50*) > (*49*) > (*48*) > (*47*). Upon addition of detergent, the resulting distribution changes depending on the surfactant used. For SDS, the apportionment of dimers

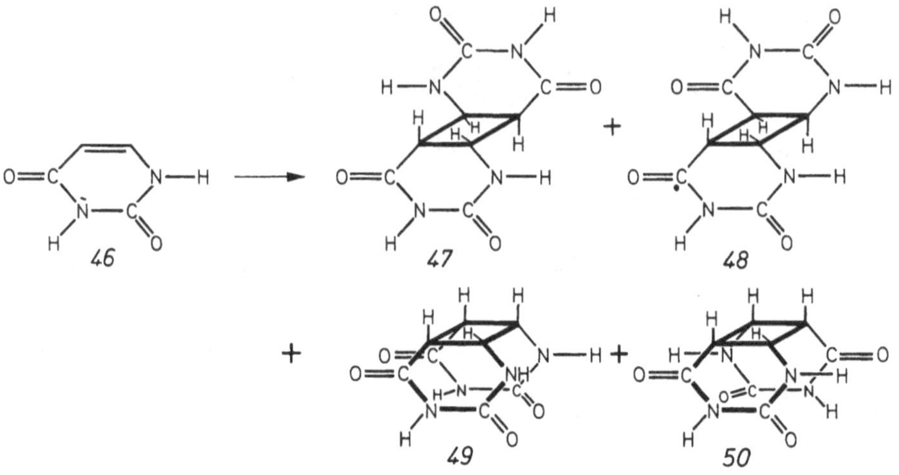

(Scheme XVII)

is (*50*) > (*47*) > (*48*) > (*49*), while in HDTCl the order is (*50*) > (*48*) > (*47*) > (*49*). The factors responsible for the dramatic changes in the relative yields of dimers is not yet understood.

The rate and yields of dimerization for isophorone (*51*)[48] has been found to be enhanced in micellar microemulsions relative to homogeneous solutions (Scheme XVIII). In homogeneous solutions, the ratio of head-to-head/head-to-tail (HH/HT) dimers increases with increasing polarity, but in micelles polarity effects alone cannot fully explain the observed ratios. The change in the regioselectivities and high yields is attributed to the localization of the substrate in the vicinity of the Stern layer where polarity and orientational effects are quite strong.

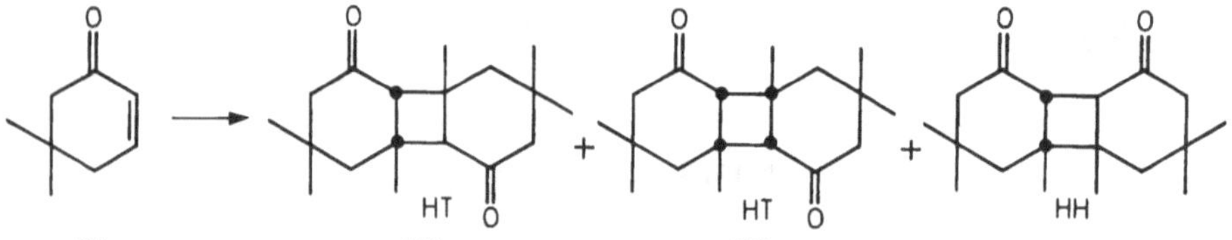

51 52 53 54

	HT	HH
Cyclohexane	10	90
SDS/butanol/H_2O	95	5

(Scheme XVIII)

While the irradiation of substituted cyclopentenones in aqueous environment gives cyclobutane dimers with a ratio of head-to-head (HH) to head-to-tail (HT) of about one, the same photoreaction in potassium decylanoate (KDC) gives the cyclobutane dimer which is almost exclusively HH (Scheme XIX)[49]. The dependence of the HH/HT ratio as a function of surfactant concentration shows a dramatic increase in the ratio at the reported CMC value. The regioselectivity is attributed to the carbonyl groups being oriented in the Stern layer with the olefin extending into the micellar core. Thus, an increase in the HH/HT ratio is observed over the random orientation of ketones in homogeneous solutions. The increase in the quantum yield for the reaction upon addition of surfactant is attributed to the high local concentrations within the micelle.

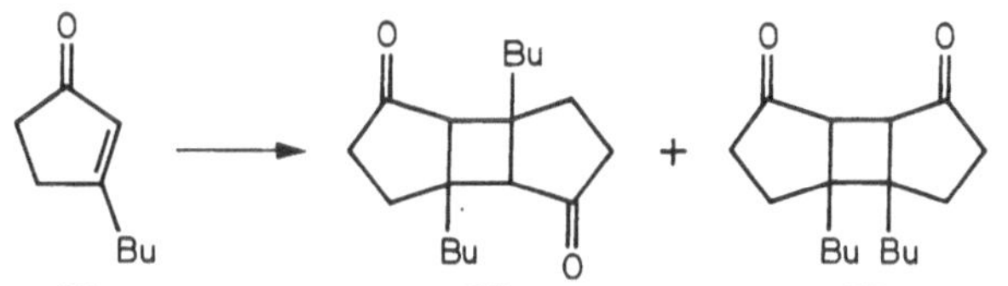

	55	*56*	*57*	
Homogeneous C_6H_6		91	9	1% yield
Micelles (KDC)		2	98	100% yield

(Scheme XIX)

Similar results have been reported for the photocycloaddition of olefins to cyclopentenones [50]. For example, irradiation of 3-n-butylcyclopentenone, (*55*) in the presence of various olefins ($\sim 10^{-2}$–10^{-1} M) in potassium dodecanoate (KDC) gave good to excellent yields of addition products (Scheme XX). The ratio of *58* to *59* isomers increases for all cases in the presence of KDC (Table 6). The change in the

Table 6. Product Distribution from the Photoaddition of Olefins to Cyclopentanones. (See Scheme XIX)

Olefin	Medium	% Yield	% *58*	% *59*
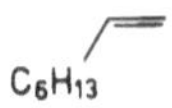	KDC	99	88	12
	Acetonitrile	92	63	37
	Methanol	96	62	38
	Diethyl ether	99	57	43
	Cyclohexane	92	53	47
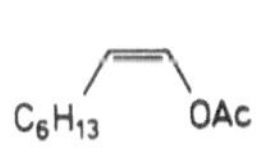	KDC	68	70	30
	Methanol	65	0	100
	Cyclohexane	53	0	100

regioselectivity is again explained by the orientation of the carbonyl within the Stern layer (Fig. 20). The observed ratios are seen to be a micellar effect and not a polarity effect by comparison of the (*58*)/(*59*) ratios of the micelle with that of methanol. The largest effect on (*58*)/(*59*) ratio and rate enhancement is observed for those olefins possessing polar groups (e.g., OAc) wherein the polar group favors the orientation away from the micellar core.

The photodimerization of coumarin (*60*) is affected by the presence of micelles (Scheme XXI) [52]. Alterations in the nature of the micelle by employing anionic (SDS), cationic (HDTBr) and non-ionic (Triton X-100) micelles slightly affect the relative quantum efficiencies of the cycloaddition and supports the contention that

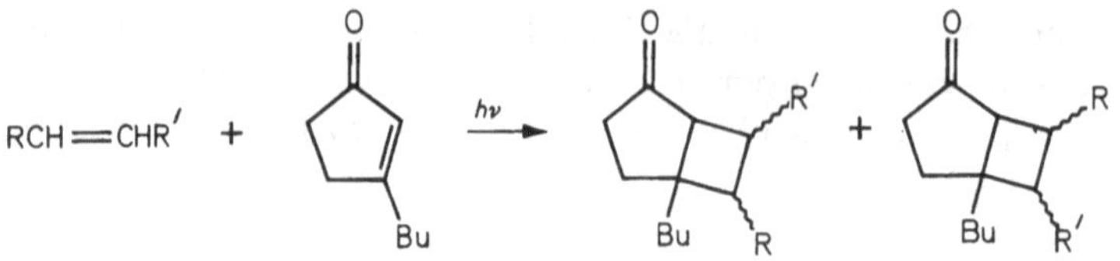

(*Scheme XX*)

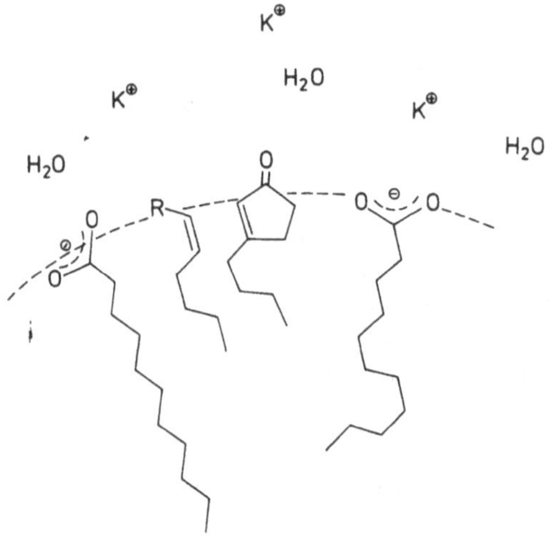

Fig. 20. The orientation of 3-butyl-cyclopentenone (*47*) and the olefin in the Stern Layer of the micelle accounts for the observed regioselectivity of the photocycloaddition

the reaction site is in the micelle core. It was further demonstrated that the observed dimerization is not caused by aggregation of the coumarin in water, and that polarity effects play an important role.

The photodimerization of aromatic hydrocarbons is also enhanced in micellar solutions over homogeneous solutions. Acenaphthylene (*63*), irradiated at the same

Table 7. Product Yields and Ratios for Acenaphthylene Dimerization

Solvent	Conc. (mM)	% Yield of 64 + 65	Ratio 64:65
Benzene	1340	9.8	22.3 :1
	100	8.2	14.4 :1
	24	8.1	3.5 :1
1% aq. PBC-34	4.9	97.3	0.77:1
5% aq. PBC-34	19	94.5	0.80:1
	9.7	93.3	0.78:1
	1.9	91.2	0.13:1
10% aq. PBC-34	28	96.0	0.88:1
	19	95.1	0.93:1
	1.9	94.4	0.06:1
5% aq. SDS	9.7	96.5	2.7 :1

bulk concentration in micelle and benzene yielded 96% of dimer in micelle, but no detectable amount in benzene (Scheme XXII)[51a)]. Furthermore, the ratio of the cis-cis (*64*) is slightly dependent on the surfactant used (Table 7).

The photocycloaddition of acrylonitrile to acenaphthylene is found to be competitive with photodimerization in the non-ionic detergent Nippol PBC-34, while

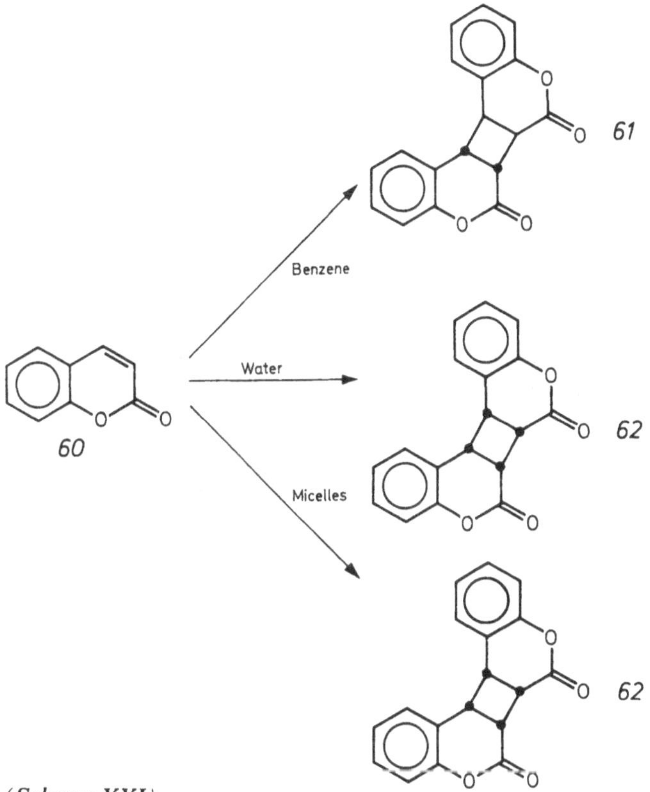

(Scheme XXI)

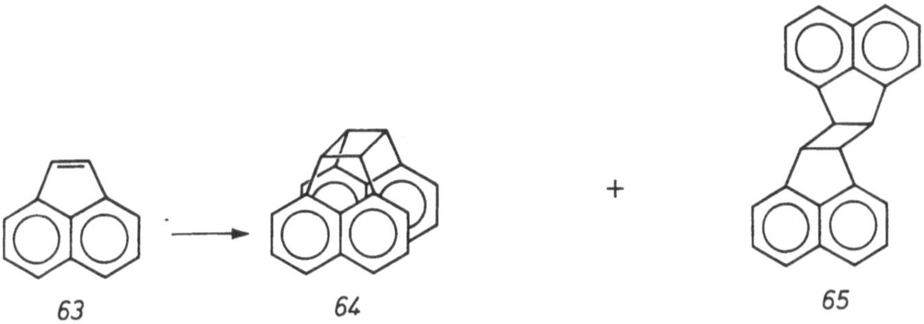

(Scheme XXII)

under the same concentrations in benzene, only dimers are detected (Scheme XXIII)[51b].

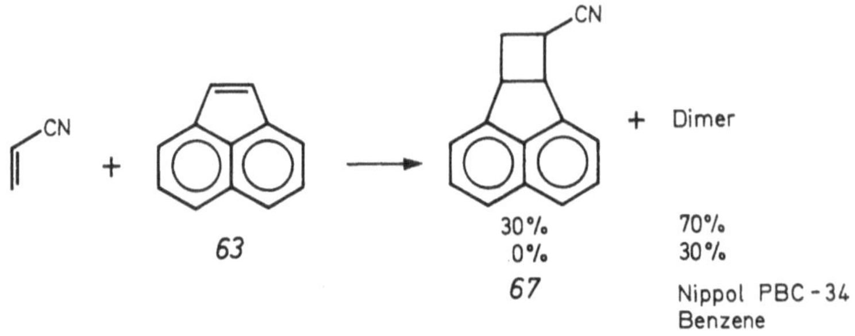

(Scheme XXIII)

Micelles also influence the dimerization of substituted anthracenes[53]. For example, 9-hydroxy-anthracene (*68*) when irradiated in SDS yields a 30–60% increase in the HH/HT ratio observed in homogeneous solutions (Scheme XXIV). Since the HH dimer is equally unstable in both micellar solutions and homogeneous solutions, the increase in the ratio is attributed to the proximity of the hydroxyl groups to the Stern layer and the orientation of the anthracene chromophores toward the micelle core thereby increasing the initial yield of the HH dimer.

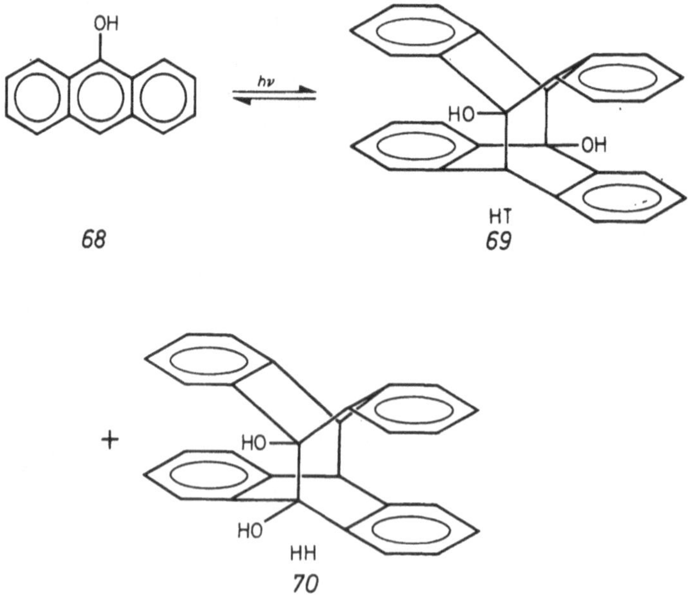

(Scheme XXIV)

3.4 Photooxidations

Photooxidation reactions involving oxidation of substrates by molecular oxygen have been extensively studied in micellar systems. SDS solutions are excellent environments for the photooxidation of benzyl sulfide to benzyl sulfoxide [54]. In SDS, the reaction occurs with a higher yield of products, decreased degradation of the phenothiazine sensitizer, and fewer by-products compared to the reaction in organic solvents. The mechanism in SDS appears to be different from the reaction in organic solvents. In the micelle the first intermediates formed are the phenothiazine cation radical and superoxide ($O_2^{\bar{\cdot}}$). Electron transfer between these two ions leads to the formation of singlet oxygen ($^1O_2^*$) which oxidizes the sulfide. It is known that anionic micelles can enhance photoionization and stabilize cation radicals, especially of phenothiazine dyes [55].

Micelles affect the competition between photooxidation mechanisms involving singlet oxygen or superoxide as intermediates. In the photooxidation of protoporphyrin (*69*) two types of products are observed: two isomeric 1,4-(*70*) and two isomeric 1,2-addition (*71*) products (Scheme XXV) [56-58]. The relative yields of the two products vary dramatically upon going from organic solvents to SDS or Brij 35 micellar solutions. Both products can be formed by a mechanism involving singlet oxygen in organic solvents. However, the increase yield of the 1,2-addition product in micelles is proposed to be due to its formation by a non-singlet oxygen mechanism which is catalyzed by the micelle (Scheme XXVI). The new mechanism appears to involve the formation of porphyrin cation radical and superoxide.

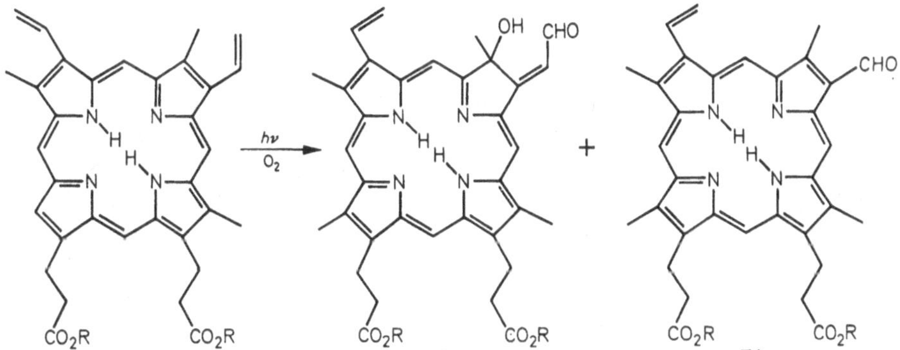

	69	70	71
		1,4-addition	1,2-addition
Methylene chloride		97	3
Brij 35 micelles		58	42

(Scheme XXV)

Competition between Type I and Type II photooxidations are affected by micellar media. Type I photooxidation involves initial quenching of the sensitizer excited state by substrate, while Type II photooxidations involve initial quenching of the sensitizer excited state by oxygen. Since, competition between Types I and II photooxidations are altered by the concentration of the substrate, local concentration effects in micelles play an important role. The photooxidation of tryptophan and tryptamine sensitamine sensitized by hematoporphyrin in aqueous and micellar solutions can occur by both a Type I and Type II mechanism [59,60]. Incorporating HMP into micellar system increases the Type II pathway for tryptophan, since it is not solubilized in the micellar phase. However, for tryptamine which is solubilized inside the micelle, micellar systems favor the Type I pathway for the photooxidation.

$$P^* + O_2 \nearrow P + {}^1O_2^* \rightarrow \text{products of type } 70$$
$$P^* + O_2 \searrow P^{\cdot +} + O_2^{\cdot -} \rightarrow \text{products of type } 71 \text{ (catalyzed by micelles)}$$

(Scheme XXVI)

Pyrene can sensitize the photooxidation of 1,3-diphenylisobenzofuran (*72*) in DTAC micellar solutions [61]. The reaction involves sensitization of singlet oxygen by pyrene which diffuses into another micelle and reacts with (*72*). Indole and tryptophan, which also react with singlet oxygen, quench the above reaction in ethanol solutions. However, in micellar systems they enhance the rate of reaction. Because of the high local concentrations of the quencher, the pyrene excited state is quenched by indole and tryptophan which leads to the photooxidation of (*72*) by a Type I process.

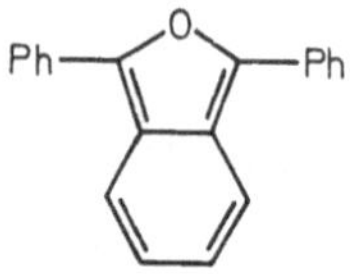

72

Micelles have been used as membrane models to study the photooxidation of biological compounds. The ability of promazine and other drugs to photosensitize both singlet oxygen and free radicals was enhanced in micellar solutions, compared to organic solvents [62]. SDS micelles were used to measure the rate of reaction of bilirubin with singlet oxygen in aqueous environments [63]. In neutral aqueous micellar solutions chlorophyll can efficiently sensitize singlet oxygen formation. The quantum yield for diffusion of singlet oxygen out of these membrane type environments into the aqueous phase was 0.7–0.85 [64].

Micelles are useful environments for studying the behavior of singlet oxygen in aqueous media. The photooxidation of micelle solubilized (*72*) with either water or micellar solubilized sensitizers is the most frequently studied reaction [65–71]. The lifetime of singlet oxygen in D_2O (30–53 μs) and H_2O (3–4 μs) was calculated by measuring the rate of reaction of (*72*). This rate can be measured by either a steady state quantum yield experiment or by time resolved laser flash photolysis. The time resolved method was used to calculate the rate constants for quenching of singlet oxygen in H_2O and D_2O by a variety of substrates [72].

These studies also indicate that singlet oxygen generated in either the aqueous phase or in a micelle can diffuse in and out of micelles easily. Therefore, a reaction of singlet oxygen can occur in a micelle other than the one in which it was generated. In the photooxidation of protoporphyrin IX, where the sensitizer is also the substrate, only part of the reaction is quenched by a water soluble quencher (sodium azide)[56]. This result indicates that the reaction is occurring through an intramicellar and an intermicellar pathway (Scheme XXVII). The intermicellar pathway is not quenchable because the singlet oxygen generated inside the micelle reacts before it could diffuse into the aqueous phase.

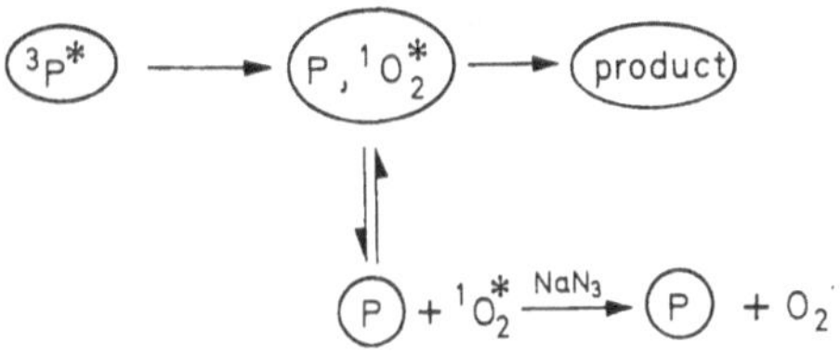

(*Scheme XXVII*)

3.5 Miscellaneous Photochemical Reactions

4-Methoxy-1-nitronaphthalene (*73a*) and 1-nitronaphthalene (*73b*) undergo photochemical aromatic substitution reactions with cyanide (Scheme XXVIII). A two-fold increase in the quantum yield for the reaction is observed for (*73a*) when the reaction occurs in HDTCl compared to aqueous solution[73]. However, a 6800-fold catalytic increase in quantum yield is observed for (*73b*). SDS micelles decrease the quantum yield compared to aqueous solutions. The higher local concentration of cyanide near the HDTCl micelles can explain a least partially the increase in quantum yield. However, the 6800-fold increase for (*73b*) is also due to a polarity effect on the reaction. This was demonstrated by an increase in the quantum yield of the reaction with decreasing polarity.

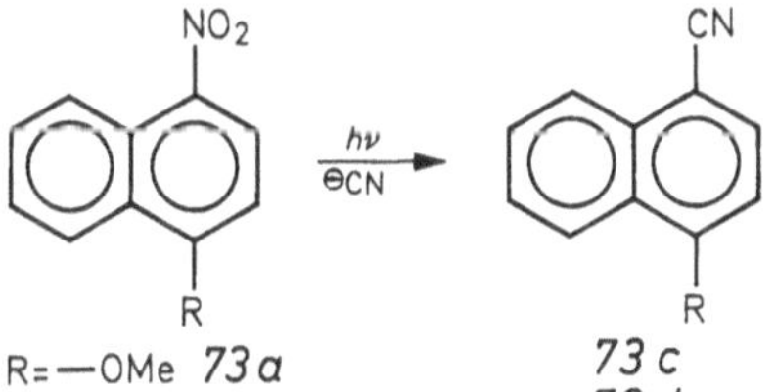

R=—OMe *73a*
R=H *73b*

73c
73d

(*Scheme XXVIII*)

The photohydrolysis of 3,5-dinitroanisole (*74*) to 3,5-dinitrophenoxide ion (*75*) is catalyzed by hydroxide ion and studied in both aqueous and micellar systems (Scheme XXIX)[74]. The quantum yield for the reaction decreased in tetradecyltrimethylammonium chloride (TTACl) micelles compared to aqueous solution. The results are explained by the decrease triplet lifetime of 3,5-dinitroanisole in the non-polar micellar environment compared to the aqueous phase.

Several photochemical reactions have been used to measure the microviscosity of micellar systems. The photolysis of potassium 6-nitrododecanoate (KND) in potassium dodecanoate micelles occurs through an intermediate alkoxyl radical which can disproportionate to give a ketone (K), or it can undergo γ-hydrogen abstraction to give an hydroxy ketone product (HK) [75]. A plot of (HK)/(K) *vs.* 1/viscosity gives a straight line for a variety of organic solvents. From the ratio of (HK)/(K) products the viscosity of the micelle was calculated to be 10 cP. Presumably, the high viscosity prevents movement necessary to achieve γ-hydrogen abstraction as is observed for type II ketone photochemistry.

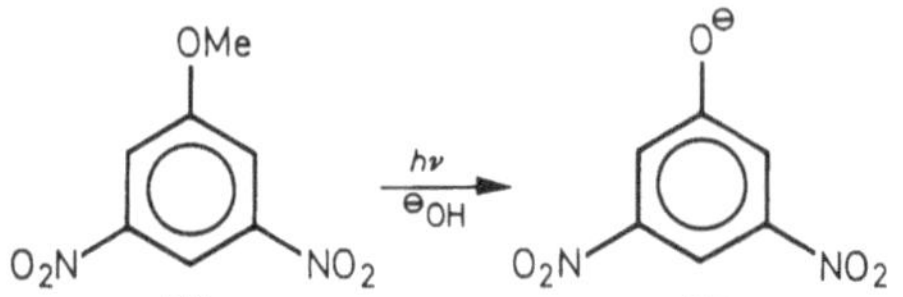

74 *75*

(Scheme XXIX)

A surfactant stilbene (*76*) has been synthesized as a viscosity probe of micellar systems [76]. The stilbene surfactant undergoes photochemical trans-cis isomerization with quantum yields very similar to stilbene. Like stilbene, the quantum yield (Φ_{t-c}) of (*76*) decreases with viscosity while the quantum yield of fluorescence (Φ_{ft}) increases. For example, Φ_{t-c} decreases from 0.5 in CH_2Cl_2 to 0.39 in HDTBr, while the Φ_{fl} increases from 0.04 in CH_2Cl_2 to 0.18 in HDTBr. This probe has also been used to study bilayer vesicle systems. Cyanine dyes become more photostable and have higher Φ_{ft} in SDS micelles than in organic solvents [77]. This is also due to an increase in viscosity.

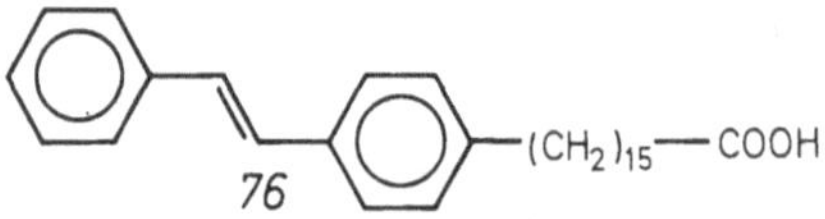

76

A photoresponsive surfactant (*77*) containing an azobenzene moiety has been used to control micellar catalysis [78]. The proton abstraction from benzoin which can be monitored by the disappearance of 2,6-dichlorophenolindophenyl is catalyzed by cationic micelles such as hexadecylpyridinium bromide (HDPB). If the reaction occurs in micelles containing both (*77*) and HDPB, the reaction is slower than in pure

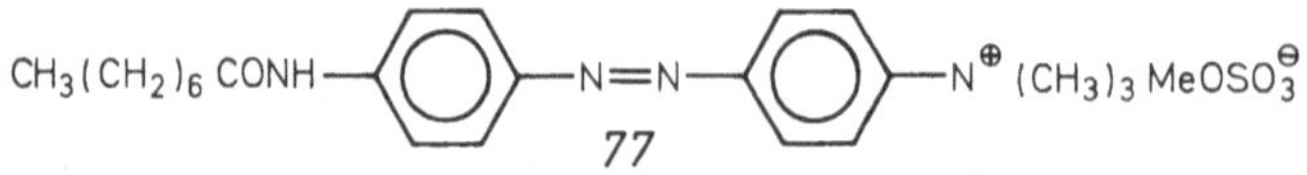

77

HDPB micelles. However, if (*77*) is irradiated to yield cis-(*77*), then the rate of reaction increases by 3.2-fold. The result is due to the ability of trans-(*77*) to break up the micelle more than cis-(*77*) can.

4-Nitrophenylnitromethanide ion (*79*) formed from 4-nitrophenylnitromethane (*78*) will undergo photochemical rearrangement and cleavage to yield 4-nitrobenz-

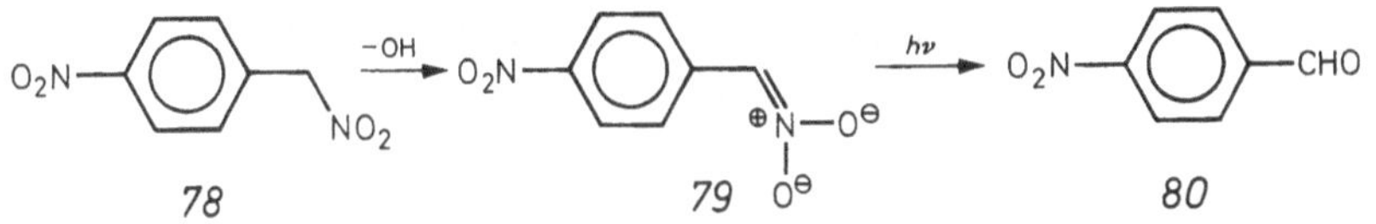

78 79 80

(Scheme XXX)

aldehyde (*80*) (Scheme XXX)[79]. The quantum yield is increased by a factor of twenty in HDTBr *vs.* aqueous solution. The decrease in observed pK_a of (*78*) explains in part the micellar effect.

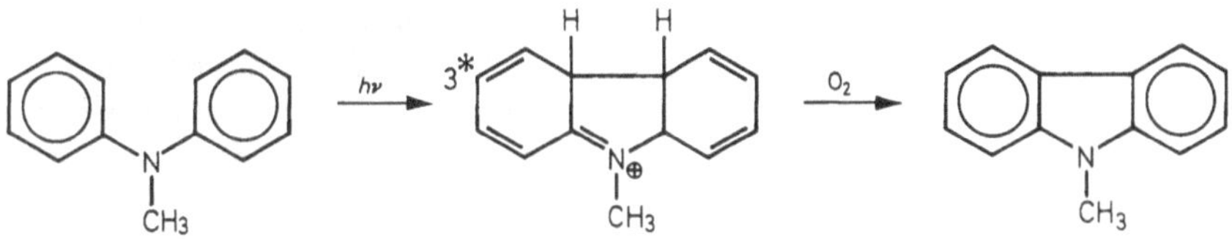

81 82 83

(Scheme XXXI)

The photocyclization of N-methyldiphenylamine (*81*) to N-methylcarbazole (*83*) has been examined as a probe for the kinetics of oxygen quenching in micellar systems (Scheme XXXI)[80]. Studies of the oxygen quenching of a triplet intermediate (*82*) of the reaction indicated that differences in rates of oxygen quenching could be explained by differences in oxygen concentration in hexane compared to SDS and HDTBr solutions. This result lends further support to the fact that oxygen can diffuse in and out of a micelle at close to diffusion control.

The photohydrolysis of m-methoxybenzylacetates occurs readily in micellar systems with few of the side products that are observed in dioxane/H_2O mixtures which were used previously for the reaction[81]. The chemiluminescence of hydrophobic 9-methyleneacridans is increased dramatically in micelles compared to aqueous solution due to the decreased polarity in the interior of the micelles[82].

3.6 Proton and Electron Transfer Reactions

Several other photochemical reactions have been studied in micellar systems. These include excited state acid-base chemistry, photoionization and electron transfer. However, since these reactions typically do not produce permanent products, and since they have been reviewed previously, they will be discussed only briefly in this review.

Many aromatic amines, phenols and carboxylic acids have different pK_a's in the excited state than in the ground state[83-85]. Therefore, at certain pH values a compound can undergo acid-base reactions in the excited state which do not occur in the ground state. For example, 1-naphthol loses a proton in the excited state leading to fluorescence from the protonated form only [86]. In micelles the reaction is hindered such that fluorescence from the protonated form is also observed. Decreased equilibrium in the excited state because of increased viscosity and decreased concentration of base (i.e., H_2O) appear to explain the results.

Photoionizations are catalyzed by anionic micelles [4,5,7,55,87-93]. Studies of pyrene, phenothiazine and tetramethylbenzidene indicate that upon excitation the photoejection of an electron into the aqueous phase occurs in higher yield and at several eV lower energy than in hydrocarbon solvents. The increased yield of photoionization is due to the stability of the cation radical of the aromatic hydrocarbon in the anionic micelle, the increased rate of escape of the electron from the micelle due to repulsion of the anionic head groups, and the decreased rate of back reaction of the electron and the cation radical due to head group repulsion. The lower threshold energy is due to the stability of cation radical and free electron in the anionic micellar systems compared to hydrocarbon environment. For example, phenothiazine (PT) will undergo photoionization in both SDS micelles and in ethanol [94]. The quantum yield of photoionization ($\Phi_I = 0.1$). The PT cation radical has a half-life of 100 ns in methanol, while in SDS the half-life is several milliseconds.

Photochemical electron transfer reactions have been examined in micellar systems as probes for the diffusion and location of quenchers, and as environments for solar energy storage [2,3,90,95,96]. The relative rates of quenching will depend on the location of the donor and acceptor (Scheme XXXII). For example, the rate of quenching of a hydrophobic donor located inside the micelle by Cu^{2+} is much faster in anionic micelles compared to cationic micelles. Similarly a hydrophobic excited state is quenched faster by a hydrophobic donor or acceptor than by a hydrophilic one in micellar systems.

$$D^* + A \rightarrow D^{+\cdot} + A^{-\cdot}$$

$$A^* + D \rightarrow D^{+\cdot} + A^{-\cdot}$$

$$D^{+\cdot} + A^{-\cdot} \rightarrow D + A$$

(Scheme XXXII)

After photochemical electron transfer occurs, the charge on the D or A is altered such that hydrophobic quenchers can become hydrophilic or *vice versa*. This phenomenon combined with charge repulsion can lead to the location of D^+ and A^- in different sites from D and A. This separation of D^+ and A^- can lead to a decreased rate of back electron transfer compared to homogeneous solution (Scheme XXXII). This phenomenon has been used to stabilize high energy intermediates produced in solar energy storage. For example, singlet excited pyrene (Py) is quenched by dimethylaniline (DMA) to yield pyrene anion radical and dimethylaniline cation radical [97]. In cationic micelles DMA^+ is expelled from the micelle, leading to stabilization of Py. The half-life Py in HDTBr micelles is 500 µs compared to 67 µs in SDS and 6 µs in methanol.

4 Concluding Remarks

Micelle aggregates can influence both photophysical and photochemical properties of solvated molecules. Some of the observed effects include the increase in the local concentration of the solute, the inclusion of the solute into a "super cage" from which the exit rate is remarkably slow, a ten-fold increase in the microscopic viscosity, the orientation of the solute in the micelle core with respect to the polar head group and the non-polar backbone, and the stabilization of charged species within the micelle due to electrostatic interactions. Inclusion of compounds in micelles has led to photoreactions with high selectivity as well as enhanced yields in comparison to homogeneous solution. The application of this knowledge to other systems and environments should prove to provide the bases for many future investigations.

5 Acknowledgements

The authors thank the National Science Foundation and the Army Research Office for their generous support of this research.

6 References

1. For recent reviews on micelles, see: (a) Tanford, C.: "The Hydrophobic Effect", Wiley and Sons: New York, 1973; (b) Mittal, K. L.: "Micellization, Solubilization, and Microemulsions", vol. 1 and 2; Plenum Press: New York, 1977; (c) Fendler, J. H.; Fendler, E. J.: "Catalysis in Micellar and Macromolecular Systems", Academic Press: New York, 1975; (d) Fisher, L. R., Oakenfull, D. G.: Chem. Soc. Rev. *6*, 25 (1977); (e) Menger, F. M.: Acc. Chem. Res. *12*, 111 (1979); (f) Fendler, J. H.: "Membrane Mimetic Chemistry", Wiley, N.Y., 1982; (g) Lindman, C., Wennerström, H.: "Micelles. Amphiphile Aggregation in Aqueous Solution" and Eicke, H.-F.: "Surfactants in Nonpolar Solvents. Aggregation and Micellization"; Topics in Current Chemistry, Vol. 87, Springer-Verlag, Berlin Heidelberg New York 1980
2. Whitten, D. G., Russell, J. C., Schmehl, R. H.: Tetrahedron *38*, 2455 (1982)
3. (a) Turro, N. J., Gratzel, M., Braun, A. M.: Angew. Chem. Intern. Ed. Engl. *19*, 675 (1980); (b) Thomas, J. K.: Chem. Rev. *80*, 283 (1980); (c) Thomas, J. K.: Acc. Chem. Res. *10*, 133 (1977)
4. Lindig, B. A., Rodgers, M. A. J.: Photochem. Photobio. *31*, 617 (1980)
5. Thomas, J. K., Greiser, F., Wong, M.: Ber. Bunsenges. Phys. Chem. *82*, 937 (1978)
6. Thomas, J. K., Almgren, M. in "Solution Chemistry of Surfactants", vol. 2, Mittal, K. L. ed.; Plenum Press: New York, 1979, pg. 559;
   Kalyanasundaram, K.: Chem. Soc. Rev. *7*, 453 (1978)
7. (a) Schore, N. E., Turro, N. J.: J. Am. Chem. Soc. *96*, 306 (1974);
   (b) Schore, N. E., Turro, N. J.: J. Am. Chem. Soc. *97*, 2488 (1975);
   (c) Chiang, H.-C. Lukton, A.: J. Phys. Chem. *79*, 1935 (1975);
   (d) Turro, N. J., Aikawa, M., Yekta, A.: J. Am. Chem. Soc. *101*, 772 (1979);
   (e) Weber, G.: Ann. Rev. Biophys. Bioeng. *1*, 553 (1972)
8. (a) Forster, Th., Selinger, B. K.: Z. Naturforsch. *A19*, 38 (1964);
   (b) Kalyanasundaram, K. *et al.*: J. Am. Chem. Soc. *95*, 3915 (1975);
   (c) Dorrance, R. C.: J. Chem. Soc., Faraday Trans. *70*, 1572 (1974)
9. Turro, N. J.: "Modern Molecular Photochemistry"; Benjamin/Cummings, 1978, p. 52
   (a) Eliel, E. L.: "Stereochemistry of Carbon Compounds", McGraw Hill, New York, 1962
   (b) Seeman, J. I.: Chem. Rev. *83* (2), 83 (1983)
   (c) Turro, N. J.: "Modern Molecular Photochemistry", Benjamin/Cummings, 1978 pages 526–528

10. (a) Quinkert, G. *et al.*: Tetrahedron Letters 1963, 1863;
(b) Engle, P. S.: J. Am. Chem. Soc. *92*, 6074 (1970);
(c) Robbins, W. K., Eastman, R. H.: ibid. *92*, 6077 (1970)
11. Turro, N. J., Kraeutler, B.: J. Am. Chem. Soc. *100*, 7432 (1978)
12. (a) Turro, N. J., Cherry, W. R.: ibid. *100*, 7431 (1978);
(b) Turro, N. J. *et al.*: ibid. *103*, 4574 (1981)
13. Turro, N. J., Kraeutler, B., Anderson, D. R.: ibid. *101*, 7435 (1979)
14. Turro, N. J., *et al.*: ibid. *103*, 3886 (1981)
15. (a) Turro, N. J. *et al.*: ibid. *103*, 3892 (1981);
(b) Turro, N. J. *et al.*: ibid. *103*, 4574 (1981)
16. Kraeutler, B., Turro, N. J.: Chem. Phys. Lett. *70*, 270 (1980)
17. Kraeutler, B., Turro, N. J.: ibid. *70*, 266 (1980)
18. Turro, N. J., Chow, M.-F., Kraeutler, B.: ibid. *73*, 545 (1980)
19. Turro, N. J., Weed, G. C.: J. Am. Chem. Soc. *105*, 1861 (1983)
20. (a) Turro, N. J.: Pure Appl. Chem. *53*, 259 (1981);
(b) Turro, N. J.: Tetrahedron *38*, 809 (1982)
21. Hayashi, H., Sakaguchi, V., Nagakura, S.: Chem. Lett. 1980, 1149
22. Buchachenko, A. L.: Russ. J. Phys. Chem. *51*, 1445 (1977)
23. Turro, N. J., Kraeutler, B.: Acc. Chem. Res. *13*, 369 (1980)
24. Turro, N. J. et al.: Tetrahedron Letters *23*, 3223 (1982)
25. Turro, N. J., Anderson, D. R., Kraeutler, B.: ibid. *21*, 3 (1980)
26. Turro, N. J., Baretz, B. H.: J. Am. Chem. Soc. *105*, 1309 (1983)
27. Turro, N. J., Mattay, J., Lehr, G. F.: "Inorganic Reactions in Organized Media", ACS Symp. Series 177, Holt, S. L. ed. Washington, D.C. 1982
28. Lehr, G. F., Turro, N. J.: Tetrahedron *37*, 3411 (1981)
29. Hutton, R. S. et al.: J. Am. Chem. Soc. *101*, 2227 (1979)
30. Turro, N. J., Mattay, J.: ibid. *103*, 4200 (1981)
31. Turro, N. J., Tung, C.-H.: Tetrahedron Letters *21*, 4321 (1980)
32. Turro, N. J., Kiu, K.-C., Chow, M.-F.: Photochem. Photobiol. *26*, 413 (1977)
33. Worsham, P. R., Eaker, D. W., Whitten, D. G.: J. Am. Chem. Soc. *100*, 7091 (1078)
34. Breslow, R., Kitabatake, S., Rothbard, J.: ibid. *100*, 8156 (1978)
35. Czarniecki, M. F., Breslow, R.: ibid. *101*, 3675 (1979)
36. Mitani, M. et al.: Tetrahedron Letters (1979) 803
37. Breslow, R., Rajagopala, R., Schwarz, J.: J. Am. Chem. Soc. *103*, 2905 (1981)
38. Scaiano, J. C., Abuin, E. B.: Chem. Phys. Lett. *81*, 209 (1981)
39. Lougnot, D. J., Jacques, P., Fouassier, J. P.: J. Photochem. *17*, 75 (1981)
40. Braun, A. M. et al.: J. Am. Chem. Soc. *103*, 7312 (1981)
41. Sakaguchi, Y., Nagakura, S., Hayashi, H.: Chem. Phys. Lett. *72*, 420 (1980)
42. Sakaguchi, Y. et al.: ibid. *82*, 213 (1981)
43. Scaiano, J. C., Albuin, E. B., Stewart, L. C.: J. Am. Chem. Soc. *104*, 5673 (1982)
44. Kano, K., Matsuo, T.: Chem. Lett. (Chem. Soc. Japan) 1973, 1191
45. Tanimoto, Y., Itoh, M.: Chem. Phys. Lett. *83*, 626 (1981)
46. Tanimoto, Y., Udagawa, H., Itoh, M.: J. Phys. Chem. *87*, 724 (1983)
47. Fendler, J. H., Bogan, G.: Photochem. Photobiol. *20*, 323 (1974)
48. Fargues, R. et al.: Nouv. J. Chem. *3*, 487 (1979)
49. Lee, K. H., de Mayo, P.: Photochem. Photobiol. *31*, 311 (1980)
50. de Mayo, P., Sydnes, J.: J. Chem. Soc., Chem. Commun. 1980, 994
51. (a) Nakamura, Y. et al.: J. Chem. Soc., Chem. Commun. 1977, 887; (b) Nakamura, Y., Imakura, Y., Morita, Y.: Chem. Lett. 1978, 965
52. Muthurama, M., Ramamurthy, V.: J. Org. Chem. *47*, 3976 (1982)
53. Wolff, T.: J. Photochem. *16*, 343 (1981)
54. Hovey, M. C.: J. Am. Chem. Soc. *104*, 4196 (1982)
55. Moroi, Y., Braun, A. M., Gratzel, M.: ibid. *101*, 567 (1979)
56. Cox, G. S., Krieg, M., Whitten, D. G.: ibid. *104*, 6930 (1982)
57. Whitten, D. G., et al.: Ber. Bunsenges., Phys. Chem. *82*, 858 (1978)
58. Horsey, B. E., Whitten, D. G.: J. Am. Chem. Soc. *100*, 1293 (1978)
59. Sconfienza, C., van de Vorst, A., Jori, G.: Photochem. Photobio. *31*, 351 (1980)

60. Rossi, E., Van de Vorst, A., Jori, G.: ibid. *34*, 447 (1981)
61. Miyoshi, N., Tomita, G.: ibid. *29*, 527 (1979)
62. Moore, D. E., Burt, C. D.: ibid. *34*, 431 (1981)
63. Matheson, I. B. C.: ibid. *29*, 875 (1979)
64. Kraljic, I., Barboy, N., Leicknam, J. P.: ibid. *30*, 631 (1979)
65. Lindig, B. A., Rodgers, M. A. J.: J. Phys. Chem. *83*, 1683 (1979)
65. Gorman, A. A., Lovering, G., Rodgers, M. A. J.: Photochem. Photobio. *23*, 399 (1976)
67. Gorman, A. A., Rodgers, M. A. J.: Chem. Phys. Lett. *55*, 52 (1978)
68. Matheson, I. B. C., Lee, J., King, A. D.: ibid. *55*, 49 (1978)
69. Usui, Y., Tsukada, M., Nakamura, H.: Bull. Chem. Soc. Japan *51*, 379 (1978)
70. Matheson, I. B. C., Massoudi, R.: J. Am. Chem. Soc. *102*, 1942 (1980)
71. Miyoshi, N., Tomita, G.: Z. Naturforsch. *33B*, 622 (1978)
72. Lindig, B. A., Rodgers, M. A. J.: Photochem. Photobio. *33*, 627 (1981)
73. Hautala, R. R., Letsinger, R. L.: J. Org. Chem. *36*, 3762 (1071)
74. Bonilha, J. B. S. et al.: J. Phys. Chem. *83*, 2463 (1979)
75. Law, K.-Y., de Mayo, P.: J. Chem. Soc., Chem. Commun. 1978, 1110
76. Russell, J. C. et al.: J. Am. Chem. Soc. *102*, 5678 (1980)
77. Humphry-Baker, R., Gratzel, M., Steiger, R.: ibid. *102*, 847 (1980)
78. Shinkai, S. et al.: Tetrahedron Letters 1981, 1409
79. Yamada, K. et al.: Bull. Chem. Soc. Japan *51*, 2447 (1978)
80. Roessler, N., Wolff, T.: Photochem. Photobio. *31*, 547 (1980)
81. Wolff, T.: J. Photochem. *18*, 285 (1982)
82. Shinkai, S. et al.: Bull. Chem. Soc. Japan 1981, 1523
83. Bagno, O., Soulignac, J. C., Joussot-Dubien, J.: Photochem. Photobio. *29*, 1079 (1979)
84. Klein, U. K. A., Hauser, M.: Z. Phys. Chem. NF *96*, 139 (1975)
85. Klein, U. K. A., Hauser, M.: ibid. *90*, 215 (1974)
86. Selinger, B. K., Weller, A.: Aust. J. Chem. *30*, 2377 (1977)
87. Gratzel, M.: in "Micellization Solubilization, and Microemulsions", K. L. Mittal, ed. 1977, Plenum Press: New York vol. 2, pp. 531–548
88. Gratzel, M., Thomas, J. K.: J. Phys. Chem. *78*, 2248 (1974)
89. Wallace, S. C., Gratzel, M., Thomas, J. K.: Chem. Phys. Lett. *23*, 359 (1973)
90. Gratzel, M.: "Topics in Surface Chemistry", ed. E. Kay and P. S. Bagus, Plenum: New York 1978, pg. 103
91. Thomas, J. K., Piciulo, P.: J. Am. Chem. Soc. *100*, 3239 (1978)
92. Gratzel, M., Kalyanasundaram, K., Thomas, J. K.: ibid. *96*, 7869 (1974)
93. Iwaoka, T., Kondo, T.: Chem. Lett. 1978, 731
94. Alkaitis, S. A., Beck, G., Gratzel, M.: J. Am. Chem. Soc. *97*, 5723 (1975)
95. Gratzel, M.: in: "Micellization, Solubilization, and Microemulsions", vol. 2, Mittal, K. L., ed. Plenum Press: New York, 1977, pg. 531
96. Kiwi, J., Gratzel, M.: J. Am. Chem. Soc. *100*, 6314 (1978)
97. Katusin-Razem, B., Wong, M., Thomas, J. K.: ibid. *100*, 1679 (1978)

Arylated Phenols, Aroxyl Radicals and Aryloxenium Ions Syntheses and Properties

Karl Dimroth

Fachbereich Chemie der Universität, D-3550 Marburg, Hans-Meerwein-Straße, FRG

Table of Contents

1 Introduction

In spite of the immense quantity of literature available on alkyl-substituted phenols including their syntheses, chemical and physical properties, their great importance as intermediates in the synthesis of natural products, as well as their many applications in science or industry, much less is known about aryl-substituted phenols and their derivatives. Even in recent comprehensive reviews on phenols [1,2] no chapter specifically concerning arylated phenols has been included. What may be the reasons for this strange gap?

1) Contrary to the easy synthesis of alkylated phenols, e.g. via the electrophilic alkylations of phenols using alkyl halides, alkenes or other alkyl derivatives, there is no universal method for the preparation of arylated phenols. Radical or other arylation procedures usually lead to a mixture of difficult to separate isomers. Aryl-substituted phenols have, therefore, to be prepared by special routes.

2) Whereas the alkyl phenols can, in most cases, be easily characterized using chemical or physical methods, aryl substituted phenols are often more difficult to characterize. When 2,4,6-triphenylphenol is substituted by only one new substituent, e.g. by a bromine atom, seven isomeric monobromo derivatives have to be considered. Two bromo atoms in the same aryl-substituted phenol can give rise to forty isomeric dibromo derivatives. When the less symmetrical 2,3,5-triphenylphenol is substituted, the number is even greater (11 mono- and 65 dibromo isomers). It is sometimes rather difficult to establish the constitution of such a reaction product.

3) The nomenclature of arylated phenols is often complicated and several versions are found in the literature. In the subject registers of Chemical Abstracts it is sometimes difficult to find a particular compound, e.g. biphenylols are registered instead of phenylphenols. In the most recent subject index 2,4,6-triphenylphenol is found under [1,1′:3′,1″-terphenyl]-2′-ol, 5′-phenyl. 4-Methyl-2,6-diphenylphenol now is named [1,1′:3′,1″-terphenyl]-2′-ol, 5′-methyl[1]. These are relatively simple examples. Derivatives with substituents in the aryl ring positions have more complicated systematic names. Even an expert in this field may have difficulties in finding a compound or in understanding its true structure from the name found in the index. This of course, complicates any comprehensive review of these compounds.

4) The chemical reactions of aryl substituted benzenes and phenols are usually quite different from those of the alkyl substituted ones. The easy attack at an α-hydrogen atom of a primary or secondary alkyl group giving rise to an oxydized or otherwise substituted derivative is impossible with aryl substituents. t-Butyl groups behave quite differently to aryl groups. Whereas the aryl substituents are strongly bonded to the arene ring, t-butyl substituents are often easily substituted by other groups. The bromination of 1,3,5-tri-t-butylbenzene with iron as catalyst, gives rise to 1-bromo-3,5-di-t-butylbenzene [3]. The bromination of 2,4,6-tri-t-butylphenol in tetrachlorobenzene at 20 °C gives 2-bromo-4,6-di-t-butylphenol via substitution of a t-butyl group by a bromine atom. Many other substitutions of this kind are known. Bromination of 2,4,6-tri-t-butylphenol in acetic acid and methanol at −10 °C,

---

[1] In this review we use phenol as a basis substance and the aryl rests as substituents in the same manner as is used with alkylated phenols.

however, leads to 4-bromo-2,4,6-tri-t-butyl-2,5-cyclohexadienone [4]. Under similar conditions 2,4,6-triphenylphenol is substituted in the 4-phenyl ring to yield 4-(4′-bromophenyl)-2,6-diphenylphenol [5].

5) Substantial differences are found, when the two radicals, 2,4,6-tri-t-butyl-phenoxyl (*2*) and 2,4,6-triphenyl-phenoxyl (*5*) are compared. The former was independantly discovered by Cook [6] and by Müller and Ley [7], who have studied in detail its chemical and physical properties. It is the prototype of the otherwise rather unstable phenoxyl intermediates and is stabilized by the bulky t-butyl groups. The radical is a deep blue crystalline and persistant compound. Air gives the 4,4′-bis-(2,4,6-tri-t-butylcyclohexa-2,5-dienone)peroxide. No dimer *3* is known. The second phenoxyl derivate *5* which is stabilized by strongly electron delocalizing aryl groups, is deep red, somewhat depending on the solvent [8], and equilibrates with its colorless dimer *6*. It is absolutely inert to oxygen [9]. Its oxidation potential (see 4.3.1.3) is much higher than that of *2* [10]. Since the aryl residues can be easily converted into electron attracting or donating groups, the variability of these radicals is very large.

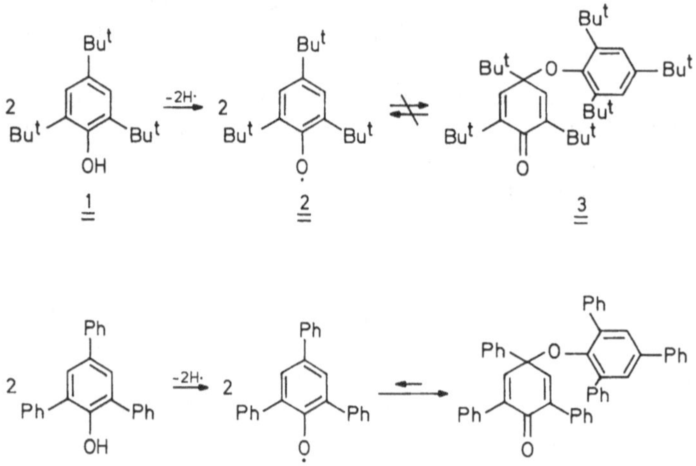

This review includes many hitherto unpublished results concerning aryl substituted phenols, their synthesis and their chemical and physical properties, which will supplement the many available publications of t-butyl substituted aroxyls by Cook [11] and Müller [12] or other authors [13–18] and the many reviews of this type of sterically hindered phenoxyls [19].

It is hoped that further study of the neglected class of arylated phenols and their oxidation products will lead to a revival in their research and practical applications.

2 Syntheses

The simplest route is by a direct substitution of a hydrogen atom in an arylated benzene for an hydroxyl- or an acyloxy group, which can be easily hydrolysed to the corresponding phenol. Unfortunately there are important obstacles to such a synthesis:

i) Only a few of the arylated benzenes are commercially available, e.g. biphenyl or 1,3-, 1,4-diphenylbenzenes and 1,3,5-triphenylbenzene. All the other arylated benzenes, especially those with substituted phenyl rings in a precise position, have to be prepared by special procedures which are often cumbersome and expensive. 1,3,5-Triphenyl-benzene is easily available by the condensation of three Mol acetophenone [20], an inexpensive by-product of the phenol synthesis from cumene.

ii) Contrary to biphenyl, where only three position isomeric phenols exist, the higher phenyl substituted benzenes can give rise to a large number of phenols. No clear-cut route exists which allows the substitution of just one of the many aromatic hydrogen atoms leading to a particular phenol.

iii) When more than one of the isomers is formed, which is normally the case, the separation and the exact determination of the structure can be rather difficult. It is, therefore, not surprising that nearly all phenol synthesis from an arylated benzene are accomplished with biphenyl because in this case only three substitution products are possible.

Similar problems arise when conventional syntheses of phenols from aromatic hydrocarbons are applied, e.g. sulfonation, nitration, halogenation etc., followed by the conversion of the introduced functional group into the phenolic hydroxyl group. An important exception has been observed with 1,3,5-triphenylbenzene. When it is nitrated, contrary to an earlier statement [21], the nitro group attacks one of the three equivalent positions of the central ring giving rise to 2,4,6-triphenylnitrobenzene in excellent yield [22]. Reduction, diazotation and hydrolysis of the nitro group then leads to 2,4,6-triphenylphenol (*4*) [23]. Bromination [24] and chlorination [25] of 1,3,5-triphenylbenzene give also the 1-halogeno-2,4,6-triphenyl derivatives in nearly quantitative yields. Sulfonation, however, leads mostly to a mixture of sulfonic acids [26].

Only one other synthesis from this review, the aldol self-condensation of cyclohexanone, followed by dehydrogenation of the initial products, is of preparative importance. It has even been used in the large scale industrial syntheses of 2,6-diphenylphenol (see 2.4). All the other routes mentioned in this account, are primarily of scientific interest and are not useful for larger scale preparative work. They are, however, of interest insofar as they emphasize differences between the chemical properties of alkyl and aryl substituted phenols.

Finally, in sect. 2.9, the conversion of 2,4,6- or higher substituted pyrylium salts into the corresponding nitrobenzene derivatives through reaction with nitromethane [27] will be discussed in more detail. Since pyrylium salts of any desired complexity can be easily prepared [28], the conversion of the nitroarenes leads to a large number of phenols of the desired constitution. This method is by far the most versatile in the preparation of a desired arylated phenol. These syntheses, however, involve many steps.

2.1 Hydroxylation and Acyloxidation of Arenes

In spite of the essential difficulties mentioned above concerning the direct substitution of a particular hydrogen atom of multi phenylated benzenes giving rise to a well defined phenol or phenol ester, much work has been done [29]. Its emphasis, however, has been on mechanism rather than on synthesis. In this account we will only discuss some of the more promising synthetic routes of arylated phenols. An interesting example may be that of 2,4,6-triphenylphenol via direct hydroxylation or acyloxylation of 1,3,5-triphenylbenzene.

2.1.1 Hydroxylation Using Hydroxyl or Acyloxy Radicals or Cations

Hydroxylations of arenes using hydroxyl radicals have been throughly studied and excellent reviews are available [30]. Unfortunately, important difficulties prevent general applicability. Radical attack on arenes shows poor regioselectivity. The resulting hydroxylated arenes are more easily oxidized than the starting hydrocarbons, giving mixtures of higher hydroxylated compounds as well as other compounds. Strongly electrophilic trifluoroacetyl radicals or cations, produced by an homolytic or heterolytic cleavage of trifluoroperacetic acid or its peroxide in the presence of a small amount of sulfuric acid in dichloromethane seems more promising [31]. When these reagents are reacted with electron rich arenes, only ortho and para trifluoroacetates, which are not further substituted, are formed in acceptable yields. As far as can be ascertained no experiments with diphenyl or higher phenylated benzenes have been reported.

2.1.2 Electrochemical Acyloxidations

At first sight, electrochemical oxidation of arenes with controlled potential of the anode would appear to be the most promising route to the desired phenols and their derivatives. But as Eberson, Schäfer, Nyberg, et al. [32] have shown in their reviews, serious difficulties such as poor regiospecificity, a high tendency to overoxidation and low yields are encountered. On the other hand, these methods are very important in obtaining a deeper insight into the mechanism of such oxidations [33]. When electron rich arenes, such as anisole or biphenyl, are electrolysed in acetic acid/sodium acetate using a carefully controlled anodic potential, the first electron transfer involves the arene, and not, as formerly has been supposed, the acetate ion. This has been established by determining the oxidation potentials using either polarography in acetonitrile [34] or in acetic acid containing 0,5 M sodium acetate vs. a saturated calomel electrode [35]. Under the latter set of conditions anisole gives $E_{1/2} = 1.67$ V and biphenyl, $E_{1/2} = 1.91$ V. The potentials are lower than that of the acetate ion ($E_{1/2} = 2.0$ V). If an acetate radical were formed, this would be relatively unstable loosing carbon dioxide to give a methyl radical which would then attack the arene. In agreement with the electrochemical measurements no arene methyl derivatives have been observed. The now generally accepted mechanism of anodic oxidation involves an initial transfer of an electron from the arene derivative to the anode which gives rise to the arene cation radical. In some cases this cation radical can be directly observed by ESR spectroscopy. Addition to the acetate ion gives rise preferentially to the ortho- or para acetoxylated cyclohexadienyl radicals. In a second anodic electron transfer step, these radicals are converted into the corresponding cyclohexadienyl cation derivatives which are in turn deprotonated to give the acetoxylated arene derivatives (see scheme 1).

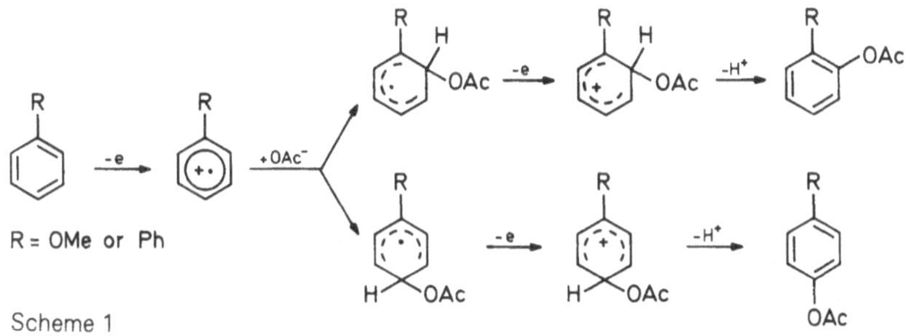

Scheme 1

The isomeric distribution of the anisole and biphenyl reactions, determined by gas phase chromatography of a small scale electrolysis in the neighbourhood of the anode has been found to be 2-, 4- and 3-acetoxyanisole, (63%, 29%, 4%, respectively) and 2-, 4- and 3-acetoxybiphenyl, (31%, 68%, <1%, respectively)[36]. Longer reaction time in the biphenyl reaction gives a mixture of 2- and 4-acetoxy biphenyl (yield 37%) as well as higher oxidized products and tarry polymers. This can be explained by looking at the half potentials of 2-acetoxybiphenyl ($E_{1/2} = 2.04$ V) or 4-acetoxy biphenyl ($E_{1/2} = 1.9$ V) which are measured polarographically. These compounds which are produced in the immediate surrounding of the anode have potentials which are not far from that of biphenyl itself ($E_{1/2} = 1.91$). They can easily be further oxidized to give a mixture of different products. The preparative value of this method in large scale syntheses of these and other phenol acetates seems, therefore, to be limited. The electrolysis in trifluoroacetic acid/sodium trifluoroacetate[37] appears more promising. The solvent mixture nitromethane/trifluoroacetic acid 2:1 using sodium trifluoroacetate as nucleophile has also been reported. Preparative yields of 65% or even 80% of phenol trifluoroacetates are claimed when benzene or some benzene derivatives are anodically oxidized[38].

2.1.3 Acyloxidation Using Chemical Oxidants

Two different mechanisms will be considered. The first strongly resembles the electrolytical route in scheme 1. Presumably an initial step gives a reversible charge transfer complex between the arene and the oxidant[39]. The subsequent electron transfer leads to the arene cation radical, probably by a chain reaction. Experiments involving strong oxidizing metal complexes, such as cobalt(III) salts (Heiba[40]), cobalt(III) trifluoroacetate (Kochi and coworkers[41]), Ag(II) (2,2'-bipyridyl)-persulfate and potassium persulfate in acetic acid, sodium acetate and barium acetate (Nyberg and Wistrand[42]), are in agreement with the mechanism presented in scheme 1. Although Nyberg and Wistrand have also investigated biphenyl as a starting material preparative yields were not reported. Liquid gas chromatography analysis shows an isomeric distribution of 2-, 4-, and 3-acetoxybiphenyl, (23%, 77%, <1%, respectively). These results are very similar to those of the anodic oxidation.

The second mechanism which cannot always be strictly differentiated from the first one[41] should be considered as a typical electrophilic substitution of the arene by the high valent metal compound. The latter is still more electrophilic with trifluoroacetate than with acetate anions. The organometallic arene intermediates were isolable in some cases[43]. In the second step, the metallic residue is substituted by the acetate or trifluoroacetate ion.

These routes have been realized with electron rich arenes and the acetates or trifluoroacetates of metals, such as lead(IV), cobalt(III), mercury(II), thallium(III), manganese(III), possibly palladium(II), etc. When these reactions are performed in the presence of cooxidents, the metal ions are reoxydized and are important catalysts[42]. With higher phenylated benzenes no experiments are reported. Biphenyl gives a 98% yield of 4-trifluoroacetoxy-biphenyl when treated with lead tetrakis-trifluoroacetate in trifluoroacetic acid[43]. It would appear worthwhile to study similar trifluoroacetoxylations using higher phenylated benzenes. 1,3,5-Triphenylbenzene would be the most interesting candidate. When biphenyl has been acetoxylated in acetic acid with palladium(II) acetate in an oxygen atmosphere for 4 h at 115°, apart from traces of 2- and 4-acetoxybiphenyl, only 3-acetoxybiphenyl is formed[44]. This result is difficult to understand. A possible explanation is a thermolytic rearrangement of the primary products to the most stable meta isomer. Preferential acetoxylations of methyl substituted arene derivatives, e.g. 1-, 2- or 3-methylanisole, using cerium(IV)-hexanitrodiammonium salt in acetic acid occurs in the side chain. When cobalt(III) acetate is used, higher temperatures (60° vs. 40°) are required[45]. In conclusion, none of the phenol syntheses via direct substitution of a particular hydrogen atom of a higher phenylated benzene by a hydroxyl- or an acyloxy-group is developed to the extent that it can be used in large scale preparations. Photochemical hydroxylations, reviewed in Ref.[46] are however even less suitable.

2.2 Arylation of Phenols and Phenol Derivatives

2.2.1 Homolytic and Arylmetal Arylations

Homolytic aromatic arylations of phenols, in particular phenolethers, by an aryl radical are strongly limited beause of position and substrate selectivities. The many modifications of the well known Gomberg-Bachman-reaction[47] are reviewed in Ref.[48]. An interesting variation, using titanium(III) sulfate as oxidant and presumably also as a complexing agent of the phenol, and the phenyldiazonium reagent in a water suspension at 2–3 °C gives a 12.5:1 mixture of 2- and 4-phenylphenol (yield 35%)[49]. 2.4-Diphenylphenol has been synthezised from 4-hydroxybiphenyl by

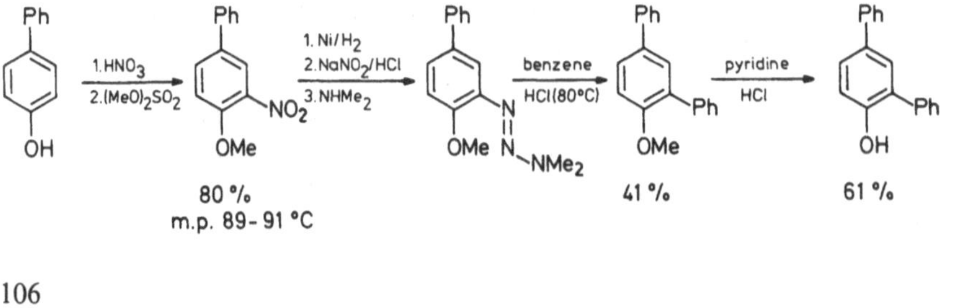

nitration, methylation, reduction of the methylether, diazotation and reaction with dimethylamine, to give initially 4-phenyl-anisyl-2-(3,3-dimethyltriazene), m.p. 78 to 79 °C. The crude product is dissolved in a large excess of benzene and heated at 80 °C 1.5 h. Hydrogen chloride is then passed through the solution. After destilling off the benzene chromatography of the resulting brown oil yields 40% of 2,4-diphenylanisole, m.p. 92–93 °C. Hydrolysis using pyridine/pyridine hydrochloride gives 61% of 2,6-diphenylphenol m.p. 90° [50].

Arylations of phenols using aryl-lead-triacetate [51] or other lead triesters, probably give aryl lead intermediates first [52]. Only alkyl substituted phenols have been reported. 2,4,6-Trimethylphenol gives the 2-arylated cyclohexa-3,5-dienone derivative. The ortho and para phenylation of phenol with lead tetraphenyl is reported in a US patent [53]. Depending on the reaction conditions, 2-phenyl-, 2,6-diphenyl- and even 2,4,6-triphenyl phenol are claimed to be the reaction products. The arylation of phenol with benzene in the presence of benzene and mercury(II) acetate, palladium(II) salts and oxygen, to give biphenyl as well as its hydroxy derivatives, is also described [54]. Triphenylbismuth carbonate or pentaphenylbismuth are usually used in the arylation of 1- and 2-naphtols as well as of alkylated phenols. 2,4-Dimethylphenol gives rise to 2,4-dimethyl-6-phenylphenol in 26% yield [55].

2.2.2 Arylation of Arylhydroxylamine Derivatives and Aryliodo- or Chlorophenols

When N-tosylated phenylhydroxylamines are treated with trifluoroacetic acid in the presence of benzene, 2- and 4-hydroxybiphenyl are found in yields of 25% and 20%, respectively. The aryloxenium ion is believed to be the intermediate [56]. No higher arylated phenol derivatives have been examined.

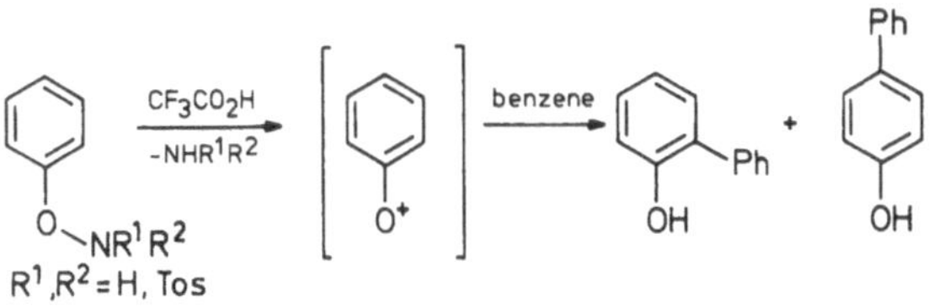

A laboratory synthesis of phenylated phenols has been developed by N. Kharasch and co-workers. They first introduced iodine in the 2-, 4- and/or 6-positions of phenol and then photolysed the iodophenols in the presence of benzene. Phenylated phenols in yields of 60–75% were found [57].

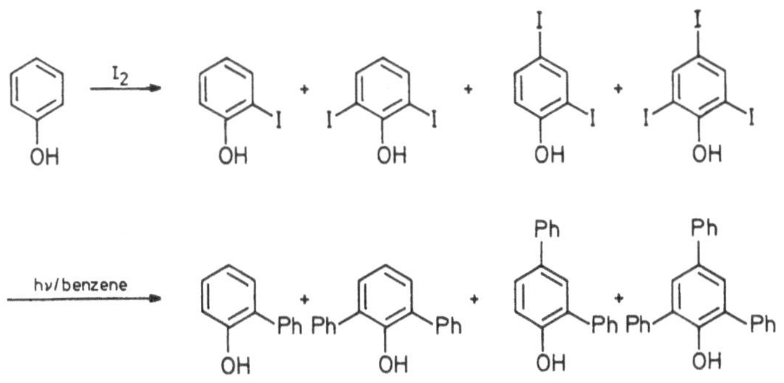

4-Cyanophenol, which is first transfered into 4-cyano-2,6-diiodo-phenol, can be photolysed with benzene to give 4-cyano-2,6-diphenylphenol in 71% yield [58]. The latter is identical with the compound prepared from 2,6-diphenyl-4-nitrophenol [59].

Another example is the arylation of 1,3,4-triphenylresorcinol which after iodation gives 6-iodo-2,4,5-triphenylresorcinol. Photolysis in benzene affords 2,4,5,6-tetraphenylresorcinol [60].

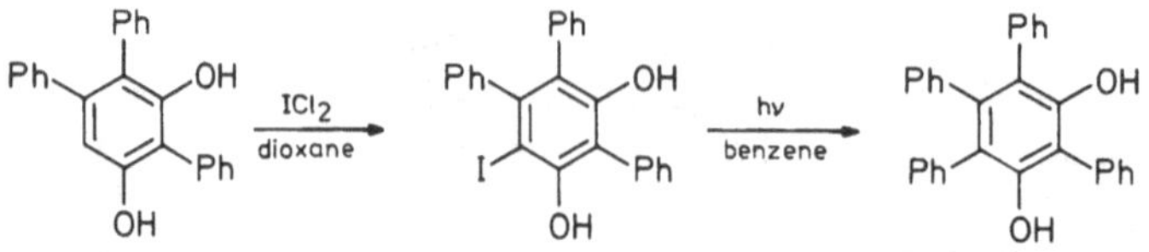

When 2-chloro-6-phenyl-phenolmethylether is transfered into the Grignard compound and then reacted with cyclohexanone, dehydrated and finally dehydrogenated, 2,6-diphenylphenolmethylether is formed [61]. The synthesis is versatile since a variety of cyclohexanone-derivatives can be used. This and other phenolethers can then easily be demethylated. One of the mildest demethylations is to react the methylether with borontribromide at low temperatures [62]. Hydroiodine acid, Lewis acids, such as aluminium-chloride [63], pyridine/pyridinium chloride, or sodium in pyridine [64] are also valuable tools in ether demethylations.

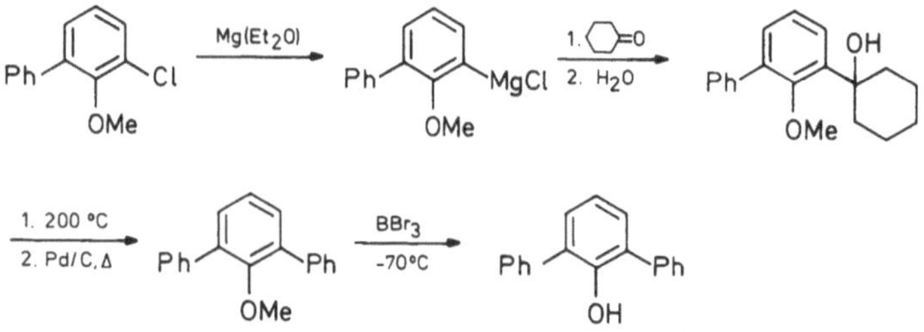

Chlorination or bromination of 4-phenyl-phenolesters is described in Ref. [65].

Direct alkylation of phenol with cyclohexene in the presence of aluminium phenolate gives 2-cyclohexenylphenol (42%) which after hydrogenation gives 2-cyclohexylphenol [66]. Dehydrogenation gives 2-phenylphenol [61]. Further arylation of 2-phenylphenol can be used to synthesize 2,6-diphenylphenol.

2.3 Transformation of Arene Substituents (NO_2 Halogens SO_3H etc) into Phenols

In the case of arylated benzenes, the usual phenol syntheses, the transformation of aryl substituents into a hydroxyl group, are very limited, unless the regiospecificity of the substitution is well known (see p. 103). Nitration of biphenyl gives good yields of 2- and 4-nitrophenyl. The ratio depends on the nitration conditions. 3-Nitro-biphenyl can be prepared by nitrating 4-amino-biphenyl and removing the amino group via diazotation and reduction.

It was previously mentioned, that the nitration of 1,3,5-triphenylbenzene should give 1-(4′-nitrophenyl)-3,5-diphenylbenzene (see p. 103). The independent synthesis of 2,4,6-triphenylnitrobenzene from 2,4,6-triphenylpyrylium salt and nitromethane, however, has shown that the nitro group goes in to the central ring of 1,3,5-triphenyl-benzene [22]. Catalytic reduction to the amine and conversion to the diazonium salt give the 1-hydroxy-, 1-fluoro-, 1-chloro-, -or 1-bromo-2,4,6-triphenylbenzene, respectively. The latter compound is identical with that found by Kohler and Blanchard [24] which was prepared in quantitative yield by direct bromination of 1,3,5-triphenylbenzene. But in his synthesis the possibility of 1-(2′-bromophenyl)-4,6-diphenylbenzene as an alternative cannot be excluded. The latter could be formed by substituting one of the six ortho hydrogen atoms of the external 1,3,5-phenyl rings and could give the same reaction products as in Ref. [24].

When biphenyl is either chlorinated or brominated in the presence of iron(III) or antimony(V)chloride, 2- and 4-halogenated biphenyls are formed. Their hydrolysis, however, requires, strong conditions such as heating in fused potassium/sodium hydroxide with copper at 250–300°. This can cause some isomerization to 3-hydroxy-biphenyl via an aryne mechanism [67].

The industrial phenol synthesis of Dow or Bayer via hydrolysis of chlorobenzene with sodium hydroxide at 360–390° is well reviewed in Ref. [1].

Lüttringhaus et al. [68] have isolated many interesting substances from the by-products (~8%) of the Bayer process. The mechanism has been fully clarified and shown to be an aryne route. When chlorobenzene or diphenylether are treated with sodium phenyl, the products are ortho metalated derivatives and benzyne. These give the same products which are formed in the industrial phenol synthesis. The most interesting compounds are 2- and 4-hydroxy biphenyl, 2,6-diphenyl- and 2,4-diphenyl-phenol [1]. For similar syntheses see [69].

Another method of transforming halogenaryls into phenols is to first make the Grignard or lithium compounds which are then oxidized by hydrogen peroxide, dibutylperoxide or other oxidants [1,70] to give mainly modest yields of the phenols. Better yields are attained, when the metalated arenes are first reacted with boric esters to give $ArB(OR)_2$. Oxidation with hydrogen peroxide then gives the phenols, e.g. 3-hydroxybiphenyl in 71% yield [71]. Another interesting route is when a halogen,

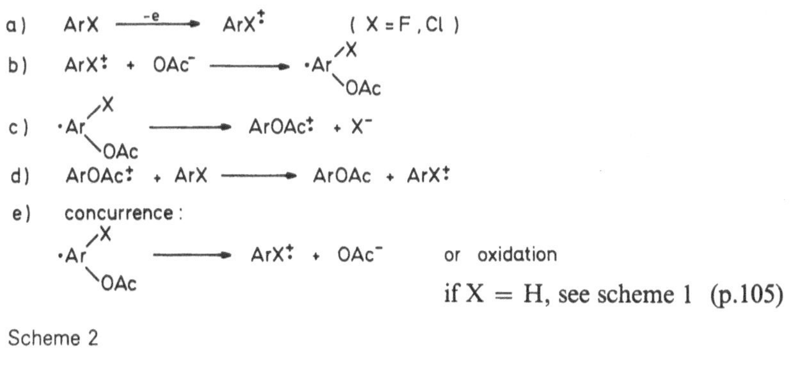

Scheme 2

preferably fluorine, is exchanged for an acetoxy group [72]. The first step of mechanism which is similar to the oxidative acetoxylation of arenes is shown in scheme 1. It is an anodic or metal induced oxidative nucleophilic substitution of the fluoro- (or chloro-) arenes in a radical chain process, called $S_{ON}2$ (see scheme 2). Though no arylated arene halides have been examined so far, the oxidation of fluorobenzene give phenol, usually with Cu(II)- or Fe(III)-$S_2O_8^{2-}$ with yields of 30%, may be interesting enough to extend this reaction to aryl substituted arylhalides [73]. Contrary to the aryne substitution, Cu(II)/O_2 is necessary for the ipso substitution [74a]. One important difference to the mechanism of scheme 1 is the influence of the good leaving group. After addition of the acetoxy ion to the radical cation according to step b), a fluoro- (or chloro-) ion instead of an acetoxy anion leaves the intermediate cation. Step c) produces the acyloxy arene cation as the chain carrier which again oxidizes the arene halide to its cation radical (step d):
Another interesting reductive radical arene substitution, the $S_{RN}1$ mechanism has been reviewed by Bunnett [75]. It has much in common with the interesting side chain radical substitions studied by Kornblum [76] and by Russell [77]. Nothing, however, is known about phenol syntheses from arylated arene halides. Sulfonation of biphenyl with either sulfuric acid or stronger sulfurating agents, gives mainly biphenyl-4-sulfonic acids together with higher substituted sulfonic acids [77b]. Alkali fusion has been reported to give 4-hydroxybiphenyl [78]. But the conditions are rather rough and no higher arylated phenols have been prepared by this route.

2.4 Aldol-Like Self-Condensation of Cyclohexanone and its Dehydrogenation

The most important industrial process for the production of 2-phenyl- and 2,6-diphenyl-phenol, is based on the long established self-condensations of cyclohexanone under either acidic, or preferentially basic conditions to give mono- or di-substituted aldol-like condensation products [79]. These easily loose water giving semicyclic or endocyclic ring double bonds. Plešec et al. have studied these reactions extensively [80].

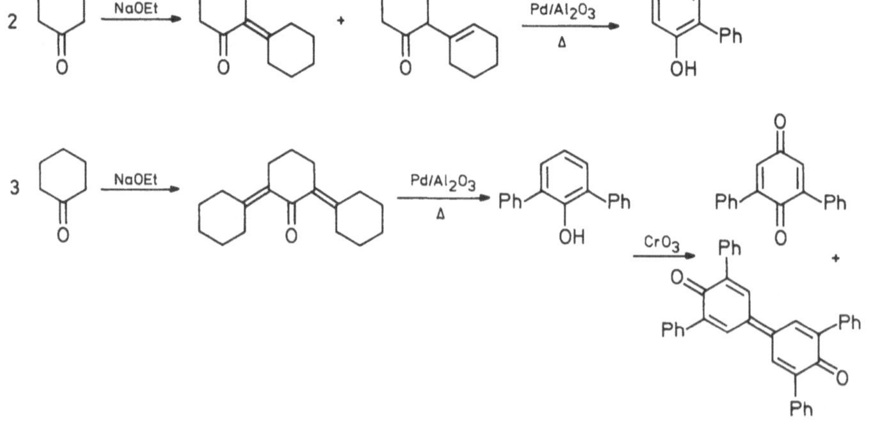

Dehydration and dehydrogenation lead to 2-phenyl- and 2,6-diphenylphenol. Oxidation of the latter with an oxidant e.g. chromiumtrioxide give rise to 2,6-diphenylquinone and 2,2′,6,6′-tetraphenyl-diphenoquinone [80]:

A commercial synthesis of 2,6-diphenylphenol is reported by Hay at General Electric, USA [81]. First, cyclohexanone was condensed with 50% sodium hydroxide at 150–190 °C giving the 2-mono- and the 2,6-disubstituted cyclohexanone derivatives. In the second step, after removal of water and sodium hydroxide, these are dehydrogenated at 300–350 °C with a palladium aluminium oxide catalyst for 20 minutes (e.g. 45% total yield of 2,6-diphenyl-phenol). It is a useful compound in technical production and has been studied by Hay and coworkers [82] (see also [83, 84]). Polymeric diphenylphenol ethers ("Tenas®") [85] or copolymers with polystyrene ("Normyl®") have been produced on a large scale by General Electric e.g. as thermoplasts [86].

Other oxidants or bulky groups (e.g. t-butyl) in the 2,6′-position of the phenol causes a C-4–C-4′-carbon bonding, which gives 2,2′,6,6′-tetraphenyl or -t-butyl-diphenoquinone. Temperatures less than 25 °C and small 2,6-substituents (e.g. Me) favour the phenyl-O/C-4 coupling, whereas C-4/C-4-coupling to dipheno-1,4-hydroquinones requires a temperature of 100 °C. Reaction conditions such as small or bulky substituents in 2,6-positions, temperature, Cu/phenol rations or pyridine/Cu rations) have been discussed by Davies et al. [87] (see also Sect. 4.2.1.5)

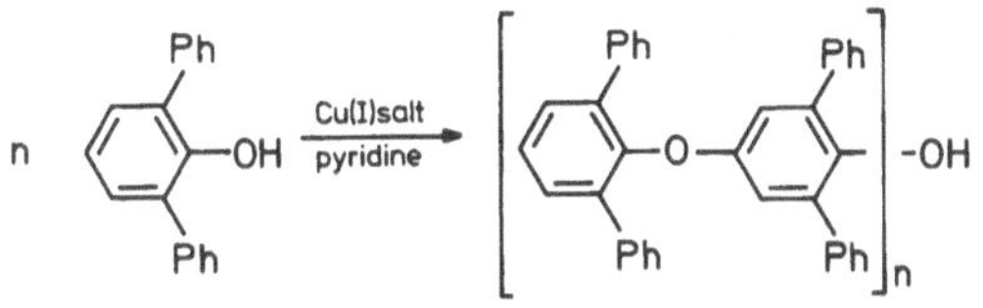

Ullmann coupling of 4-iodo-tri-methylsilyl-phenolether by hydroxy-arylation gives 4,4′-hydroxy-biphenyl in 55% yield [88].

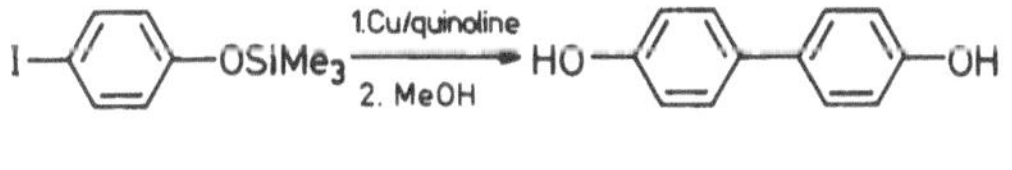

2.5 Other Routes Using Cyclohexanone and 1,3-Cyclohexadione (Dihydroresorcinol)

The α-arylation of ketones, such as cyclohexanone, can be achieved using different methods. A convenient route by Pinhey et al. [89], reacts cyclohexanone-2-carboxylic esters with aryllead triacetates in pyridine. The protection of the β-carboxylic ester prevents α,α-di- or even higher arylations in α′-positions. The ester group can be removed by basic hydrolysis and mild thermal decarboxylation or by heating in wet dimethylsulfoxide with sodium chloride (120–180 °C) [90]. Barton et al. [91] have found a similar α-arylation route using the less electrophilic triphenylbismuth carbonate. In both cases probably the lead- or bismuth-enolates, respectively, are the first intermediates. The facile cleavage of the metal aryl bond then affords the 2-aryl-2-ethoxycarbonylcyclohexanone:

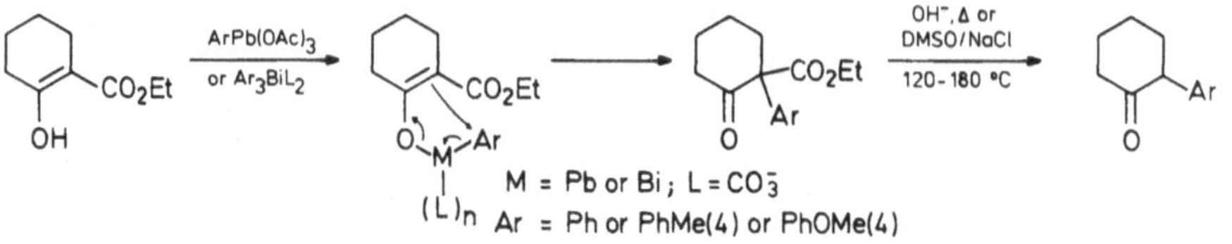

25

A useful modification, which prevents 2,2-diarylations, first introduces a 2-hydroxymethylene group [92].

The reaction with 4-tolyl or anisyl lead-triacetates in THF/pyridine at room temperature gives rise to the 2-aryl, 2-formyl-cyclohexanones which on hydrolysis afford the 2-aryl-cyclohexanones in 70–74% yield.

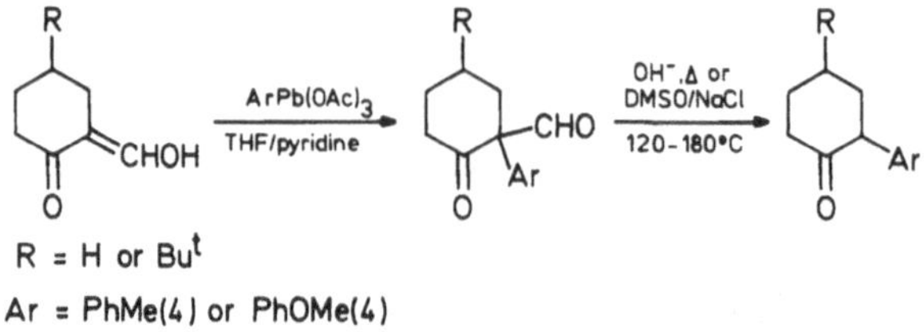

Dehydrogenation to give the corresponding arylated phenols has so far not been reported, but this should be possible using established routes.

Dihydroresorcinol or its monoethylether have been reacted with phenyl-magnesiumbromide to give 3-phenylcyclohexanone. Palladium carbon catalyst [93] in the presence of cymine (different solvents are examined) gives 3-hydroxybiphenyl (65% yield) [94].

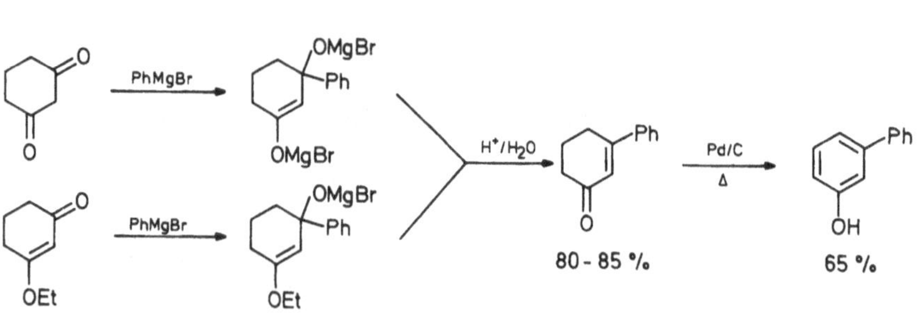

Since the arylated 1,3-cyclohexadienone derivatives are usually synthesized from smaller building blocks, their reactions will be found in Section 2.6. This is also true for the cyclohexanone derivatives.

3-Phenylphenol can be produced either by a Cu(I) catalysed phenyl-Grignard reaction [95] or by $LiCuPh_2$-addition [96] to cyclohex-2-enone and subsequent dehydrogenation. It is obtained in good yields when 3-phenyl-2-cyclohexenone is reacted, preferentially in cymene, with palladium-carbon [97]. 3-Phenyl-2-cyclohexenone can in turn be prepared from resorcino-monoethylether and phenylmagnesium bromide.

2.6 Aliphatic Building Blocks

Many arylated cyclohexanone and 1,3-cyclohexadione derivatives are synthesized through aldol condensations, in many cases in combination with Michael additions. Some examples are given below; Petrows ketone [98] via bromination at low temperature, hydrolysis and decarboxylation gives 3,5-diphenylphenol [99]:

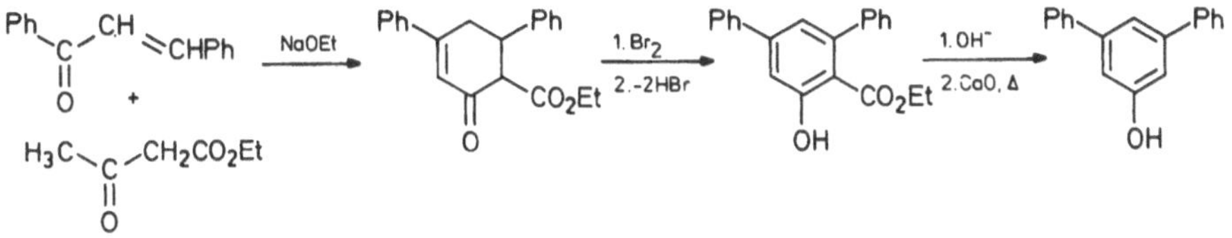

Aromatization of unsaturated ketones using bromosuccinimide and dehydrobromination is also a useful procedure [100]. Similar syntheses, giving 3-phenyl, 3,4-diphenyl- and 3,4,5-tri-phenylphenol are reported by Downes [101] and by Egli et al. [102].

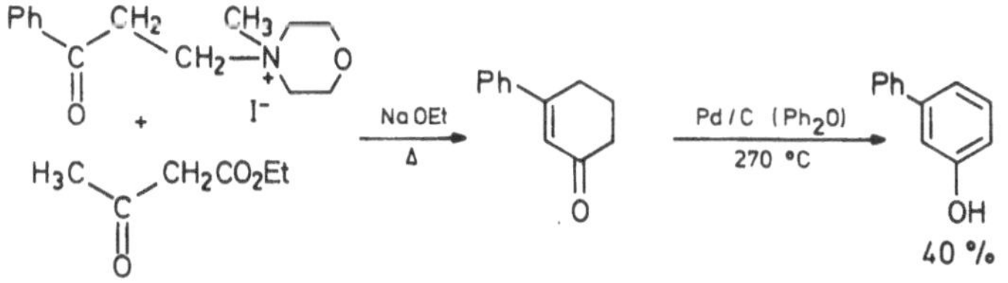

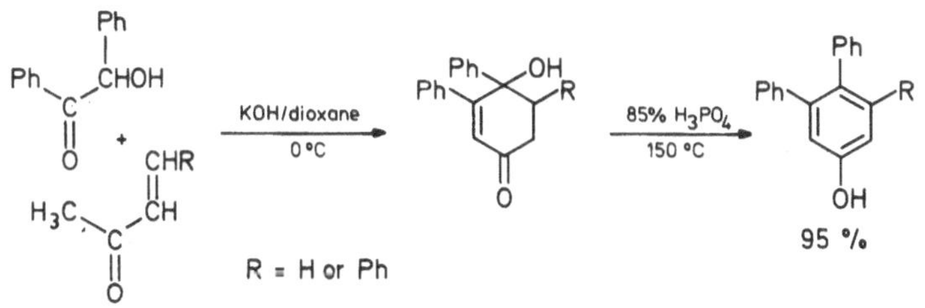

While examining the photoarrangement of 4,4-diphenylcyclohexa-2,5-dienone, Zimmerman synthezised the long sought 3,4-diphenylphenol and 2,3-diphenylphenol using β-trimethylammonium-ethyl-methyl-ketone iodide and benzylphenylketone, or benzalacetone and ethylphenylacetate [103].

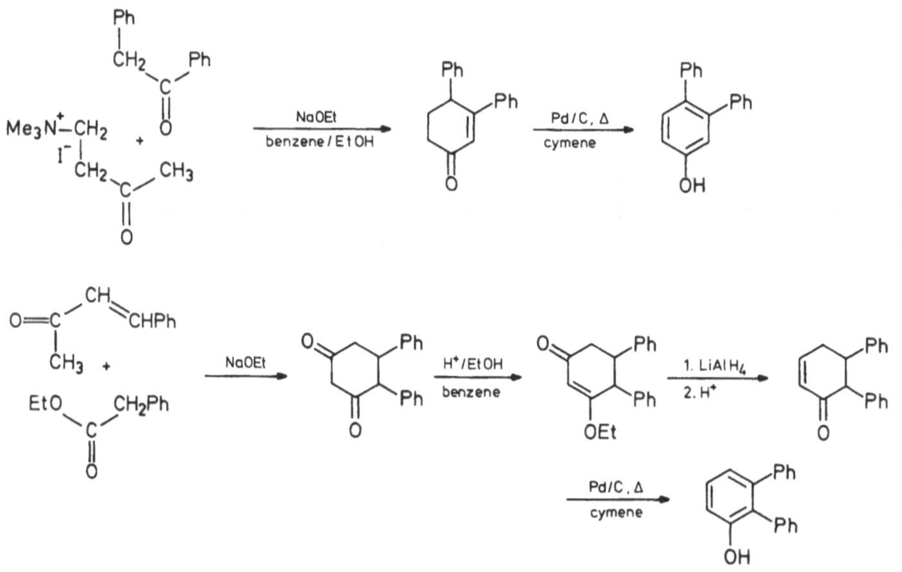

The Hill synthesis [104] of 4-nitro-2,6-diphenylphenol from nitro-malonaldehyde and dibenzylketone was the first example of a useful synthesis by which many similar 2,6-diaryl-substituted cyclohexenone- and 2,6-diarylphenol derivatives have since been prepared.

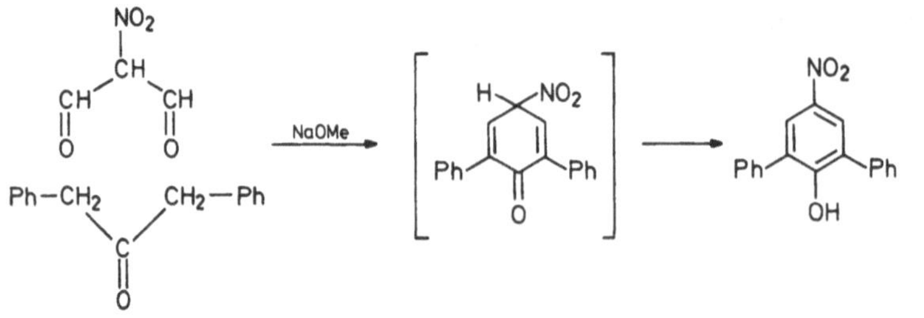

Yates and Hyre [105] synthesized the three isomeric 2,3,5,6-, 2,3,4,6- and 2,3,4,5-tetraphenylphenols (see p. 27 also) by the following route:

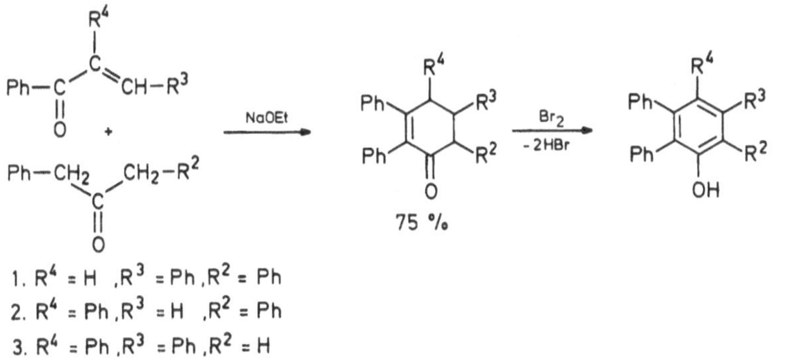

2,3,6-Triphenylphenol has been similarly prepared via the condensation of dibenzylketone with cinanamalaldehyde to give initially 2,5,6-triphenylcyclohex-2-enone. When this is treated with NBS in CCl_4 and dehydrobrominated with lutidine, 2,3,6-triphenylphenol is formed in 72–93% yield [106].

A convenient variation is the condensation of 3-phenylaryne-phenylketone and dibenzylketone to give 2,3,4,6-tetraphenylphenol in 78% yield [107].

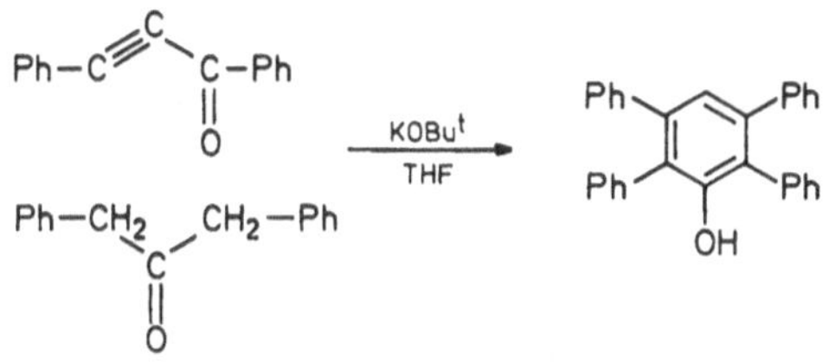

Other syntheses of arylated phenols have been developed by Suter [108], Davey [109], Harrison [110] among others. When benzalacetone is condensed with sodium malonic ester, 4-ethylcarboxy-5-phenyl-1,3-dioxocyclohexanone is formed. Hydrolysis and decarboxylation gives the 1,3-diketo derivative, the sodium salt of which on treatment with diethylsulfate yields the enolethylether. Dehydrogenation using sulfur gives 3-ethoxy-5-phenylphenol [108].

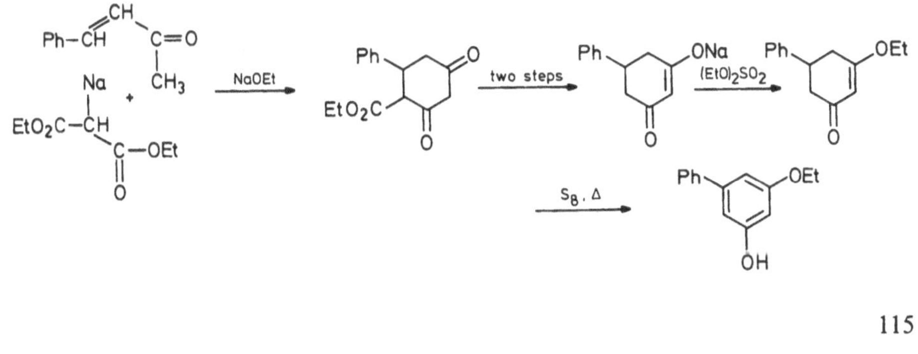

Similar syntheses of arylated 1,3-dioxo-cyclohexanone derivatives also lead to arylated phenols. The enolic methoxyl group of the enolketone is removed using lithiumaluminiumhydride. Acetone cyanhydrin transforms the enolic methoxyl group into a cyano group. Arylated cyano phenols are also available [109] by this route:

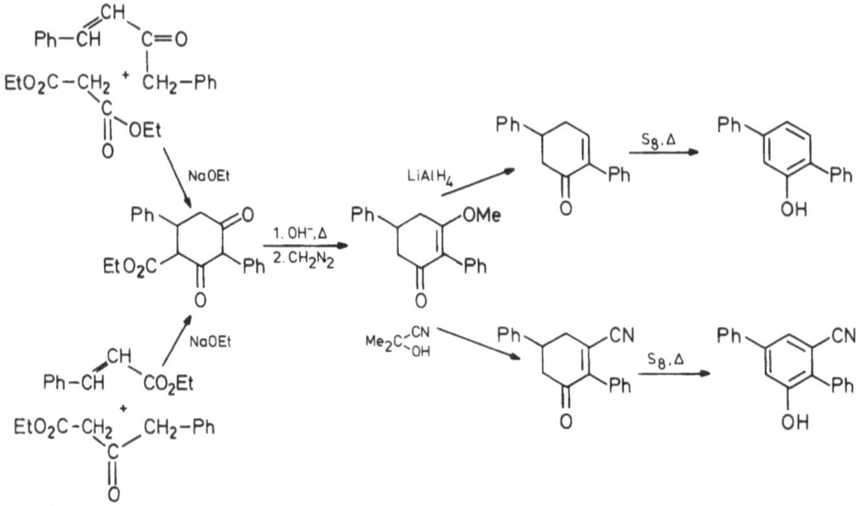

2,6- and 2,4-Diphenyl-phenols as well as 3-cyano-2,6-diphenyl-phenols can be similarly synthesized. An interesting route to 3-phenylphenol uses the reaction of methyl,1-oxo-3-phenyl-2-cyclohexene-2-methylacetate with sodium hydride in DMF at 100 °C (28–37% yield [101b)])

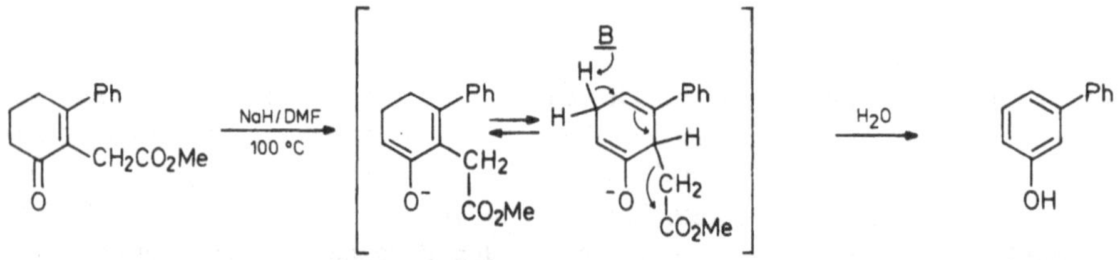

[2+4]Addition of vinylene carbonate to 1,2,3,4-tetraphenylcyclopentadieneone gives 80% of the Diels Alder addition product which upon thermolysis or photolysis yields 95% of 2,3,4,5-tetraphenylphenol [105]. Other substituted alkyl-aryl-phenols are synthesized by a similar route, using 2-alkyl-3,4,5-triphenylcyclopentadienones [110].

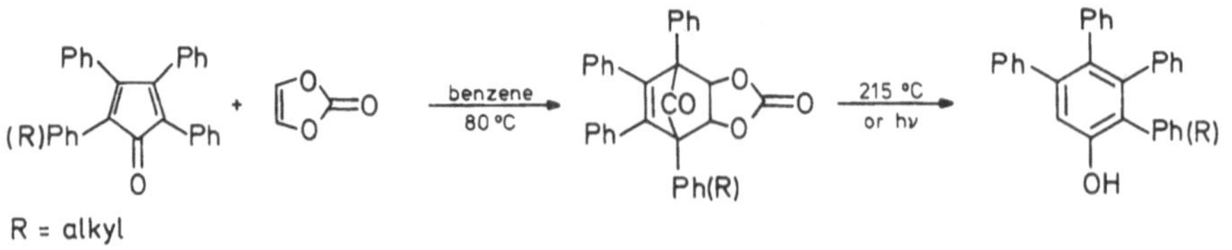

Many 4-substituted 2,6-diphenylphenols have been synthesized from Hill's 4-nitro-2,6-diphenylphenol which is also available by nitration of 2,6-diphenylphenol [50, 68]. For X-ray and spectroscopic data see Ref. [111] and [50] respectively. Reduction with thin and hydrochloric acid or even better using hydrogen and Raney nickel at room temperature gives a 80% yield of 4-amino-2,6-diphenylphenol. Diazotation of the acetate and subsequent coupling with phenols or N,N-diamino benzene, gives the 4-arylazo-2,6-diphenylphenol dyes. Substitution of the diazonium group leads to 4-dimethylamino-, 4-cyano, 4-fluoro-, 4-chloro-, 4-bromo-, 4-sulfonic acid as well as other 2,6-diphenylphenol-derivatives [50]. Oxidation of the 4-nitro-2,6-diphenylphenol with lead tetraacetate or even simple heating of the solution in benzene or acetic acid in the presence of light [112] leads to 2,6-diphenylquinone and N_2O.

2.7 Chemical, Photochemical and Thermal Rearrangements

When 2-phenylphenol is treated with aluminiumchloride at 100 °C, migration of the phenyl group to give the thermodynamically more stable 3-phenylphenol takes place [113].

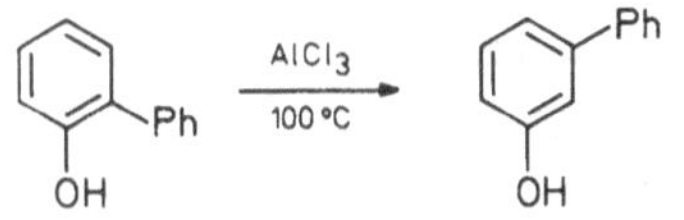

Zimmerman [103] applied the well known rearrangement of 4,4-disubstituted cyclohexa-2,5-dienones (e.g. the 4,4-diphenyl derivative tives the 3,4-disubstituted phenols [1]). Using acetic anhydride and a small amount of concentrated sulfuric acid, 97.5% of 3,4-diphenylphenolacetate is formed, which in turn is hydrolyzed to give 3,4-diphenylphenol. This is one of the easiest synthesis of this phenol. 2-Bromo-4,4-diphenylcyclohexa-2,5-dienone rearranges by a similar route to give a 49% yield of 2 bromo-3,4-diphenylphenolacetate and 47% 2-bromo-4,5-diphenylphenolacetate [114]. A [1.2]-phenyl shift of 4-methoxy-2,3,4,6-tetraphenyl-2,5-dienone in acetic anhydride and zinc chloride gives rise to 4-methoxy-2,3,5,6-tetraphenylacetate (m.p. 284 °C, 85% yield) [115].

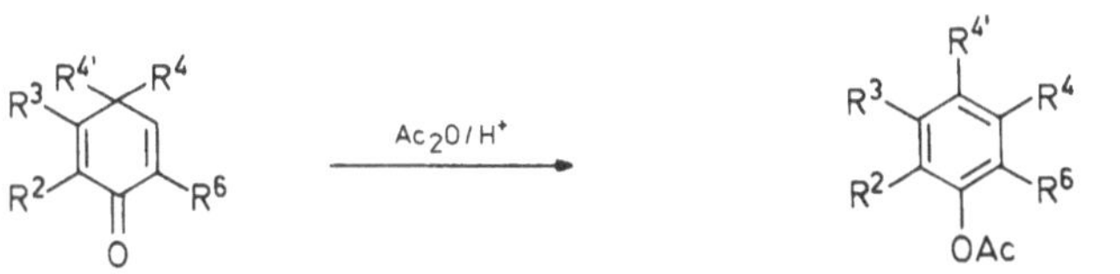

a. $R^2 = R^3 = R^6 = H$, $R^4 = R^{4'} = Ph$
b. $R^2 = R^3 = R^4 = R^6 = Ph$, $R^{4'} = OMe$
c. $R^3 = H$, $R^2 = R^4 = R^6 = Ph$, $R^{4'} = OMe$
(OAc, OH)

a. $R^2 = R^3 = R^6 = H$, $R^4 = R^{4'} = Ph$
b. $R^2 = R^3 = R^4 = R^6 = Ph$, $R^{4'} = OMe$
c. $R^3 = H$, $R^2 = R^{4'} = R^6 = Ph$, $R^4 = OMe$

Alkaline hydrolysis gives 4-methoxy-2,3,5,6-tetraphenylphenol (m.p. 247 °C). Further hydrolysis with hydrogen iodide in acetic acid affords tetraphenylhydroquinone (m.p. 300–302 °C) in quantitative yield. Oxidation with lead dioxide or alkaline potassium hexacyanoferrate (III) gives the yellow 2,3,5,6-tetraphenylquinone [115]. This can also be obtained by arylation of quinone with phenyldiazonium salt [116]. Phenylmagnesium-bromide reduces 4-methoxy-2,3,4,6-tetraphenyl-cyclohexa-2,5-dienone to 2,3,4,6-tetraphenylphenol. Such a [1.2] phenyl shift was not observed when 4-methoxy-, 4-acetoxy or 4-hydroxy-2,4,6-triphenylcyclohexa-2,5-dienone were treated with strong acids in ethanol. Only 2,4,6-triphenylphenol was isolated [5].

Photochemical rearrangements, in particular the mechanisms, of 4,4-diphenylcyclohexa-2,5-dienone have been intensively studied by Zimmerman et al. [117]. The main products are 2,3- and 2,4-diphenylphenol. Perst et al. [118] isolated 2,3,4,6-tetraphenylphenol and other substances when 2,4,4,6-tetraphenylcyclohexa-2,5-dienone was photolysed. Other photochemical rearrangements leading to 3-acetoxy-2,4,6-triphenylphenol and 2-acetoxy-3,4,6-triphenylphenol are found in Ref. [119].

Nearly quantitative yields of 2,4,5- and 3,4,5-triphenylphenol have been reported by Monahan [120]. When 1-(1,2,3-triphenyl-cycloprop-2-enyl)-3-diazopropan-2-one is heated with copper powder in benzene, the tricyclic compound A as well as 49% of the starting material is found. After UV irradiation of A in dioxane/water (60:40) under nitrogen for 22 hrs., 50% of 2,4,5-triphenylphenol was isolated. When A is heated 1 min. to 180 °C, 70% 3,4,5-triphenylphenol is found. With sodium hydroxide in refluxing dioxane A is converted into the bicyclic ketone B which on photolysis in 60% dioxane/water at 254 nm for 14 hrs gives 2,4,5- and 3,4,5-triphenylphenol in low yields (< 10%).

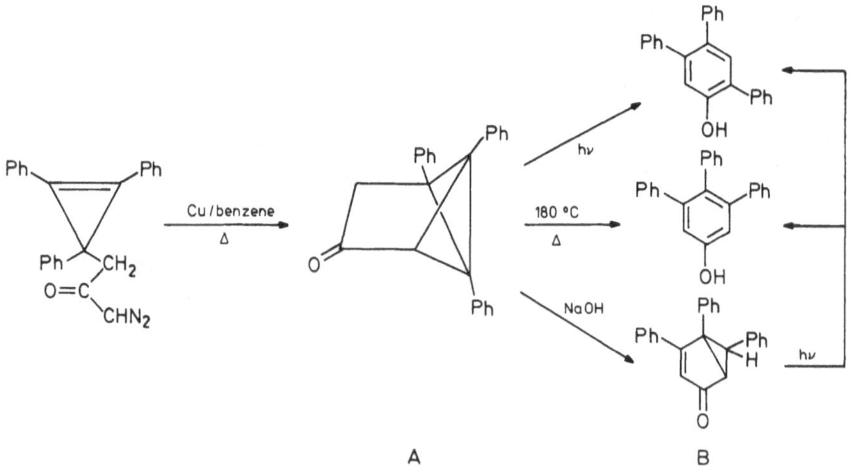

36

The addition of the diazomethane to tetraphenylcyclopentadienone in the dark gives 73–81% of the pyrazolidine C. Whereas boiling in ethanol and hydrochloric acid gives 2,3,5,6-tetraphenylphenol in 73–81% yield in tetraline at 208 °C a nearly quantitative yield is obtained. In a lower boiling solvent such as o-xylene or toluene in the absence of an acid C gives the bicyclic ketone D in 78% yield [121]. Irradiation (>280 nm) in ethanol, or thermolysis in tetraline, transforms D into 2,3,4,6-tetraphenylphenol 71–98% and 40–90%, respectively. The yields depend on the para-substituents of the phenyl-substituents (such as Me_2N, MeO, Cl or Br) [122].

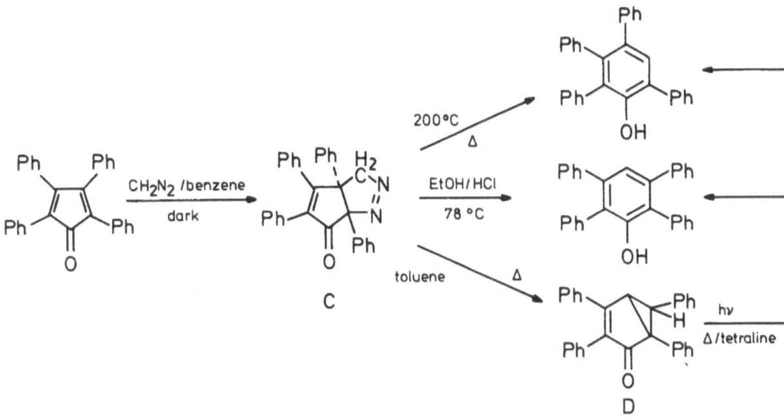

A similar thermal rearrangement has been observed in a two phase reaction when dichlorocarbene is added to tetracyclone giving E (30%). Heating of E in refluxing mesitylene or o-dichlorobenzene for 3.5 hrs gives 60% of 3-chloro-2,4,5,6-tetraphenylphenol [123].

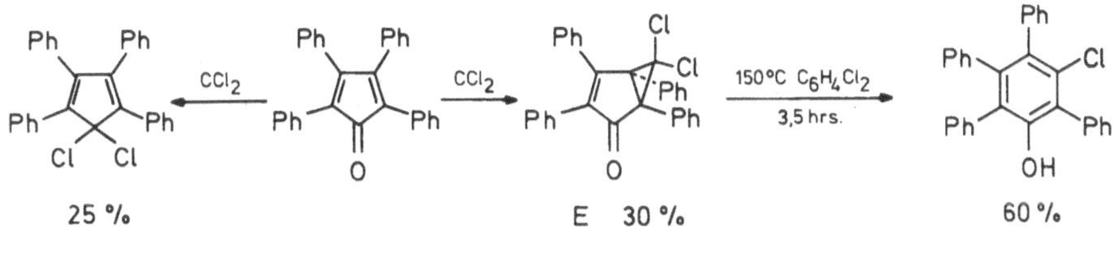

37

Pyrolysis of bis-3(1,2-diphenyl-cyclopropenyl)ether at 180° yields 2,3,4,6-tetraphenylphenol [156].

2.8 Addition of Lithium or Magnesium Organic Compounds to 1,4-Quinones

The high oxidation potential of quinones favors the oxidation of organo lithium or magnesium compounds instead of their addition to the 1,2- or 1,4-positions. Therefore, mono protected 1,4-quinones where the protecting groups are easily removed with silver fluorides such as 1-CN or 1-O-$SiMe_3$ derivatives have been developed to overcome this problem [124]. Organo lithium compounds, even with functionalized residues or aryl-magnesiumbromide, favor at low temperatures, a clean 1,2-addition to one of the carbonyl groups of 1,4-quinones.

Thus, even 2,6-dibromo-1,4-quinone adds lithium organo compounds to give 4-alkyl-4-hydroxy-2,6-dibromo-2,6-cyclohexadien-1-ones [125]. When bulky t-butyl groups are in the ortho positions of one of the carbonyl groups, such as in 2,6-di-t-butyl quinone, phenylmagnesiumbromide adds only to position 4, giving 2,6-di-t-butyl-4-hydroxy-4-phenyl-cyclohexa-2,6-dien-1-one. Reduction with zinc/hydrochloric acid gives 2,6-di-t-butyl-4-phenylphenol in good yields [126]. A similar reaction with 2,6-diphenylquinone has not been reported.

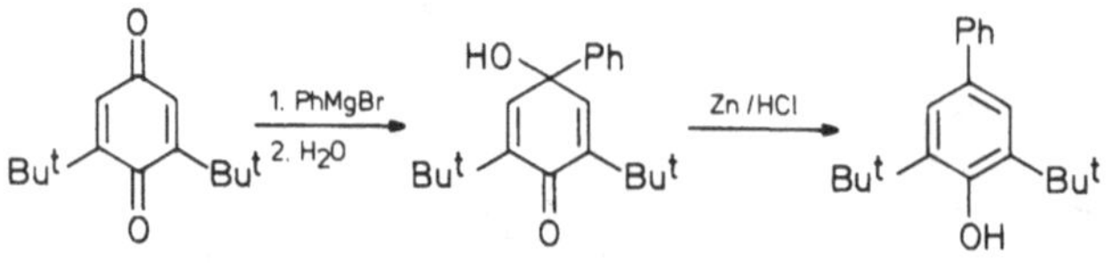

2.9 Synthesis Using Pyrylium Salts

2.9.1. Arylated Phenols via Nitromethane Condensation

2,4,6- or higher substituted pyrylium salts with aryl-, alkyl-, or mixed substituents can be converted in good yields by nitromethane in the presence of strong bases, such as potassium-t-butylate or t-amines, into the corresponding nitro arene derivatives [27]. Reduction of the nitro group, diazotation of the aniline derivative and hydrolysis in aqueous sulfuric acid affords the phenol derivatives in acceptable yields [9].

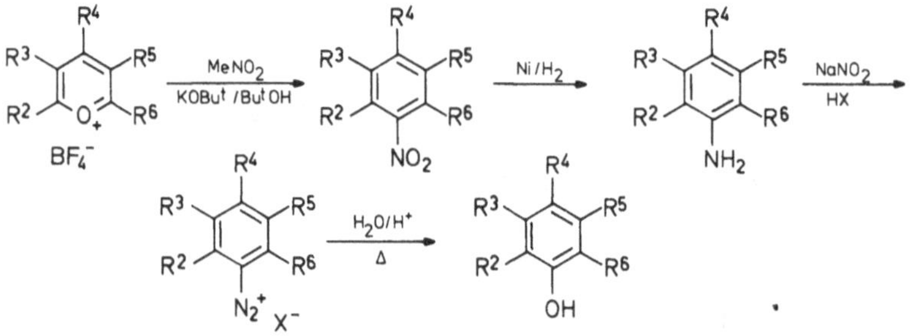

Since pyrylium salts can be synthesized by many different routes [28] this method is very versatile. Substituents in positions 2,4,6 are essential, but positions 3 and 5, can also be occupied. The pyrylium ring carbon atoms can be replaced by a ^{13}C-carbon in a definite position and each of the phenyl groups can be replaced by C_6D_5. Carbon-1 can be labelled using $^{13}CH_3NO_2$ [127]. Hydrolysis of the diazonium salts with $^{18}OH_2$ or $^{17}OH_2$ gives rise to ^{18}O- or ^{17}O-phenol derivatives, respectively [127]. There are, however, some minor limitations in the nitromethane condensation of the pyrylium salts. When the 2,6-substituents are bulky and the 4-substituent is small, the CH_2NO_2-group will add to position 4 [128]. The 4-methyl-2,6-diphenyl-pyrylium salt cannot, therefore, be converted into 4-methyl-2,6-diphenylphenol. However the

4-isopropyl (or 4-cyclohexyl-)-2,6-diphenyl-pyrylium salts gives 4-isopropyl- (or 4-cyclohexyl)2,6-diphenyl-nitrobenzene and the corresponding phenol derivatives. Contrary to 2,6-di-t-butyl substituted pyrylium salts 2-t-butyl-4,6-diphenyl-pyrylium salts and 4-t-butyl-2,6-phenyl-pyrylium salts are easily converted into their phenol derivatives [129]. Using isobutylene in sulfuric acid, Müller et al. alkylated 4-, and 2-phenyl phenol to give 2,6-di-t-butyl-4-phenyl- and 2,4-di-t-butyl-6-phenyl-phenol, respectively [130]. The prototype of arylated phenols, 2,4,6-triphenylphenol, was prepared for the first time using the above pyrylium salt method. It can also be prepared via direct nitration 1,3,5-triphenyl benzene [22,131] etc. A great advantage of the synthesis of both arylated and aryl-alkyl-substituted phenols by the pyrylium salt route, is the fact that there are no difficulties in determining the structure of the substituted compounds (see introduction p. 101). The synthesis of all seven possible monobromo isomers of 2,4,6-triphenyl-pyrylium salts and their conversion into the corresponding phenols [132] has shown that only one isomer, 4-(4'-bromophenyl)-2,6-diphenylphenol (81% yield) arrise, when 2,4,6-triphenylphenol is brominated. On the other hand, when 2,4,6-triphenylphenol is chlorinated in different solvents a difficult to separate mixture of chloro compounds is formed. Chlorination of 2,4,6-triphenylphenol-acetate, however, in tetrachloromethane gives a 92% yield of 3-chloro-2,4,6-triphenylphenolacetate, which can be hydrolyzed to the corresponding phenol. Independant synthesis, as shown below, used also in the synthesis of 3-bromo-2,4,6-triphenyphenol confirms the structure [133].

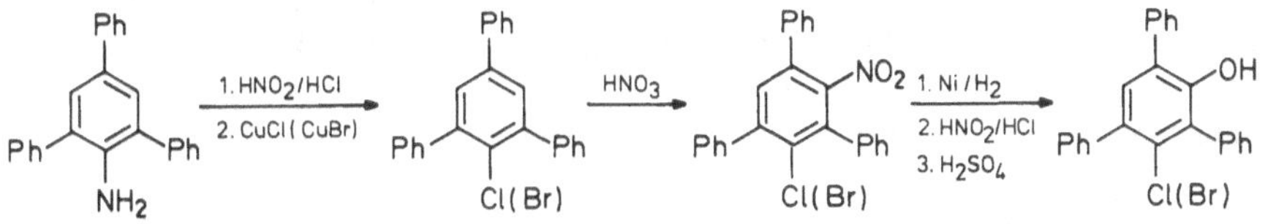

41

The easily available 4-(4'-bromophenyl)-2,6-diphenylphenol has been used to make an interesting water soluble persistant and oxygen stable phenoxyl radical [132]. It has been synthesized by reacting the (4'-bromo-4-phenyl)-phenol derivative with butyllithium in ether and dry sulfuryl chloride.

The 4'-sulfenic acid was isolated in 78% yield and characterized by its anilinium salt, m.p. 234–235 °C. Oxidation of sodium salt in 2 N sodiumhydroxide with 30% hydrogenperoxide gives the sodium salt of the sulfonic acid, which was analyzed after conversion into its S-benzyl-isothiuronium salt, m.p. 229–232 °C. Oxidation of the sodium sulfonate in acetonitrile with lead dioxide gives a red-violet solution of the water soluble phenoxyl. The intensity of the color increased with higher temperatures or by dilution, but the crystallized dimer could not be isolated when the solution was cooled down. Catalytic dehydrogenation, first observed using 4-nitromethyl-2,6-diphenylpyrane [134] is possible when the solution of the nitro-pyrane derivative in tetrachloromethane is shaken with an aqueous alkaline solution of potassium-hexacyanoferrate(III) and a small amount of the water soluble 4-(4'-phenylsodium-sulfonate)-2,6-diphenylphenol. A quantitative yield of 4-nitro-methylene-2,6-diphenylpyrane was isolated from the organic phase. No dehydrogenation is observed when the reaction is carried out in the absence of the phenolsulfonate derivative:

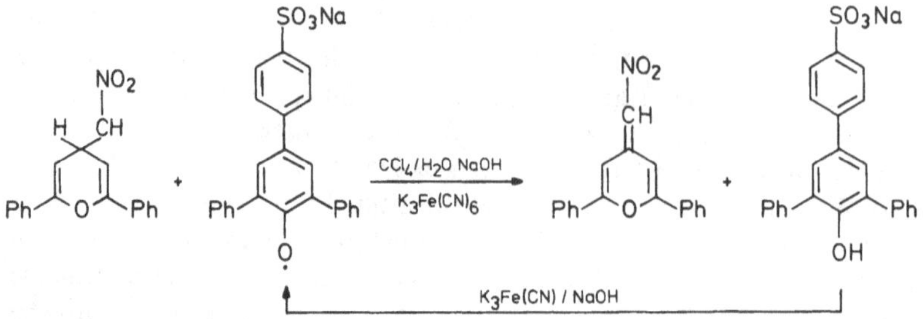

Blöcher synthesized another interesting phenolic compound [132] when he treated 2,4,6-triphenylphenol with sulfurylchloride and aluminiumchloride in carbondisulfide. Bonding of the C-4′-carbon atoms of the 4-phenyl rings gave 4,4‴-dihydroxy-3,5,3‴, 5‴-tetraphenyl-1,1′-4′,1″,4″,1‴-quater phenyl (47% yield). Alternatively, one can proceed via an Ullmann synthesis of 4-(4′-bromophenyl)-2,6-diphenyl-phenylmethylether (m.p. 85–87 °C) by first it transfering in into the Grignard compound in tetrahydrofuran, then reacting it with anhydrous copper chloride, and finally hydrolysing in pyridine hydrochloride/pyridine. Oxidation of (*1*) in benzene with an aqueous alkaline solution of potassium hexacyanoferrate(III) gives a deep green solution with strong ESR signals. Titration using hydroquinone in acetone indicates a diradical *2* content of 80%. The fact that the intensity of the green color decreases, the lower the temperature and the higher the concentration indicates an equilibration with its quinol dimer. The dimethylether of compound *3* was synthe-

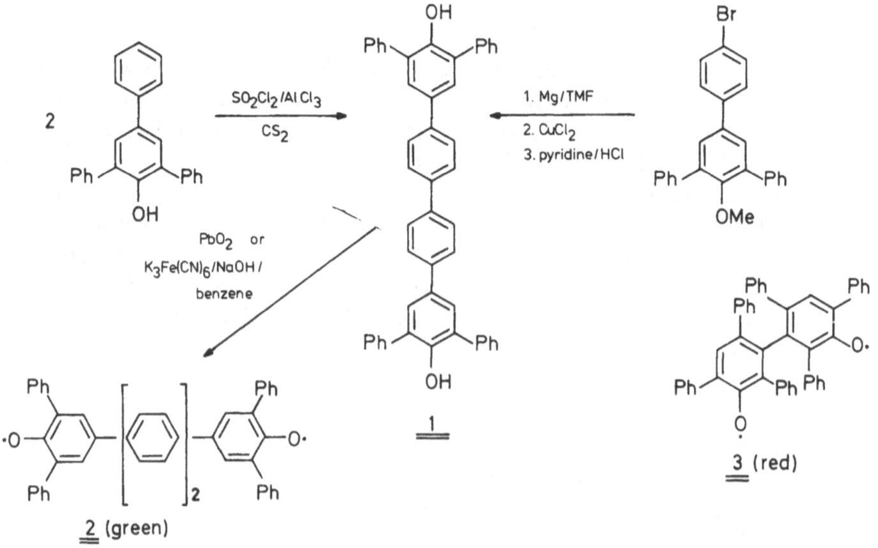

sized [132] from 3-bromo-2,4,6-triphenylanisol (m.p. 178–179°) by coupling the Grignard compound with copper lithium giving the 3,3'-dimeric dimethylether (m.p. 303–304°) in low yield (6%). After demethylation this was converted to the corresponding diphenol (72% yield, m.p. 305–306°) and the air stable cherry red diradical *3*.

4,4''-Dihydroxy-3,5,3'', 5''-tetraphenyl-1,1',4',1''-terphenyl (*4*) and its bis-phenoxyl *5* have also been synthesized by Umbach [135]. The bis-pyrylium-tetrafluoroborate was prepared by condensing terephthalaldehyde with acetophenone to give initially benzene-1,4-chalcone [136] which upon reaction with acetophenone in carbon disulfide and iron(III) chloride gave the pyrylium-bis-tetrachloroferrate, and on addition of tetrafluoroboric acid in acetic acid the bispyrylium salt (m.p. 310–325°). Condensation with nitromethane gave the 4,4''-dinitro compound (m.p. 342–344°, 85% yield). Its diamine (m.p. 308–309°, 68% yield) and bis-diazonium salt gave the diphenol *4* (m.p. 247–248 °C) in 72% yield (diacetate m.p. 268–270°). Oxidation of a benzene solution with aqueous alkaline potassiumhexacyanoferrate (III) in the usual manner gave a deep blue solution of the bis-phenoxyl radical *5* which crystallized in dark blue needles, m.p. 220–225°. The solid compound as well as its solution gave strong ESR signals. The absence of a carbonyl IR band near 1667 cm^{-1}, as well as no concentration or temperature dependent change in color intensity (λ_{max} = 610 nm, ε = 91000) show that no dimerization to the quinol derivative occurs.

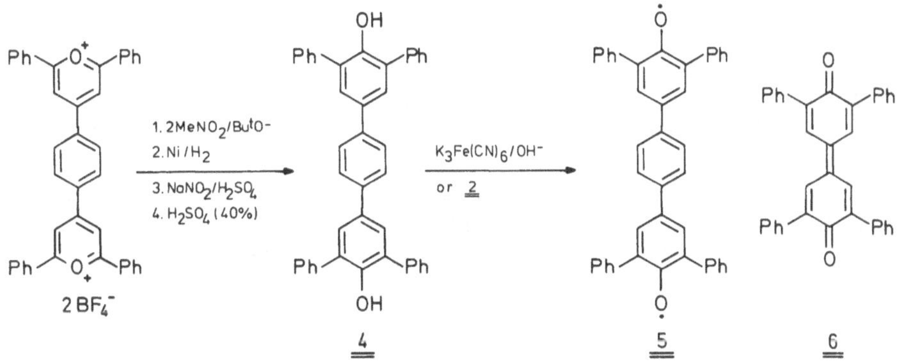

The persistant air stable biradical *2* is a stronger oxidant than *5* which is shown by the fact that the biphenol *4* gives *5* in a benzene solution of *2*. In contrast to the bis-phenoxyl radicals, pure 3,5,3',5'-tetraphenyldiphenoquinone (*6*) shows no ESR signals and is, therefore, a true quinone and not a diradical.

2.9.2 Rearrangements

Interesting rearrangements have been observed when the phenylnitromethane anion is added to 2,4,6-arylated pyrylium salts. A sensitive intermediate, probably the dihydrobenzene derivative *1* was isolated from 2,6-diphenyl-4-(p-tolyl)-pyrylium-tetrafluoroborate and phenylnitromethane in t-butanol with only one mole potassium-t-butylate. When heated in toluene, the dihydrobenzene derivative *2* which is subsequently hydrolysed and dehydrated to give the 2,3,4-triphenyl-6-(p-tolyl)-phenol *3* in 78% yield is formed probably by a [1.3]$NO_2 \rightarrow$ ONO rearrangement. On the other hand, heating of *1* in the presence of an excess of t-butylate affords 2,3,4-triphenyl-6-(p-tolyl)-nitrobenzene (*4*) by a [1.3]$NO_2 \rightarrow NO_2$ rearrangement in 81% yield. Similar rearrangements for 2,4,6-triphenyl-pyrylium-tetrafluoroborate, 4-4'-nitro-phenyl)- and (4-(4'-methoxy-phenyl)-2,6-diphenyl-pyrylium salts and phenyl-nitromethane are shown in Table 1 [137].

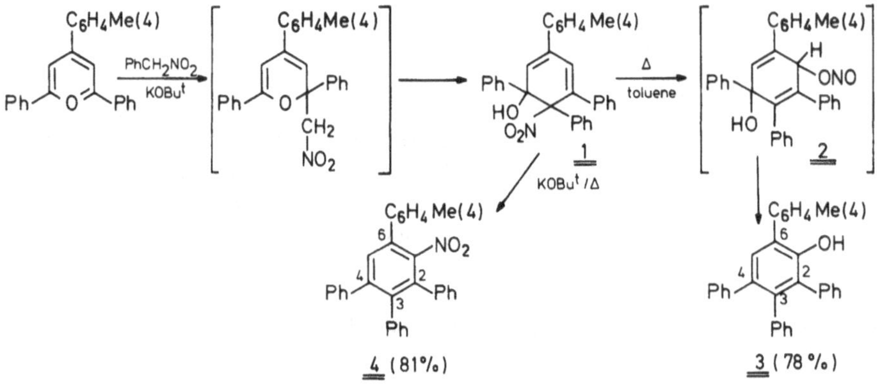

2,4,6-Triphenyl-pyrylium salts and ring substituted phenylnitromethane derivatives similarly rearrange to the phenols *6* or nitrobenzenes *7*. When the ringsubstituted phenylnitromethane is reacted with the 2,4,6-arylated pyrylium salt in t-butanol with one mole of potassium t-butylate, *5* is the suspected intermediate. It rearranges to the phenol *6* when heated in 1,2-dichlorobenzene with one mole of ethyl-diisopropylamine and to the nitro compound *7* with an excess of potassium t-butylation in t-butanol (47–81% yield [138]). Some similar examples are shown in Table 1.

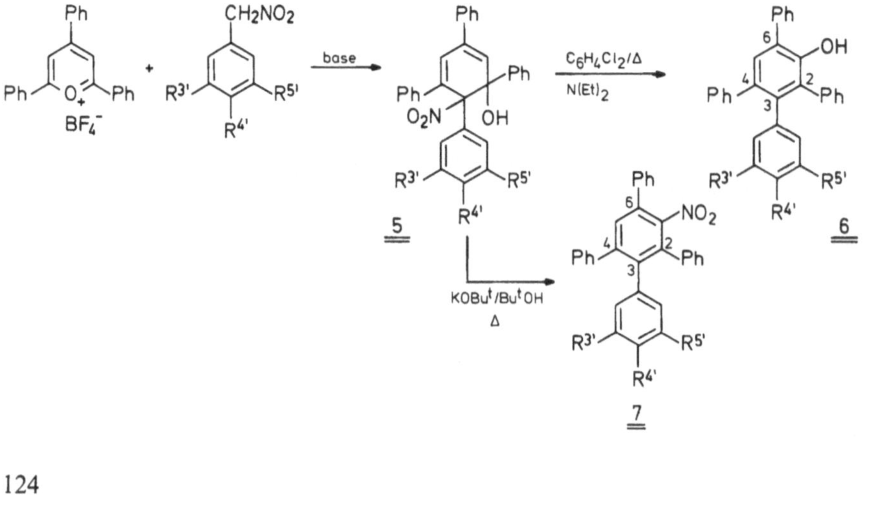

Table 1. Phenols *3* and *6* and nitrobenzenes *4* and *7* from 2,4,6-arylated pyrylium salts and nitromethyl-arenes

Phenol derivatives *3* and *6*							Nitrobenzene derivatives *4* and *7*							
R^2	R^3	R^4	R^5	R^6	yield (%)	m.p. (°C)	R^2	R^3	R^4	R^5	R^6	yield (%)	m.p. (°C)	Ref.
Ph	Ph	Ph	H	Ph	42	244–6	Ph	Ph	Ph	H	Ph	67	220	[137,115]
Ph	Ph	Ph	H	PhMe(4)	78	239	Ph	Ph	Ph	H	PhMe(4)	81	199–200	[137]
Ph	Ph	Ph	H	$PhNO_2(4)$	27	222–3	Ph	Ph	Ph	H	$PhNO_2(4)$	47	198–9	[137]
Ph	Ph	Ph	H	PhOMe(4)	55	219–20	Ph	Ph	Ph	H	PhOMe(4)	70	233–4	[137]
Ph	Ph	Ph	Ph	PhMe(4)	39	191–2	Ph	Ph	Ph	H	PhMe(4)	65	210	[137]
Ph	PhOMe(4)	Ph	H	Ph	15	199–200	Ph	PhOMe(4)	Ph	H	Ph	50	240–1	[138]
Ph	PhOCOPh(4)	Ph	H	Ph	27	230–1	Ph	PhOCOPh(4)	Ph	H	Ph	38	241–2	[138]
Ph	$PhNO_2(4)$	Ph	H	Ph	37	244–5	Ph	$PhNO_2(4)$	Ph	H	Ph	62	266–7	[138]
Ph	$PhNH_2(4)$	Ph	H	Ph	87	254–6	Ph	$PhNO_2(4)$	Ph	H	Ph	—	—	[138]
Ph	PhOH(4)	Ph	H	Ph	95	231–3 diOAc	Ph	PhOH(4)	Ph	H	Ph	90	221–2	[138]
Ph	PhOAc(4)	Ph	H	Ph	14	211–2	—	—	—		—	—	—	[138]
Ph	$RNO_2(4)$	Ph	H	Ph	40	229–30 diOAc	Ph	$R—NO_2(4)$	Ph	H	Ph	64	290–1	[138] R=Ph-(3′,5′-Ph)₂
Ph	ROAc(4)	Ph	H	Ph	82	204–6	—	—	—		—	—	—	[138] R=Ph-(3′,5′-Ph)₂
Ph	ROH(4)	Ph	H	Ph	95	182–3	—	—	—		—	—	—	[138] R=Ph-(3′,5′-Ph)₂
PhOMe(4)	Ph	H	Ph	Ph	3	193–4,5 261–3	PhOMe(4)	Ph	H	H	Ph	80	200–1	[138] Quinone (63%) 190–2 °C
PhOH(4)	Ph	H	Ph	Ph	95	225–30[a]	—	—	—		—	—	—	[138]a Di-acetat
Ph	Ph	R′	H	Ph	72	305–6[b]	—	—	—		—	—	—	[132,138] R′=Ph-(2′,4′,6′-Ph)₃-4′-OH, Oxidation red stable radical

[a] methylether
[b] see formula *10* p. 126 and *3* (radical) p. 122

The most interesting compound is the dinitro derivative *7* ($R^{3'}=R^{5'}=Ph$, $R^{4'} = NO_2$). It has been transfered by the usual route into 3-(4'-hydroxy-3',5'-diphenyl-phenyl)-2,4,6-triphenylphenol (*8*). Dehydrogenation of a benzene solution with alkaline hexacyanoferrate(III) or with lead dioxide gives a deep red solution. Its color intensity is strongly temperature dependant, indicating dimerization to a quinol derivative. It cannot be clearly decided from the rather complicated ESR-spectrum whether or not the dehydrogenation product *9* is a mono- or a diradical with twisted benzene rings. Since no carbon oxide is evolved from the stable red solution it appears that no five membered ring compound is formed as in the case when 2,4,6-triphenyl resorcinol is dehydrogenated [139].

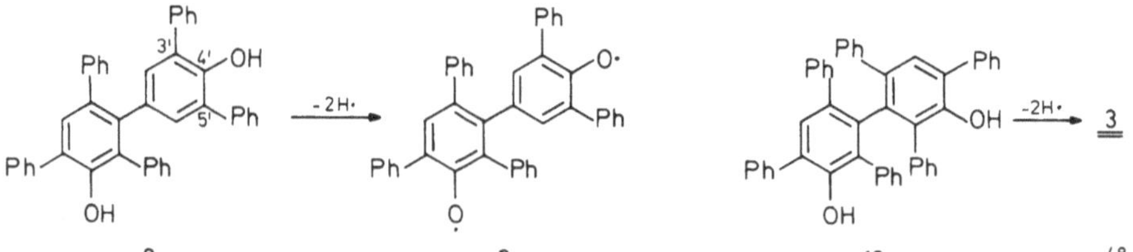

8 9 10

48

A similar compound *10* has already been mentioned p. 122.

When 2,4,6-tri- or 2,3,4,6-tetraarylated pyrylium tetrafluoroborate is treated with nitromethane under different conditions rearrangement is also observed. No rearrangement occurs when the two components are heated together with an excess of potassium t-butylate in t-butanol (route a). Under strong basic conditions the intermediate *11* probably loses the acidic proton at position 1 to give the aromatic nitro compound *12*.

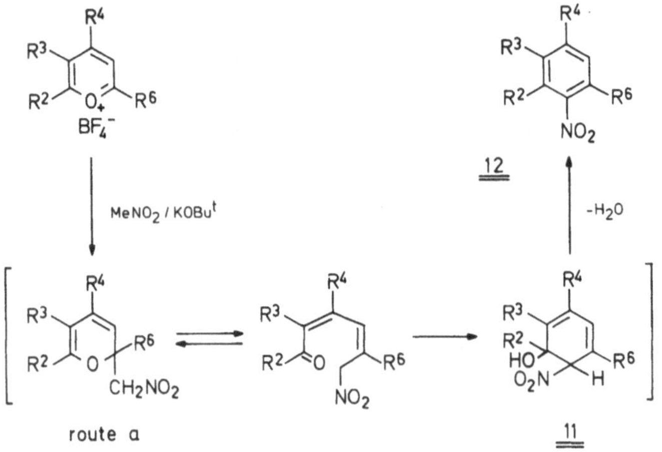

When the pyrylium tetrafluoroborate is reacted in ethanol or 1,2-dichlorobenzene with nitromethane, however, and only one mole of triethylamine is used, a much smaller yield of the "normal" nitrobenzene derivative *12* is found. Two different

Table 2. Phenols and oxidation products from the reaction of pyrylium salts and nitromethane with 1 mole NR_3 in unpolar solvents

pyrylium salt					nitro comp. *12*		phenol *14*				phenol *16*				oxidation (CrO_3)
R^2	R^3	R^4	R^5	R^6	yield (%)	m.p. (°C)	yield (%)	m.p. (°C) phenol	acetate	oxidation stabil radical	yield (%)	m.p. (°C) phenol	acetate	diphenol	m.p. (°C) quinone
Ph	H	Ph	H	Ph	~1	144.5	~1	149–50	121–2[a]	red	—	—	—	—	—
Ph	Ph	Ph	H	Ph	18.5	216–7	7.1	244–5[a]	180–1	deep red	7	272–3	253[a]	—	310
PhOMe(4)	Ph	Ph	H	Ph	8	200–1	2.3	195–196[b]	—	blue	24	193–4	—	261–2	190–2
Ph	PhOMe(4)	Ph	H	Ph	2	139–40		199–200	190–1[c]	blue	18	193–4	—	261–2	190–2

[a] 1 part pyrylium-tetrafluoroborate, 4 parts $MeNO_2$, 1 part NEt $(iPr)_2$ in 1,2-dichlorobenzene at 179 °C

[b] Lit. 105)

[c] 2,4,6-triphenyl-3-(4'-hydroxyphenyl)-phenol m.p. 231–3 °C

phenols *14* and *16*, formed by a $NO_2 \rightarrow ONO$ rearrangement are isolated instead. The yields of the three products *12*, *14* and *16* are strongly dependent on the pyrylium salt substituents (see Table 2).

The mechanism of this rearrangement is not clear. We suppose that the intermediate *11* dissociates to give an ion pair *11a*. Route b, which involves addition of the NO_2-anion leads to the nitrous ester *13* which on hydrolysis and elimination of water gives rise to the phenol *14*. Route c which involves a [1.2]phenonium rearrangement is favoured in compound *11a*, R^2 = PhOMe(4), Table 2 because when R^2 is 4-methoxyphenyl it is a better migrating group than when R^2 is phenyl Deprotonation of the cation gives *16*.

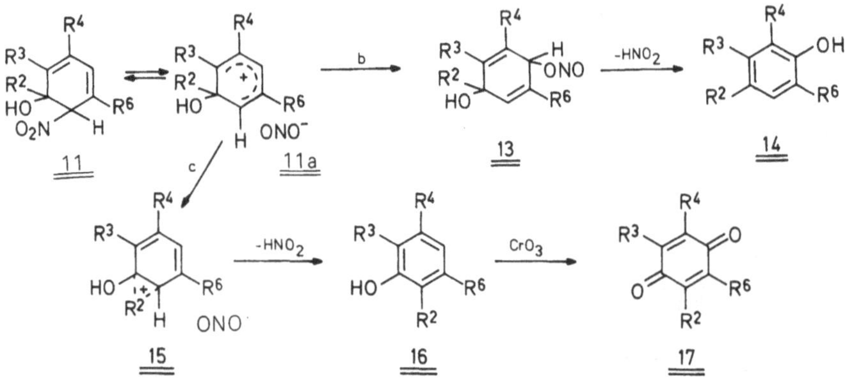

The structure of the phenols *14* were proven by the pyrylium salt synthesis using the well known route a (p. 126) and the transformation of the nitro group into a hydroxyl. The structures of *16* are confirmed by chromic acid oxidation to 2,3,4,6-substituted quinones *17*. When 2,4,6-triphenylpyrylium-tetrafluoroborate was treated using the same procedure (ethanol, one mole triethylamine, nitromethane) only 23% of the "normal" 2,4,6-triphenylnitrobenzene and 3% of 2,4,6-triphenylphenol were isolated. Heating in 1,2-dichlorobenzene and diisopropylamine for 1.5 hrs gave 20% of 2,4,6-triphenylphenol as well as traces of 2,4,6-triphenylnitrobenzene and tentatively 2,3,5-triphenylphenol. This single-step synthesis of 2,4,6-triphenylphenol from 2,4,6-triphenyl-pyrylium salt by a probable radical $NO_2 \rightarrow ONO$ rearrangement is another convenient route to this important phenol.

2.9.3 Syntheses and Properties of 2,4,6-Triarylated Benzene Derivatives with NH_2, NHR, NR_2, CH_2R and SH Substituents in Position 1

Since 2,4,6-triphenylphenol can be easily oxidized to a stable radical, we were interested in the behaviour of various substituted 2,4,6-triphenylbenzenes. When Schromm [140] oxidized 2,4,6-triphenyl-aniline by shaking a solution in benzene with lead dioxide, no radical was detected either by ESR spectroscopy or by trapping with triphenylmethyl. A red color (λ_{max} = 520 nm) possibly due to the ArNH radical was, however, observed. [140]. The addition of 70% perchloric acid changes the color to deep blue-violet. The compound was isolated when a benzene solution of 2,4,6-

triphenyl-aniline was boiled for 12 h with an excess of a saturated aqueous solution of potassium permanganate. Even better yields were possible when a boiling solution of the aniline derivative in acetonitrile and small portions of potassium permanganate were added for 4 h. Chromatography (Al_2O_3 in benzene) gave 2,4,6,2',4',6'-hexaphenylazobenzene (*1*), m.p. 204–205 °C (88% yield). Table 3 gives some other examples of similarly synthesized substituted azobenzenes. In contrast to the stable 2,4,6-tri-t-butylphenazyl radical [141], the 2,4,6-triphenyl-phenazyl radical [140] easily dimerizes to give initially the Ar—NH—NH—Ar compound which can be further dehydrogenated to give the azo compound [143,144]. This is possible because the steric hindrance of the bulky t-butyl groups is lacking. This is also true in the oxidation of 2-t-butyl-4,6-diphenylaniline. Reduction of *1* with zinc/HCl leads back to the 2,4,6-triphenylaniline, or, in the presence of acetanhydride to its monoacetate. Protonation of the azo dyes, e.g. *1*, in acetic acid with 70% perchloric acid, first gives a blue-violet color (λ_{max} = 560 nm) of high intensity due to the perchlorate *2*. Deprotonation with ether or pyridine leads back to *1*. But (in benzene/$HClO_4$) the solution has changed to brown. Two isosestic points (at 520 and 400 nm) were observed. After work-up a new perchlorate of the benzo[c]-cinnolium salt *3* was isolated. Quantitative yields from *1* by oxidation with half concentrated nitric acid in dichloromethane and perchloric acid were possible. Reduction with zinc and acetic acid gave the 5,6-dihydro derivative *4a* which can be alkylated with triethyloxoniumtetrafluoroborate, to give *4b*. *4a* and KOH affords *4c*. Further examples of related cinnolinium salts are presented in Table 3 [140].

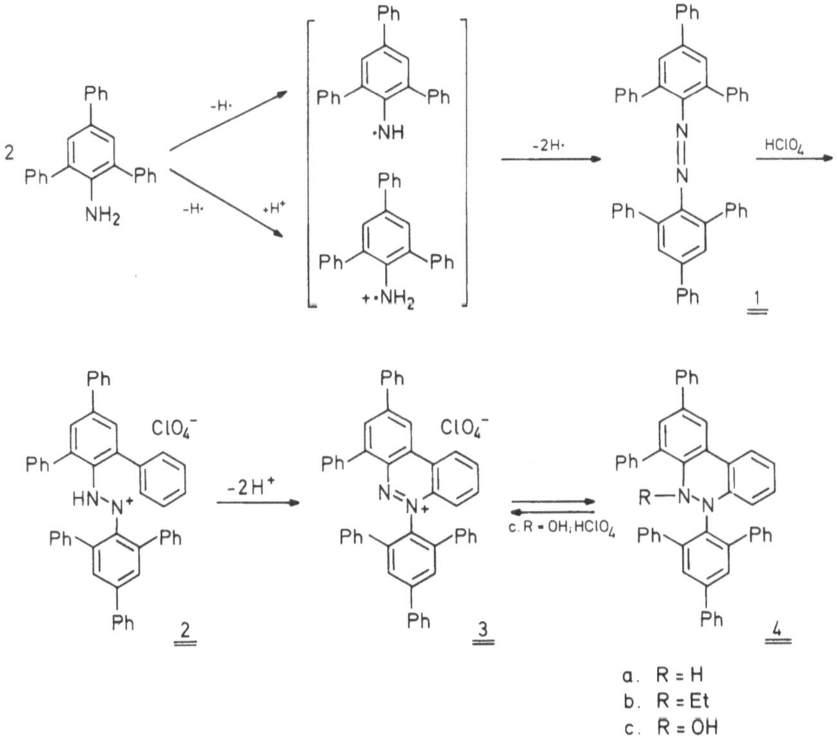

53

Table 3. Arylazocompounds *1* and aryl-benzo[c]cinnolinium salts *3* [140]

Azo-compounds *1*									cinnolinium perchlorate *3*	
$R^2 = R^{2'}$	$R^3 = R^{3'}$	$R^4 = R^{4'}$	$R^6 = R^{6''}$	m.p. (°C)	color	λ_{max}	$\varepsilon \cdot 10^2$	yield (%)	m.p. (°C)	yield
Ph	H	Ph	Ph [143]	204–205	blueviolet	518	20.7	88[a]	185–195	98
Ph	Ph	Ph	Ph	175–185	red	508	15.0	60.5	—	—
PhOMe (4)	H	PhOMe (4)	PhOM° t4)	225–226	pink	527	20.7	18	—	—
PhCl (4)	H	H	PhCl (4)	225–227	yellow-orange	460	7.2	17	—	—
PhCl (4)	H	PhCl (4)	PhCl (4)	278–279	violet-blue	525	20.4	46.5	303–306	53
Bu^t	H	Ph	Ph	173–174	red	495	21.6	81	188–191	51
Ph	H	H	Ph	233–234	orange	—	—	16.5	169–271	71
Ph	H	Cl	H	225–227	yellow-orange	460	7.2	—	—	—
Ph	H	Cl	Cl	119–120	orange	468	10.2	—	—	—
Me	H	Me	Me [143]	173–174	orange	—	9.8	—	—	—
Cl	H	Cl	Cl [144]	188–189	brownish	450	5.6	30	—	—
Ph	H	H	H	144–145	orange	433	—	17.5	—	—

[a] Cinnoliniumperchlorate *3* with 2nKOH in dioxan-acetonitrile gives 2,4-diphenyl-5-hydroxy-6-(2′,4′,6′-triphenylphenyl)-5,6-dihydro-benzo[c]cinnoline, m.p. 227–228 °C (44% yield). Perchloric acid leads back to the cinnolinium perchlorate *3*.

Condensation of N-formyl or N-acetyl-2,4,6-triphenylaniline with polymeric phosphoric acid at 150–160 °C give the phenanthridine derivatives *5a* and *b*. When iodobenzene and copper powder are heated for 36 h with 2,4,6-triphenylaniline in nitrobenze and potassium carbonate, 1,3,9-triphenylcarbazol *6* is formed.

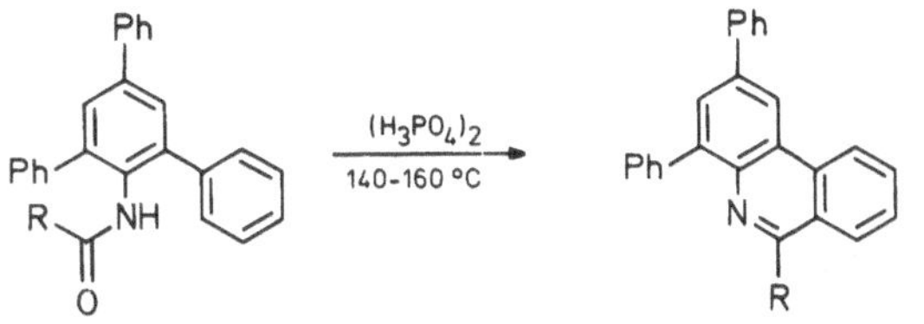

a. R = H , 61 %, m.p. 200-201 °C
b. R = Me, 79 %, m.p. 145-148 °C

5

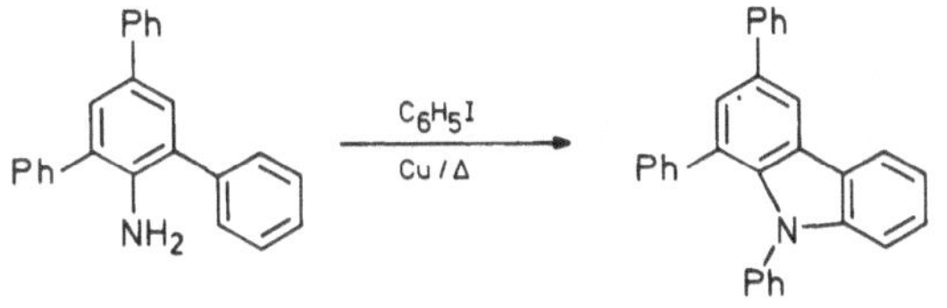

6

33 %, m.p. 211-213 °C

Many derivatives of 2,4,6-triphenylaniline, such as the azo dyes or the halogenides (F, Cl, Br) as well as the nitrile have been prepared by Saure [25] from its diazonium salt.

The nitrile has been hydrogenated to give the 2,4,6-triphenylbenzylamine which can then be transformed into the corresponding benzylalcohol and benzylchloride. The latter gives 1,2-bis-(2,4,6-triphenyl)ethane *7a* when reacted with zinc dust. No radical was, however, observed on thermolysis of either *7a* or its methyl or phenyl derivatives *7b* and *c*.

2,4,6-Triphenyl-thiophenyl (*8*) was synthesized by substituting the diazonium group of 2,4,6-triphenylaniline by sulfur dioxide/copper(I)-chloride followed by reduction of the disulfide *9* [145]. Mild oxidation gave only the disulfide *9*, (a very short lived

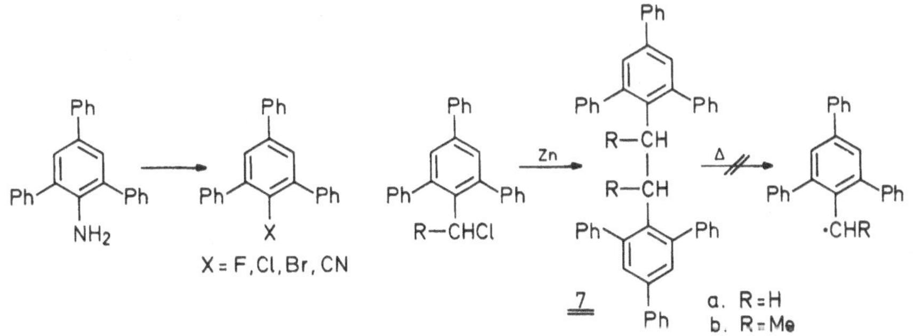

7

a. R = H
b. R = Me
c. R = Ph

S-radical is a possible intermediate). Chlorine in tetrachloromethane gave the sulfenic acid chloride *10* which can the be transformed into the trisulfide *11* as well as many other derivatives. The sulfonic acid salt of *12* was prepared by chromic acid oxidation of *8* in 45% yield. Fusion with potassium hydroxide gave 2,4,6-triphenylphenol in 78% yield.

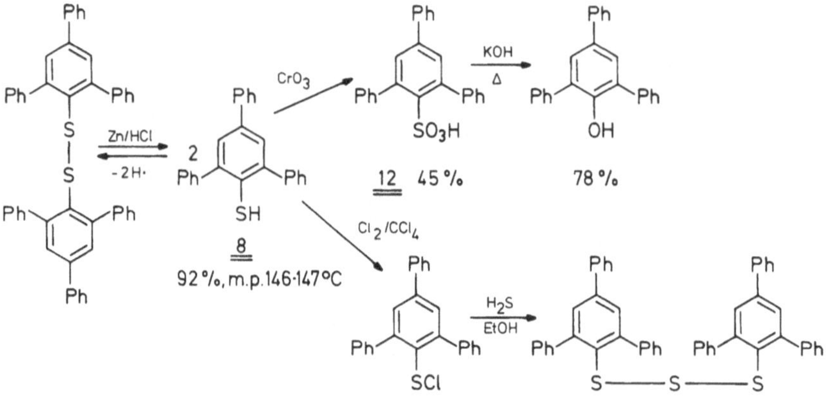

9
30 %, m.p. 247-248 °C

10
72 %, m.p. 73-75 °C

11
80 %, m.p. 183-184 °C

The instability of the 2,4,6-triphenyl-thiophenoxy radical is probably not due to stereochemical reasons but to the fact that the preferred delocalization of the sulfur unpaired electron into the aryl-system which requires a partial C—S double bond is unfavored. The S—S-dimerization giving *9* is, therefore, the prefered reaction.

2.10 Synthesis of Aryl-Cyanophenols and of 2,4,6-Tricyanophenol

4-Cyano-2,6-diphenylphenol has been prepared from 4-nitro-2,6-diphenylanisol via reduction to the 4-amino-2,6-diphenylanisol, followed by diazotation, substitution of the diazonium group by the cyano group and demethylation [59]. Bromination of 2,3,5,6-tetraphenylphenol gives the 4-bromo derivative in 90% yield. Treatment with CuCN in N-methylpyridone at 200° gives 4-cyano-2,3,5,6-tetraphenylphenol, m.p. 325° in 80% yield [147a, 168].

2-Cyano-4,6-diphenylphenol has been prepared from 2-bromo-4,6-diphenylanisol. Lithiation with n-butyllithium, carboxylation with carbon dioxide, subsequent transformation to the carboxamide and dehydration with sulfurylchlorid followed by cleavage of the anisolderivative gives the phenol-derivative with pyridine hydrochloride (m.p. 138–140 °C) in acceptable yield [146]. 2,6-Dicyano-4-phenyl-phenol has been synthesized from 4-phenylanisol. Condensation with formaldehyde gives 2,6-bis-hydroxymethyl-4-phenylanisol from which 2,6-dicyano-6-phenylphenol, m.p. 216° has been synthesized.

Finally, 2,4,6-tricyanophenol m.p. 185° has been prepared from 2,4,6-tris-carboxymethyl-anisol by an analogous route. It is a strong acid ($p_k \sim 1.0$) which can only be prepared from the pyridinium salt to give the phenol by cation exchange with lewatit S 140 [147].

3 Tables of Aryl-Phenols

3.1 Monoarylphenols (Table 4)

Table 4. Monophenylphenols

Name	m.p. (°C)	b.p. (°C)	m.p. (°C) of derivatives		Beilstein 6	Ref.
			a)	b)		
2-phenyl-phenol	59–60	287	63	31	E IV, 4579	53, 57, 148, 149, 150, 151, 152)
3-phenyl-phenol	78	300	34	29	E IV, 4597	69, 113, 149, 153)
4-phenyl-phenol	164–5	305–8	88–9	91–2	E IV, 4600	148, 152)

a) acetate, b) methylether

3.2 Diphenylphenols (Table 5)

Table 5. Diphenylphenols

No.	Name	m.p. (°C)	m.p. (°C) of derivatives				Beilstein 6	Ref.
			a)	b)	c)	d)		
1	2,3-diphenylphenol	101–2	117–8	—	—	—	—	103)
2	2,4-diphenylphenol	90–92	—	118	97.5	—	—	57, 154)
3	2,5-diphenylphenol	192–4	144	—	97.8	—	E IV, 5009	52, 153)
4	2,6-diphenylphenol	101	—	—	42	137	E IV, 5009	53, 57, 79, 109, 157)
5	3,4-diphenylphenol	105–6	130–1	118	—	—	E IV, 5008	102, 103, 154)
6	3,5-diphenylphenol	95	—	124	113–4	—	E IV, 5009	100)

a) acetate, b) benzoate, c) methylether, d) 4-nitro-derivative

3.3 Triphenylphenols (Table 6)

Table 6. Triphenylphenols

No.	Name	m.p. (°C)	m.p. (°C) of derivatives		Beilstein 6	Ref.
			a)	b)		
1	2,3,4-triphenylphenol	—	—	—	—	unknown
2	2,3,5-triphenylphenol	162–3	—	—	—	138)
3	2,3,6-triphenylphenol	163–4	197–20	—	—	106, 107)
4	2,4,5-triphenylphenol	93–4	153–5	—	—	120)
5	2,4,6-triphenylphenol	149–50	121–2	107–8	E IV, 5141	52, 53, 131, 138)
6	3,4,5-triphenylphenol	229–30	191–2	122	E IV, 5141	102, 120)

a) acetate, b) methylether

3.4 Tetraphenylphenols and Pentaphenylphenol (Table 7)

Table 7. Tetraphenylphenols and pentaphenylphenol

No.	Name	m.p. (°C)	m.p. of derivatives			Ref.
			a	b	c	
1	2,3,4,5-tetraphenylphenol	183–4	—	—	—	105,110)
2	2,3,4,6-tetraphenylphenol	244–5	181	200	179–80	105,107,110,118,121,137,155,156)
3	2,3,5,6-tetraphenylphenol	277–8	228–30[d]	269	226	105,115,122,156)
4	2,3,4,5,6-pentaphenylphenol	267	317[e]	—	178–9	115)

[a] acetate, [b] benzoate, [c] methylether, [d] 4-bromo-derivative, [e] subl. 285 °C

3.5 Special Arylphenols (Table 8)

4 Physical and Chemical Properties of Arylated Phenols

4.1 IR-, UV- and NMR-Data

Aryl substituents in either the 2- or 2,6-position cause a downfield shift of the OH stretching frequency in CCl_4 due to the formation of an intramolecular hydrogen bond to the aromatic ring [170)] (phenol 3605 cm^{-1}, 2-phenylphenol and 2,6-diphenylphenol 3555 cm^{-1}; 2,4,6-triphenylphenol 3553 cm^{-1}). Under the same conditions t-butyl-phenols give rise to higher frequencies (2,6-di-t-butylphenol and 2,4,6-tri-t-butylphenol 3645 cm^{-1}, compared with phenol 3612 cm^{-1}) [171)].

The bulky 2,6-di-t-butyl groups, and to a lesser extent the 2,6-diphenyl groups, diminish the solubility in aqueous sodium hydroxide. In addition, the acidity of the phenols is strongly influenced by electron donating or attracting substituents, especially in positions 2, 6, and 4. Only a few exact p_K values of arylated phenols are known. Because of their insolubility in water the acidity constants cannot be compared with the p_K of water soluble phenols. Electrochemical measurements in acetonitrile/tetraamoniumchloride by titration with tetramethylammonium hydroxide in acetonitrile afforded the relative "p_K" values of 2,4,6-tri-t-butyl-phenol, 2,4,6-triphenylphenol and 4-cyano-2,6-diphenylphenol of 7, 5.2, and 2, respectively [147)]. The influence of hydrogen bonds on the acidity of substituted phenols in DMSO solution is discussed in Ref. [147)].

The ^1H- and ^{13}C—NMR spectra of arylated phenols give little structural information due to the fact that their chemical shifts fall in the low region of aromatic compounds.

4.2 Substitution

2,4- and 2,6-diarylphenols easily react with electrophiles to give either the 6- or 4-substituted diarylphenol derivatives in excellent yields. For 4-nitro, 4-chloro-, 4-bromo-, and many other 4-substituted 2,6-diphenylphenols see table 8. 4,4′-Thio-

Table 8. Derivatives of arylated phenols and their dehydrogenation products

No.	R^2	R^3	R^4	R^5	R^6	m.p. (°C) phenol	m.p. (°C)		dehydrogenation		Ref. and remarks
							acetate methyl-ether*)	method of synthesis	radical stable(st) instable(inst)	dimer m.p. (°C)	
1	Ph	H	H	H	Ph	101	103	—	—	polymer or quinone 136–7	table 5,4, [157] for 2,6 PhR(4), R = MePh, OMe
2	Ph	H	Me	H	Ph	—	—	L	—	—	[158,160] and CH_2, CHR—, —CR_2—
3	Ph	H	Bu^t	H	Ph	96–7	145	A	green(st)	dimer 184–92	[129] very weakly dissoziated
4	Ph	H	OMe	H	Ph	67.5–8	98	C	green(st)	158–9	[59,160] also OEt (4)
5	Ph	H	OAc	H	Ph	166–7	125–6	C	green(st)	165–70	[59]
6	Ph	H	S—S	H	Ph	—	166–7	C	—	—	[59] $C_{4,4'}$
7	Ph	H	SPh	H	Ph	—	146–7	C	violet(inst)	—	[59]
8	Ph	H	SO_2Cl	H	Ph	—	133 and 152	C	—	—	[59]
9	Ph	H	SO_2Ph	H	Ph	148	161–2	C	—	—	[59]
10	Ph	H	$SO_3H \cdot H_2O$	H	Ph	126–31	—	C	blue(inst)	—	[59]
11	Ph	H	NO_2	H	Ph	136–7	114 152–3*)	C	—	quinone 136–7	[59,104,112]
12	Ph	H	NH_2	H	Ph	149–50	125	C		quinone 136–7	[59]
13	Ph	H	NMe_2	H	Ph	110–2	103	C	deep red(inst)	—	[59]
14	Ph	H	NMe_3BF_4	H	Ph	188–9	231–4	C	—	—	[59]
15	Ph	H	F	H	Ph	92–93.5	88–91	C	—	dimer	[59]
16	Ph	H	Cl	H	Ph	—	112–3	C	—	polymer quinone	[59]
17	Ph	H	Br	H	Ph	40–5	89	C	—	quinone 136–7	[24,59]
18	Ph	H	CN	H	Ph	165–6.5	90–4	C	green(st) strong oxid.	dimer	[59,146]

Table 8 (continued)

No.	R^2	R^3	R^4	R^5	R^6	m.p. (°C) phenol	m.p. (°C) acetate methyl-ether*)	method of synthesis	dehydrogenation: radical stable(st) instable(inst)	dehydrogenation: dimer m.p. (°C)	Ref. and remarks
19	Ph	H	Ph	H	Ph	148–9	121–2 107–8*)	A	red(st)	dimer 142 dec.	Table 6 (5), D, ^{13}C, ^{17}O, ^{18}O labelled derivatives [127]
20	Ph	H	PhMe(4).	H	Ph	226,5–7,5	—	A	red orange(st)	—	[145]
21	Ph	H	Ph—Ph(4)	H	Ph	158–60	—	A	green blue(st)	125–30	[163]
22	Ph	H	$Ph(Ph)_n$ $Ph(Ph)_2OH$	H	Ph	—	—	A	—	—	see p. 122; a = 1, 2
23	Ph	H	PhCN(4)	H	Ph	177–9	178–8.5	K	red(st)	142–5	[59,146]
24	Ph	H	$PhCO_2Me(4)$	H	Ph	153–7	—	K			[165]
25	Ph	H	$PhCO_2Ph(4)$	H	Ph	154–6	173–5	K			[165]
26	Ph	H	$PhCONH_2(4)$	H	Ph	217–8	—	K			[165]
27	Ph	H	PhCOPh(4)	H	Ph	154–6	173–5	K	red violet(inst)		[165]
28	Ph	H	PhC(OH) $Ph_2(4)$	H	Ph	182–7	—	K	red violet(inst)		[165] $HClO_4$ → blue carbenium salt
29	Ph	H	PhC(OMe) $Ph_2(4)$	H	Ph	175–80	—	K			[165] $HClO_4$ → blue carbenium salt
30	Ph	H	$PhCHPh_2(4)$	H	Ph	172–3	155–8	K	first red violet blue 172–3 O_2 sensitive		[165] O, C(4')-diradical (?)
31	Ph	H	PhOMe(4)	H	Ph	159–62	167–8	A	deep blue 92–95	183–4	[166]
32	Ph	H	PhO— $Ph(Ph)_3OH(4)$	H	Ph	—	—		deep blue		[133,8]
33	Ph	H	$PhNMe_2(4)$	H	Ph	143–4		K	green(inst)	quinonimine derivative	[167]
34	Ph	H	$(Ph—S)_2(4)$	H	Ph	197–202	190–2	K	green(inst)	—	[132,165]
35	Ph	H	$PhSO_2H(4)$	H	Ph	80 (dec)	234–6**	K			[132]**) anilinium salt
36	Ph	H	$PhSO_2H(3)$	H	Ph	70 (dec)	212–5**)	K			[132]**) anilinium salt
37	Ph	H	$PhSO_3H(4)$	H	Ph	248–50	233–4***)	K			[132,165]***) thiuronium salt
38	Ph	H	$PhSO_3H(3)$	H	Ph	285–300	172–8***)	K			[132,165]***) thiuronium salt

39	Ph	H	PhCl(4)	H	Ph	145–6	135	A	red violet		[132,168]
40	Ph	H	PhCl(3)	H	Ph	122–3	96–97	A	red violet	107–44	[132]
41	Ph	H	PhBr(4)	H	Ph	145–6	134–5 85–7*)	A	red violet(inst)		[132,133,5]
42	Ph	H	PhBr(3)	H	Ph	127–7.5	115–6	A	red violet	117–42	[132,5]
43	Ph	H	PhBr(2)	H	Ph	114–5	147–8	A	orange	122–34	[132]
44	Ph	H	Ph	H	Bu^t	68–9	111–3	A	green(inst)	105–10	[129]
45	Bu^t	H	Ph	H	Bu^t	101–2	—**)	H	violet	162–72	[161] **) benzoate m.p. 99–100 °C
46	Ph	H	Bu^t	H	Bu^t	57–8	—	A	green	dimer	[162]
47	PhMe (4)	H	Ph	H	Ph	126.5–7	—	A	red violet(st)		[145]
48	PhMe (4)	H	Ph	H	PhMe (4)	89–90	—	A	red violet(st)		[145]
49	PhMe (4)	H	PhMe	H	PhMe (4)	138–9	—	A	red violet(st)		[145]
50	$PhBu^t$ (4)	H	$PhBu^t$(4)	H	$PhBu^t$ (4)	272–3	180 162–3*)	A	violet(st)	180–5	[164]
51	$PhBu^t$ (4)	H	PhBr(4)	H	$PhBu^t$ (4)	266	238–9	H	violet(inst)	180–90	[164]
52	Ph—Ph(4)	H	Ph	H	Ph	129–30	—	A	red(st)	118–23	[163]
53	Ph—Ph(4)	H	Ph	H	Ph—Ph(4)	200–1	—	A	red(st)	175–85	[163]
54	Ph—Ph(4)	H	Ph—Ph(4)	H	Ph	156–60	—	A	blue green(st)	143–50	[163]
55	Ph—Ph(4)	H	Ph—Ph(4)	H	Ph—Ph(4)	216–8	—	A	green(st)	172–3	[163]
56	Ph-1-napht (4)	H	Ph	H	Ph	193–4	—	A	red(inst)		[163]
57	PhCl (4)	H	Ph	H	Ph	132–3	—	A	orange red(st)		[145]
58	PhCl (4)	H	Ph	H	PhCl (4)	66–7	144	A	orange red(st)		[166]

Table 8 (continued)

No.	R^2	R^3	R^4	R^5	R^6	m.p. (°C) phenol	m.p. (°C)		dehydrogenation		Ref. and remarks
							acetate methyl-ether*)	method of synthesis	radical stable(st) instable(inst)	dimer m.p. (°C)	
59	PhCl (4)	H	PhCl(4)	H	PhCl (4)	169–70	—	A	orange red(st)		[145]
60	PhBr (4)	H	Ph	H	Ph	165–6	134–5	A	red violet(st)		[132,133]
61	PhBr (3)	H	Ph	H	Ph	157–8	113–4	A	red violet(st)		[132,5]
62	PhBr (2)	H	Ph	H	Ph	50–5	140–3	A	red violet(st)		[132]
63	PhBr (4)	H	Ph	H	PhBr (4)	77–8	—	A	red(st)		[168]
64	PhBr (4)	H	PhBr(4)	H	PhBr (4)	159	—	A	yellow(inst)	202–5	[168]
65	PhO Me(4)	H	Ph	H	PhO Me(4)	159–62	122–4	A	blue, 92–95		[166]
66	PhO Me(4)	H	PhOMe(4)	H	PhO Me(4)	134–7	—	A	deep bluegreen	107–112	[166]
67	PhO Ph(4)	H	PhOPh(4)	H	PhO Ph(4)	160–3	—		deep blue	100–5	[8]
68	Ph	Ph	Ph	H	Ph	244–5	181–2 179–80*)	A	red(st)	132 dec.	[5,110]
69	Ph	Ph	Ph	Ph	Ph	267	317 200–2*)	A	red(st)	170 dec.	[115]
70	Ph	R	Ph	H	Ph	—	—	B	—	—	see Table 1 for different R and Ph
71	Ph	CN	Ph	H	Ph	122	141–2	D	yellow(st)	135–90	[168]
72	Ph	CN	Ph	CN	Ph	279	—	D	yellow(st)	—	[168] strong oxidant dimer
73	Ph	CN	Ph	Cl	Ph	267	—	D	yellow(st)	—	[168]

74	Ph	CN	PhBr	Ph	Ph	159–61	—	D			[168]
75	Ph	Cl	Ph	H	Ph	180–1	144	D	pink(st)	130–40	[5]
76	Ph	Br	Ph	H	Ph	177–8	152	D	pink(st)	135–145	[5]
77	Ph	NO_2	Ph	H	Ph	208–10 165–73*)	—	E	red(st)	—	[8]
78	Ph	NO_2	Ph	H	Ph	173–80	—	F	(inst)	—	[8]
79	Ph	OH	Ph	H	Ph	167–8	150–60**)	D	(inst)	—	[5,8] m.p. 208–209 °C
80	Ph	OAc	PhCl(4)	H	Ph		198***)	D	—	—	[166]***) diacetate
81	Ph	OAc	PhCl(3)	H	Ph		153***)	D	—	—	[166]***) diacetate
82	PhCl (4)	OAc	PhCl(4)	H	PhCl (4)		163***)	D	—		[166]***) diacetate
83	Ph	OAc	PhBr(4)	H	Ph		191***)	D	—		[166]***) diacetate
84	Ph	OH	Ph	OH	Ph	167	150–160***)	D	(green) (inst)		[166]***) diacetate (2,4,5-triphenyl-resorcinol see [169])
85	Ph	CN	CN	H	Ph	205	—	D	$K_3Fe(Cr)_6$/NaOH not dihydro-genated		[168] $Pb(OAc)_4 \rightarrow 2$ quinol-acetate m.p. 130–140 °C
86	Ph	CN	CN	CN	Ph	268–73	—	D	not dihydrogenated		[168,146]
87	Ph	Ph	OMe	Ph	Ph	247	282–3	G	yellow(st)	crist.	[115]
88	Ph	Ph	OH	Ph	Ph	300–2	304–6	G	yellow quinone	306–7	[115]
89	Ph	H	CH_2(Ph3,5-Ph)$_2$OH(4)	Ph	Ph	203–5	220–1	L	orange(inst)		[159,165]
90	Ph	OH	Ph	Ph	H	208–9	222–9 160–1***)	p. 116 p. 108	hydroxy quinone		[169]***) diacetate
91	Ph	OH	Ph	Ph	Ph	251–2			(diradical) deep red	—CO tetracyclone	[60]

A) Pyrylium salt method; B) Pyrylium salt rearrangements; C) From 4-diazonium salt; D) From quinol derivatives; E) Nitration ($HOAc/HNO_3$); F) Reduction; G) Rearrangement of the quinol derivative; H) Alkylation; I) From compound *59* with NaOPh; K) From compound *41* with LiBu etc.; L) From compound *1* with CH_2O, or COR_2, $LiAlH_4$

bis (2,6-diphenylphenol) has been prepared from 2,6-diphenylphenol by the addition of sulfur in xylene and sodium hydroxide [172]. The reaction of 2,6-diphenylphenol with formaldehyde or with ketones is described in Ref. [159,165]. When 2,4,6-triphenylphenolacetate (TPPOAc) reacts with chlorine, substitution in the central ring gives 3-chloro-2,4,6-triphenylphenolacetate. Bromine and TPPOH in acetic acid, however, give 4-(4′-bromo-phenyl)-2,6-diphenylphenol. In both cases the yields are nearly quantitative. The 4-(4′-bromophenyl)-2,6-diphenylphenol and its methylether are valuable compounds in the synthesis of 4-(4′-substituted phenyl)-2,6-diphenylphenols after first reacting them with lithium-butyl in THF and than with electrophiles. For examples see Table 8 [132,135].

4.3 Dehydrogenation

4.3.1 One Electron- or Hydrogen Atom Abstraction

4.3.1.1 Phenoxyl Radicals and Their Dimers

Whereas the dehydrogenation or oxidation of phenol and most of its alkyl- or halogeno-substituted derivatives give rise to many different compounds, including polymers [173], the 2,4,6-tri-t-butyl- or -tri-phenyl-phenols yield a deeply colored persistant phenoxyl radical. Various oxidants, such as lead- or manganese dioxide, silveroxide, cerium(IV) sulfate, chromium(III) salts, chromium(VI)oxide, or potassium permanganate etc. have been used. One of the prefered one electron acceptor oxidants is potassium hexacyanoferrate(III) in aqueous alkaline solution, in which only the phenolate is soluble. When the phenol derivative (*1a*) in a water immiscible solvent, such as benzene, carbondisulfide, tetrachloro- or dichloromethane is shaken with the aqueous phase, the phenolate *2a* is oxidized to the phenoxyl *3a*. Since this is insoluble in the aqueous phase, it immediately moves into the organic phase, where it can more or less equilibrate with its dimer *4a*. In the aqueous phase, therefore, the concentration of the phenolate *2* is always much higher than that of the phenoxyl *3*. For this reason the oxidation potential of hexacyanoferrate(III) may be considerably lower than the oxidation potential of the phenolate/phenoxyl system.

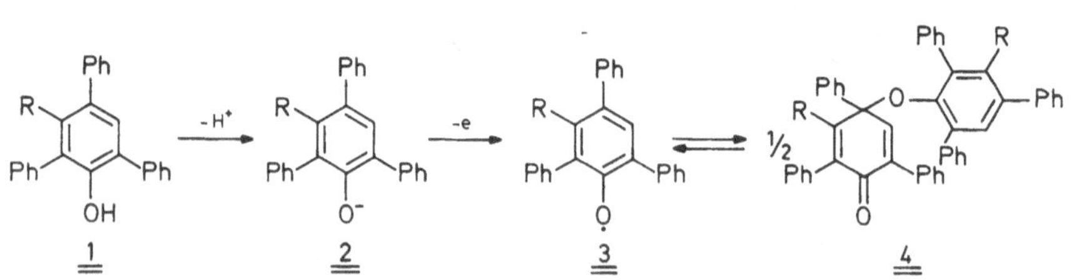

1 2 3 4

a. R = H
b. R = Br

The structure of the dimer *4* has been fully established by X-ray analysis of the bromo analogue *4b* [174]. The IR-spectrum of the dimers such as *4* in CCl_4 shows the characteristic double bands of the CO group near 1666 cm^{-1} [175] and another band at 1201 cm^{-1} due to the phenolether. Substitution of ^{16}O by ^{18}O in *1a* shifts the C=O absorption of *4a* from 1665 and 1642 cm^{-1} to 1620 and 1595 cm^{-1} and the ether absorption from 1201 cm^{-1} to 1185 cm^{-1} [127]. This is in full accordance with structure *4*.

A very useful method in transforming phenols, even unstable ones, into phenoxyl radicals is flash photolysis [176,177]. When 2,4,6-triphenylphenol (*1a*) in benzene was photolysed the radical *3a* as well as its dimer *4a* were observed. The K-value $3a \rightleftarrows 4a$, determined by flash photolysis, agrees well with the K-value of the colorimetric determination, mentioned below. Flash photolysis can also be used to evaluate the rates of dimerization or dissociation of $3a \rightarrow 4a$ (k being in the order of 0.3 sec $mole^{-1}$), as well as determining the activation parameters [177].

By using Ziegler's colorimetric method [178] to determine the dissociation constant of 2-triphenylmethyl and its dimer, the dissociation constant of the 2,4,6-triphenylphenoxyl dimer under various conditions has been measured.The values presented in Table 9 were found by evaluating the extinction coefficients of the maximum at ~530 nm of the radical depending on the concentrations [133].

Table 9. Dissociation constants K at 20° and dissociation heats of $4a \rightleftarrows 3a$ in various solvents

Solvent	K (20 °C) $mol \cdot l^{-1}$	dissociation heat ($kcal \cdot mol^{-1}$)
trichloromethane	$22 \cdot 10^{-5} \pm 6 \cdot 10^{-5}$	—
acetonitrile	$11 \cdot 10^{-5} \pm 3 \cdot 10^{-5}$	—
benzene	$4 \cdot 10^{-5} \pm 3 \cdot 10^{-5}$	—
carbondisulfide	$3 \cdot 10^{-5} \pm 1 \cdot 10^{-5}$	9.1 ± 1
tetrachloromethane	$1 \cdot 10^{-5} \pm 0.5 \cdot 10^{-5}$	14.6 ± 1

Substitution of phenyl- by biphenyl groups in positions 2 and/or 6 have only a small influence on the dissociation constants in benzene. Substitution in position 4, however, distinctly raises the K values as shown in Table 10 [163].

4-Methyl-, 4-methoxy, and 4-cyano-2,6-diphenylphenoxyls are almost fully dimerized. Since in most cases the dissociation rate of the dimers into radicals in very large, the radical content of the solutions can easily be titrated using a solution of hydroquinone in acetone. The bleaching of the colored radical gives the exact end

Table 10. Dissociation constants K of biphenyl substituted phenoxyls

R^2	R^4	R^6	K (20°) $mol \cdot l^{-1}$
Ph	Ph	Ph	$4 \cdot 10^{-5}$
Ph—Ph	Ph	Ph	$6 \cdot 10^{-5}$
Ph—Ph	Ph	Ph—Ph	$6 \cdot 10^{-5}$
Ph	Ph—Ph	Ph	$13 \cdot 10^{-5}$
Ph—Ph	Ph—Ph	Ph	$13 \cdot 10^{-5}$
Ph Ph	Ph—Ph	Ph—Ph	$13 \cdot 10^{-5}$

point of the titration. In some cases, where the dissociation rate of the dimer is low, such as with the dimer of 3-chloro-2,4,6-triphenylphenoxyl, a typical Schmidlin effect is observed. Addition of a small amount of a phenol of a lower oxidation potential, which gives a persistant colored radical, allows a relatively quick titration with the hydroquinone solution.

4.3.1.2 UV/Vis and ESR Spectra

The deeply colored, red, violet or blue arylated phenoxyls show three or four distinct maxima in the visible region, the wave lengths of which are dependent on the solvent. Addition of methanol shifts the two highest maxima to longer wave lengths. Since the extinction coefficients, especially of the long wave maxima, depend strongly on the concentrations and temperatures, table 11 gives only the absorption maxima of some stable arylated phenoxyls at 20° in benzene.

Table 11. Absorption maxima (nm) of some arylated phenoxyls in benzene at 20° [163]

R^2	R^3	R^4	R^5	R^6	very low	Extinction			Ref.
						middle	high	very high	
Ph	H	Ph	H	Ph	735	524	381	343	[163]
Ph—Ph	H	Ph	H	Ph	760	546	405	—	[163]
Ph—Ph	H	Ph	H	Ph—Ph	790	562	418	351	[163]
Ph	H	Ph—Ph	H	Ph	710	575	391	—	[163]
Ph—Ph	H	Ph—Ph	H	Ph—Ph	780	572	399	—	[163]
Ph	Ph	Ph	Ph	Ph	780	563	>400	—	[113]

The electron spin resonance spectra of the arylated phenoxyls give very many narrow lines caused by the coupling of the unpaired electron with the many aromatic protons of the substituents. By a systematic study, in which the H-atoms were substituted by D-atoms, ^{12}C- by ^{13}C-, and finally ^{16}O by ^{17}O, all the coupling constants of the complicated ESR spectra of 2,4,6-triphenylphenoxyl were determined. Since all details of the results are reviewed [179] no further comment shall be added in this account. The ^1H-coupling constants agree well with those of an ENDOR spectroscopy reported by Hyde [180].

4.3.1.3 Oxidation Potentials

When the colors of the phenoxyls are very different, as for example of 2,4,6-tri-t-butylphenoxyl (blue) and 2,4,6-triphenylphenoxyl (red orange) qualitative experiments using equimolar amounts of the phenol and the phenoxyl of the two different compounds in an organic solvent show which of the phenoxyls is the stronger oxident:

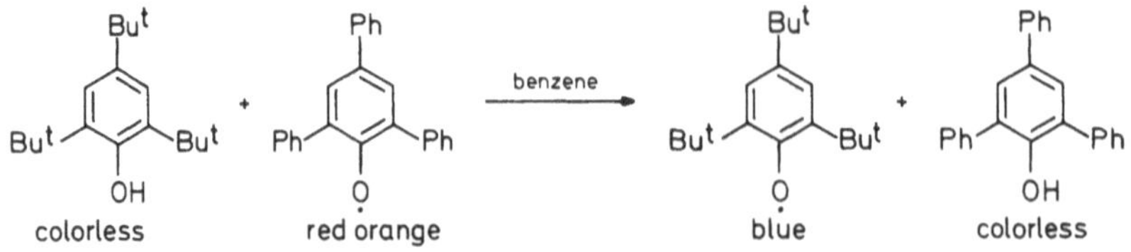

Table 12. Polarometric half wave potentials of arylated phenols

Substance	$E_{1/2}$ (mV)	Substance	$E_{1/2}$ (mV)
4-phenyl-2,6-dicyanophenol	> +589	2,4,6-tris-biphenylphenol	+ 5
2,4,6-triphenyl-2,6-dicyanophenol	+512	4-biphenyl-2,6-diphenylphenol	+ 4
3-chloro, 5-cyano-2,4,6-triphenylphenol	+401	2,6-bis-(biphenyl)-4-phenylphenol	0
2-cyano-4,6-diphenylphenol	+382	2,4,6-triphenylphenol	0
4-cyano-2,6-diphenylphenol	+338	2-biphenyl-4,6-diphenylphenol	— 9
4-(4′-bromophenyl)-3-cyano-2,6-diphenylphenol	+241	2,4,6-tris-(4′-phenoxyphenyl)-phenol	— 32
3-cyano-2,4,6-triphenyl	+222	4-fluoro-2,6-diphenylphenol	— 44
2,4,6-tris-(4′-bromophenyl)phenol	+195	2,6-bis-(4′-methoxy phenyl)-phenol	— 55
2,3,4,5,6-pentaphenylphenol	+155	2,4,6-tris-(4′-methoxy phenyl)-phenol	— 87
3-chloro-2,4,6-triphenylphenol	+136	4-t-butyl-2,6-diphenylphenol	— 91
4-(4′-cyanophenyl)-2,6-diphenylphenol	+ 83	6-t-butyl-2,4-diphenylphenol	— 99
2,6-bis-(4′-bromophenyl)-4-phenylphenol	+ 46	4,6-di-t-butyl-2-phenylphenol	—135
2,3,4,6-tetraphenylphenol	+ 27	2,6-di-t-butyl-4-phenylphenol	225
4-(4′-bromophenyl)-2,6-diphenylphenol	+ 25	2,4,6-tri-t-butylphenol	270
2,6-diphenyl-4-(4′-phenylsulfonic acid)-phenol	+ 16	3,4,5,6-tetrachloro-1,2-hydroquinol	—290

Many similar experiments have been performed. The absorption maxima of the two radicals must, however, be rather different for evaluating the equilibrium constants using colorimetry [181]. These experiments give only a qualitative scale of the oxidative power of the different arylated phenols. It agrees well with the known influence of electron donating or accepting substituents, studied in similar substituted hydroquinone/quinone series.

More exact results have been found using electrochemical measurements. Steuber [10] has used the polarography of the arylated phenolates in acetonitrile/water 10:1 and tetramethylammonium-hydroxide and -chloride. Since phenoxyls of a high potential react with mercury, giving a solid Hg(I)-phenolate (mercuri (I)-2,4,6-triphenylphenolate m.p. 183–185° [133]) a special rotating graphite index electrode has been developed. The half wave potentials of many arylated phenolate/phenoxyls have been measured using an aqueous silver/silverchloride reference electrode ($E_0' = +203$ mV against a normal hydrogen electrode), bridged by a lithium chloride 90% acetonitrile solution. They are presented in Table 12, the potential of 2,4,6-triphenylphenolate/phenoxyl as reference value = 0 mV [10].

Since these potentials have been determined in dilute solutions, the equilibrium of the phenoxyls *3* with their dimers *4* has been neglected. In each case it has been proven that the equilibrium between the phenoxylate/phenoxyl has been reached. To compare these potentials with other oxidants, the half wave potential of 3,4,5,6-tetrachloroquinolate/semiquinone under the same conditions (giving —290 mV) has also been measured. Similar oxidation potentials have been found, when lithium-phenolates (prepated from the phenols with phenyllithium) have been titrated with gold(III)chloride in acetonitrile by potentiometry, using a platinum indicator- and a silver/silverchloride reference electrode [182]. The potentials presented in Table 12, however, cannot be compared with the normal potentials of redox systems in aqueous solutions, since the solvent, hydrogen bonds, p_K-values, and liquid/liquid potentials all influence the values in an unpredictible way. The oxidation potentials of the arylated phenoxyls in Table 12 are amohg the highest for organic compounds in organic solutions. This can be shown by a simple experiment: When 2,3-dichloro-5,6-dicyanoquinone (DDQ) in benzene, one of the quinones with the highest oxidation potential [183], is mixed with 2,4,6-triphenylphenol (1:1 or 1:2) the phenol is not dehydrogenated to its phenoxyl. 2,4,6-Triphenylphenoxyl, however, mixed in benzene with 2,3-dichloro-5,6-dicyanohydroquinone (1:1 or 2:1). Under the same conditions dehydrogenates the hydroquinone to give the yellow quinone derivative.

Other careful electrochemical measurements of the oxidation potentials of 2,4,6-tri-t-butylphenol and 2,6-di-t-butyl-4-methylphenol in acetate buffered ethanol or acetonitrile have been measured by Mauser et al. [184]. They determined the static potentials using a boron carbide indicator and a mercury/mercury-acetate reference-electrode. Since in this case the oxidation of the phenols and not the phenolates to the phenoxyls has been determined the oxidation potentials cannot be compared with those in Table 12. For other electrochemical oxidations of phenols in buffered aqueous solutions using a graphite electrode see Ref. [185,186].

4.3.1.4 Chemical Reactions of Phenoxyl Radicals

Two different persistant phenoxyl in organic solvents can react to give mixed dimers [187]. At raised temperatures these partly dissociate to give the original two radicals. By means of the IR spectra of ^{18}O labelled 2,4,6-triphenoxyl it is found that in both examples *1* and *2* quinols are formed, in which the 2,4,6-triphenoxy (TTPO) part is the unchanged ether component [188].

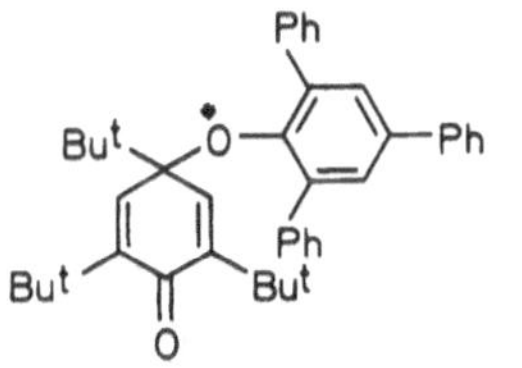

1

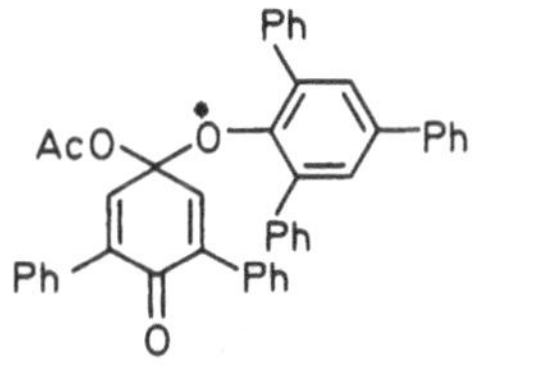

2

66

When 2,4,6-triphenylphenoxyl is heated for several hours in non-reactive solvents, such as benzene, self condensation to a higher molecular phenol is observed. Pyridine, but also visible light catalyse this reaction preferently at the 4'-positions of the 2,4,6-phenyl groups to give TPPO-ethers *3*:

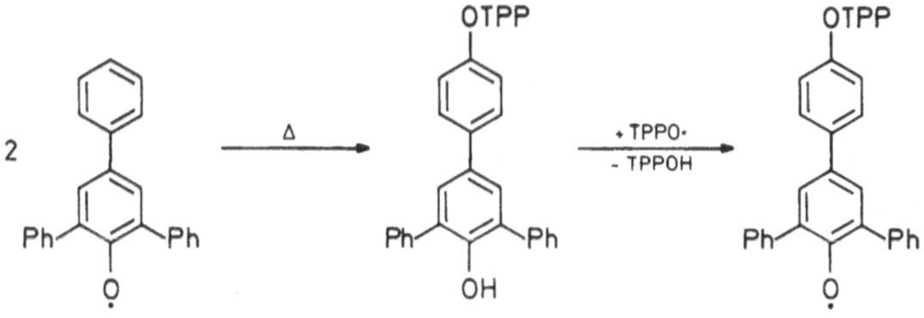

3 4

red (TPPO·) colorless blue

Since the phenol *3* has a lower oxidation potential than TPPOH it is dehydrogenated by the red TPPO· to give a blue solution of *4* [164].

To prevent the self condensation of TPPO· the three 4'-positions of the 2,4,6-phenyl rings of TPPOH were protected using t-butyl groups. This is achieved by reacting TPPOH with t-butylchloride and aluminium chloride [189]. The new phenoxyl *5* is less sensitive than TPPO· when it is heated in organic solvents. But after refluxing in 1,2-dichloroethane *5* also self condenses to give the phenol *6*. Dehydrogenation of *6* again gives a persistant violet-red phenoxyl [164].

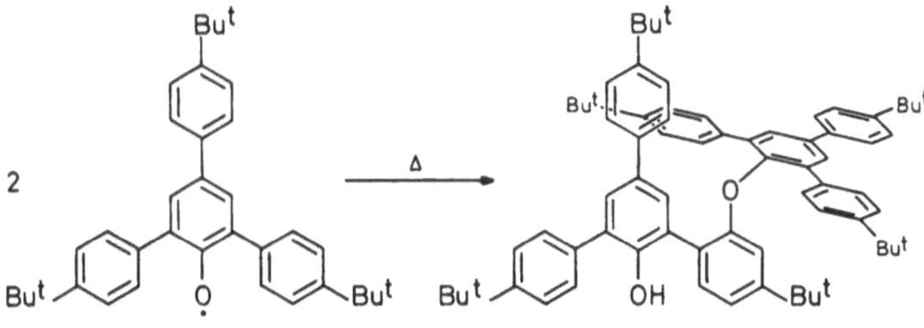

5 6

Dehydrogenation using the arylated phenoxylsTPPO· or *5*, therefore, should be carried out at room temperature to avoid side reactions (which are also observed in dehydrogenations with tetrachloroquinone [183]) and 2,3-dichloro-5,6-dicyanoquinone at higher temperatures). A very convenient procedure, which is especially useful in the dehydrogenation of hydrazo compounds *7* to azocompounds *8*, is the two-phase reaction of the organic substance in an organic solvent, e.g. dichloromethane, with an aqueous alkaline solution of potassium hexacyanoferrate(III). Addition of only 1–2% or even 1–2‰ of TPPOH or the phenol of *5* and shaking the two phases for several minutes at room temperature gives the 1,2-diazenes in almost quantitative yield. The work-up is simple, since almost no oxidant contaminates the organic phase. Even sensitive 1,2-diazenes, which at elevated temperatures easily split into nitrogen and organic radicals, have been prepared by this method.

$R^1-NH-NH-R^2 + 2TPPO\cdot \rightarrow R^1-N=N-R^2$

7 *8*

$R^1=R^2=Ph, Me_2CN; C_6H_{10}CN$

$R^1=Ph; Ph(NO_2)_2$ (2,4); $R^2=CPh_3$.

N-Acylhydrazines have to be dehydrogenated with two moles of TPPO· in an organic solvent since the dehydrogenated products are soluble in the aqueous alkaline phase. Diisopropylhydrazine gives bis-isopropylenediazane [189].

The first step of these dehydrogenations is probably an electron transfer of one electron from the unbonded electron pair of the nitrogen atom to the phenoxyl radical to give the nitrogen cation radical and the phenolate. Such a route can be shown with bis-(1,4-dimethylamino)-benzene *9* to yield Wurster's radical cation *10* [9].

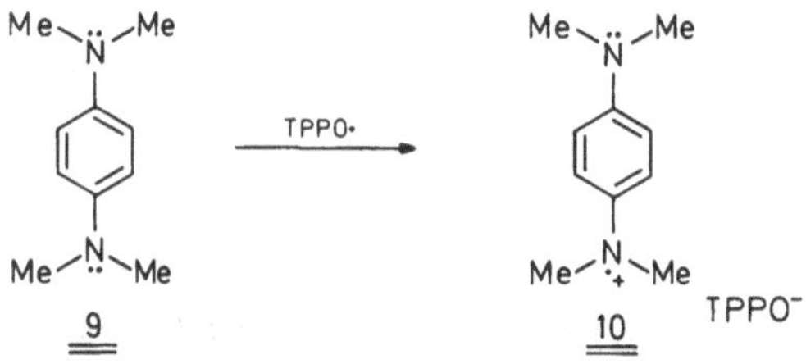

Benzylaniline *11* is substituted via a similar mechanism in the para position of the N-phenyl ring to the TPPO-ether and dehydrogenated by an excess of TPPO·

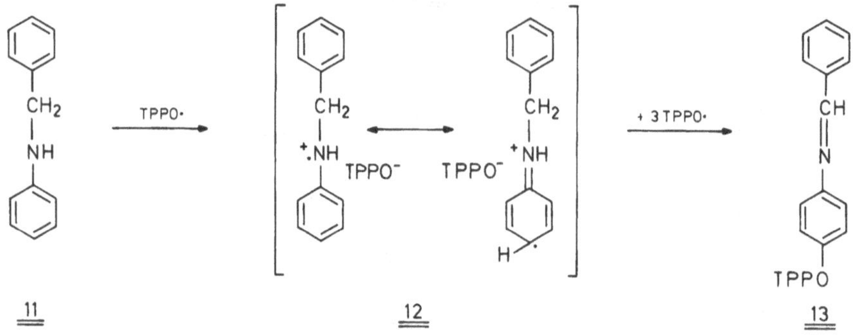

to *13* [190]. The radical cation *12* is the probable intermediate which is then substituted by the addition of the TPPO· radical.

No formation of an ether substitution product is observed when the para-anilino position is substituted, e.g. when N-benzyl-p-toluidine is dehydrogenated with TPPO·. Substitution of the CH_2-bonded phenyl ring does not occur [190].

Dehydrogenation of thiophenols with TPPO gives disulfides. Thiophenoxy radicals, even in the case of 2,4,6-triphenylthiophenol, have not been observed by ESR spectroscopy [145].

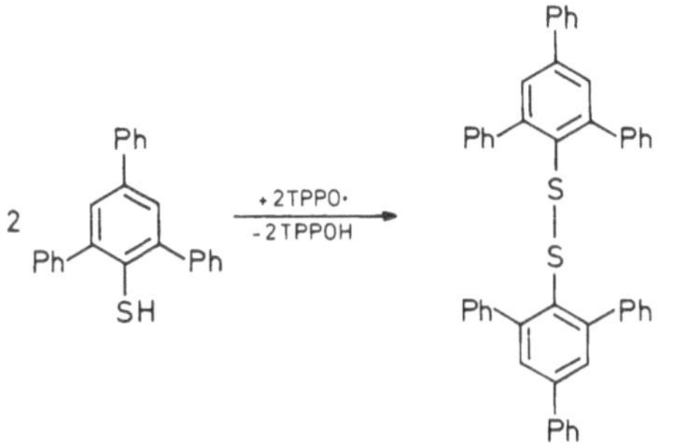

Hydrogen abstraction from C—H bonds using TPPO· or its tris-4'-t-butylated derivative *5* is only successful in special cases. Some examples which lead to arylethers of TPPO· or *5*, have been mentioned before, such as the self condensation of TPPO. or of *5* at higher temperatures as well as the formation of *13*. Another example is the reaction of triphenylmethane and TPPO· in CCl_4. A light sensitive colorless compound m.p. 145–146°, has been isolated in 80% yield. The IR spectra (no C=0 absorption at 1660 cm^{-1} but a strong absorption at 1200 cm^{-1} which points to an ether bond) disfavors a quinole derivative and supports structure *14*. Heating *14* in dichlorobenzene gives a red solution of the two radicals triphenylmethyl (which can be reacted with oxygen to give the crystalline peroxide) and TPPO·. When *14* is heated in dichlorobenzene in the presence of water, triphenyl-carbinol and TPPOH have been isolated [133].

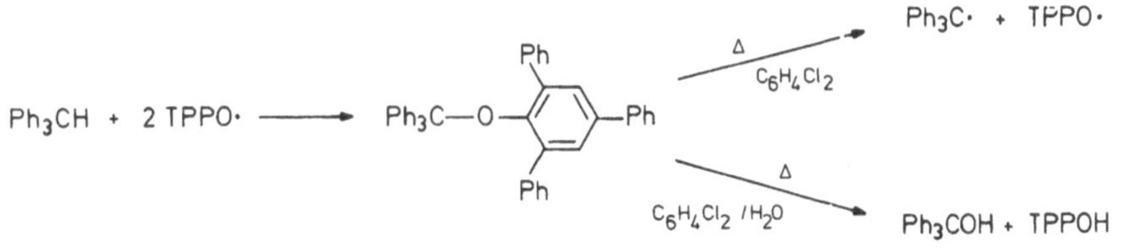

14

Whereas former experiments, which describe the dehydrogenation of cyclohexane to cyclohexadiene and benzene by heating with TPPO·. [133] could not be reproduced, naphthaline has been isolated when 1,2,3,4-tetrahydronaphthaline and 2 moles of

TPPO· have been reacted in CH_2Cl_2 at room temperature. The isolation of naphthaline and TPPOH, however, is only successful after work-up by chromatography on aluminiumoxide.

This is a radical mechanism which is similar to Criegee's reaction of tetraline and 2,3-dichloro-anthracene-1,4,9,10-diquinone [191]. He isolated the 1-tetraline-hydroquinone ether derivative. This intermediate decomposes at the m.p. (242–244°) to give 1,2-dihydro-naphthaline and the 1,4-hydroquinone derivative. A similar route using tetraline and TPPO·, therefore, seem to give the sensitive intermediate diether *15*, which decomposes with heat or Al_2O_3 to give naphthaline and TPPOH [164]:

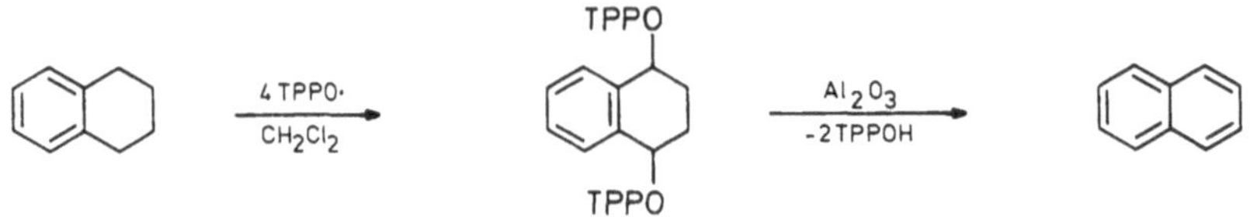

15

When 1,4- or 1,2-dihydronaphthaline in CH_2Cl_2 was reacted in a two phase oxidation with TPPO· at room temperature by adding only very small amounts of TPPOH to avoid an excess, the intermediate *17* was isolated. The elimination of TPPOH to give naphthaline is caused by melting *17*, as well as by acids, phenols or even alcohols. 1,4-Dihydronaphthaline reacts much faster than 1,2-dihydronaphthaline. Since in the former compound the two hydrogen atoms in the 1,4-position are more activated than the hydrogen atoms in position 3 and 4 of the latter compound, the abstraction of a hydrogen atom in the former is favored. In both dihydronaphthalines the allyl radical *16* will be the intermediate (not detected by ESR) which is quickly trapped by TPPO· to give *17*. It decomposes with acid or on heating to give TPPOH and naphthaline:

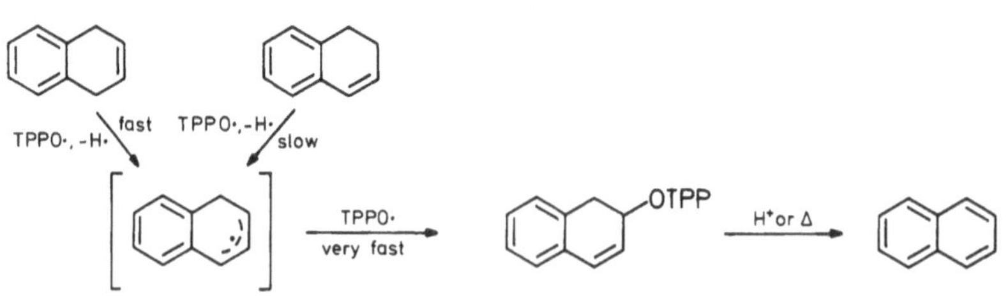

16 17

Such a radical mechanism was supported when fluorene (in CCl_4) was reacted with TPPO· [164]. After 24 h the radical color dissappeared and *18a* (m.p. 178–179°) was isolated. Similarly, *18b* (m.p. 229–230°) was found by reacting fluorene with 2,4,6-tris-(4'-t-butyl-phenyl)-phenoxyl. Both ethers are very stable. Since no 9,9-

difluorene was found, the hydrogen abstraction of fluorene to give the fluorene radical must be slower and the addition of the phenoxyl very fast:

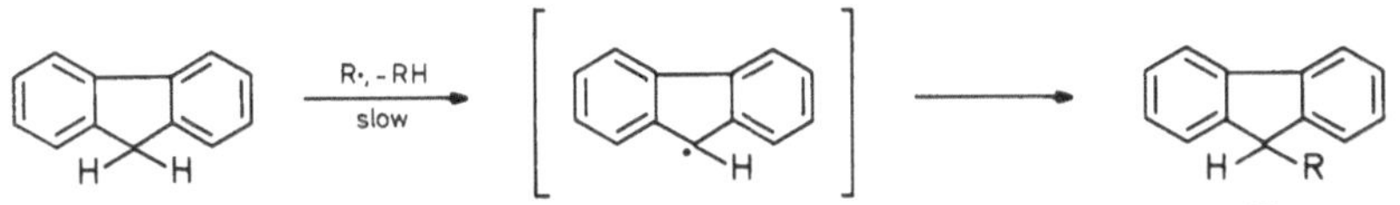

18

a. R = TPPO
b. R = tris-(2,4,6-4',4'',4'''-t-butylphenyl)-phenoxyl

By a similar route, 9,10-dihydro-anthracene gives 94% of the sensitive monophenoxyl ether (m.p. 202°) *19* (R see above formula b). It decomposes in the presence of light or traces of acids to give ROH and anthrance:

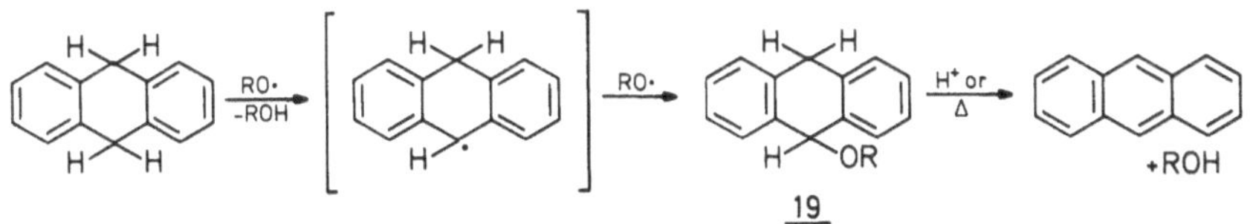

19

Both 9,9-di-deuterofluorene and 9,9,10,10-tetradeuteroanthracenes show a large kinetic isotope effect [164] $k_H/k_D \sim 6$–8. This is further evidence in favor of the proposed mechanism, i.e. hydrogen abstraction is the slow step and the radical/phenoxyl addition is fast [192].

TPPO· is light sensitive but its photochemistry remains to be studied[2]. Since Hageman [193] has found that irradiation of 2- or 2,6-diphenylquinone in acetonitrile gives the 4-hydroxydibenzofuranes *20* through an intramolecular hydrogen transfer, a ring clousure similar to that when TPPO· gives a dibenzofurane derivative and TPPOH does seem possible. When the 2- and 2,6-diphenylquinones are irradiated in methanol or acetic acid, not only the furane derivatives *20* but also cyclobutane derivatives formed via a [2 + 2]addition are observed [194].

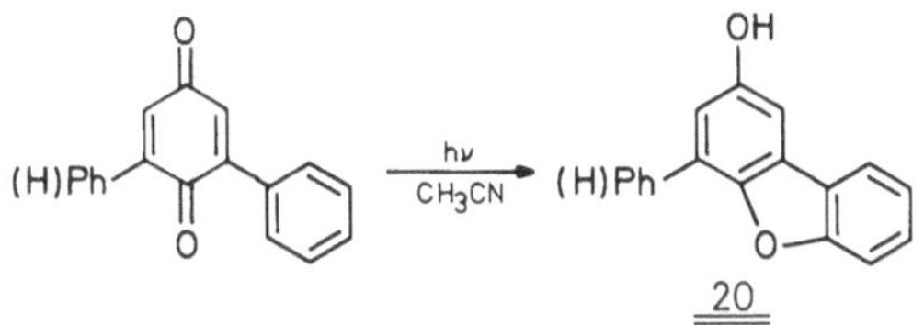

20

2 For another light induced reaction see formula *3*, p. 145

4.4 Two Electron- or Hydrogen Atom Abstraction

4.4.1 Quinones, Diphenoquinones and Phenolethers (Oxidative Coupling Reactions)

The oxidation of phenols to various compounds has been extensively covered in the literature [195]. Their importance in preparative, mechanistic as well as biochemical synthesis has been clearly established. In this account we shall limit our discussion to 2,6-diphenylphenol (*1a*) not only because it is the most important example, but also because it sometimes behaves quite differently to either 2,6-dimethyl- or 2,6-di-t-butylphenol. Three oxidation products *3*, *5* and *6*, as well as the polymer ethers *7*, are possible depending on the oxidants or reaction conditions, e.g., solvent, concentration, temperature and catalyst. The mechanism of the different pathways are by no means fully understood. Route i) which leads to 2,6-diphenylquinone is easily realized when *1a* is oxidized with bromine [68], chromic acid or in Teubers quinone synthesis [196] using Fremy's salt (potassium nitroso disulfonate). In the reaction with CrO_3 *3* is formed as well as the diphenoquinone *5* [80]. When other high valent metal oxides are used *3* is also formed [195]. In some cases the quinole derivative *2a* which equilibrates with *2b* is a possible intermediate in the transformation of *1a* to *3*. With PbO_2 in acetic acid, 4-acetoxy-2,6-diphenylphenol is isolated, and

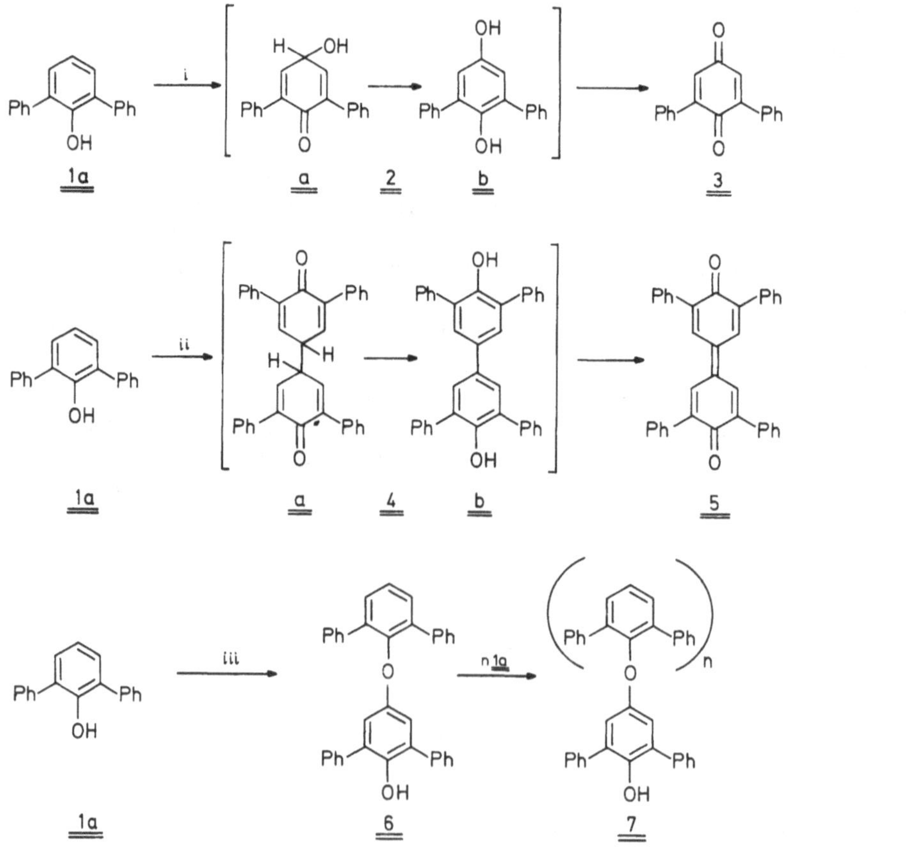

on further oxidation yields exclusively *3*. In formic acid at 30° and 4 mole PbO_2 *1a* gives *3* in 80% yield, whereas 2 mole PbO_2 in acetic acid at 25° gives only 30% of *3* and 65% of *5* [197a)]. Oxidation of *1a* with Ag_2O, MnO_2 and PbO_2 in apolar solvents gives mainly polymeric phenol ethers, such as *7* [197b)].

Route ii), the oxidative coupling of the 4,4,-carbon atoms of two molecules of *1a*, probably leads initially to the intermediate *4a* which then rearranges to become *4b*. This is further oxidized to give *5*. An intermediate, resembling *4b*, has been isolated in the oxidation of 2,6-di-t-butylphenol in CS_2 or CH_3CN with manganesium(III)-trisacetylacetate [198)]. A two phase oxidation of *1a* in benzene and aqueous alkaline $K_3Fe(CN)_6$ gives a 50:50 mixture of *5* and the polymeric ether *7*. 2,6-Diphenyl-phenoxyl is the probable reactive intermediate [59)]. Route iii) is realized by the oxidation of *1a* with oxygen, catalysed by cobalt disalylalethylenediimine ("salcomine"). It gives all three products *3*, *5*, *7*. The product yields depend strongly on the reaction conditions [199)]. Another important class of catalyst used in similar O_2-oxidations are Cu(I)-salts complexed with diethylamine [79)] or in the presence of pyridine [87)]. PbO_2 can also oxidize *1a* to give phenolethers *7* of different molecular weight [61,200)]. These polymeric ethers were, for a long time, important in industry. However, the much cheaper 2,6-dialkylphenols, often in combination with styrene, have replaced them as starting materials for important glassy polymers (see also page 111). A large number of patents describe the different products of mostly 2,6-alkylated phenols to give the phenolethers, such as *7*. The exact oxidation mechanism is, however, by no means fully understood. Phenoxyls or phenolates of Cu(I), Cr(III), Pb(II) and other metals, more or less complexed with amines, are regarded as intermediates [200)]. The coupling of the phenoxy-oxygen with an aromatic carbon atom to give a diarylether has also been observed using 2,4,6-triphenylphenoxyl (see Sect. 4.3.1.4). In the presence of pyridine, light or on heating in organic solvents they give rise to the dimer (or even polymer) ethers by the coupling of the oxygen of C-1 with the carbon at C-4′ of the para phenyl ring. Substitution of the three 4′-positions of the 2,4,6-phenylrings by t-butyl groups delays such a coupling, but does not prevent it. Heating in appropriate solvents leads to O—C coupling products to the 2′-position of one of the 2,6-phenylrings (see formula *6*, p. 145).

4.4.2 Aroxenium Ions[3]

For a long time it was assumed that a two step oxidation of phenols (*1b*), (R = various substituents) leads to aroxenium ions *2* as reactive intermediates. The mesomeric formulae of *2* indicate the reactive positions where a nucleophile can attack.

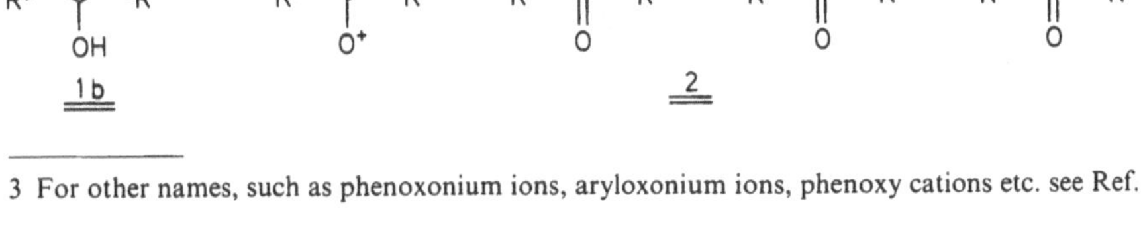

3 For other names, such as phenoxonium ions, aryloxonium ions, phenoxy cations etc. see Ref. [201a)]

Electrochemical and chemical oxidation of phenol derivatives can lead to *2*. Electrochemical methods of substituted phenols, such as 2,4,6-tri-t-butylphenol by fast response potentiostates or by cyclic voltammetry in acetonitrile containing 10% water and 0.1 M tetramethylammonium hydroxide, proceed via a one electron oxidation. The stable blue 2,4,6-tri-t-butylphenoxyl has been isolated. A second irreversible oxidation step via anodic transfer of another electron followed by the reactive intermediate reacting chemically (ECE or EEC- mechanism), has been observed at higher potentials in unbuffered acetonitrile solution and 0.5 M sodium perchlorate [202]. These observations are similar to the results of Steuber and Dimroth [10], who using polarography with a graphite electrode in acetonitrile under basic conditions found that 2,4,6-triphenylphenolate (*3*) looses one electron to give the stable phenoxyl radical *4* (see Sect. 4.3.1.3 and Table 12). At a higher potential, however, anodic oxidation of 2,4,6-triphenylphenol *5* in acetic acid/water 9:1 and 5% sodium acetate gives 2-acetoxy-2,4,6-triphenylquinol *7* by a two electron transfer. The intermediate 2,4,6-triphenylphenoxenium ion *6* has been trapped using the acetate ion. This is in agreement with an EEC mechanism.

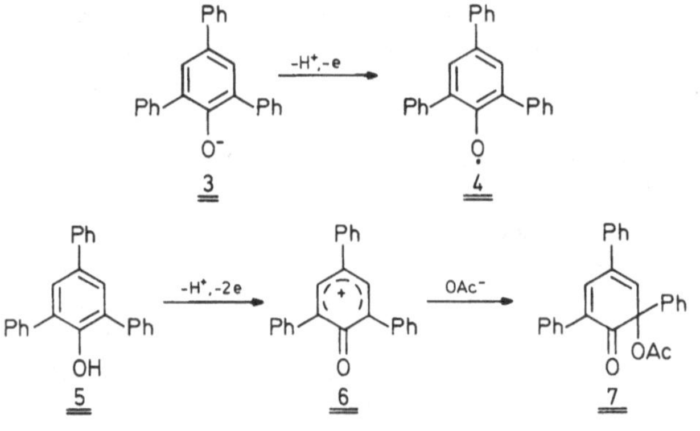

Table 13 gives the half wave potentials of two electron oxidations of some aryl and cyano-substituted phenols, measured against an aqueous Ag/AgCl electrode (conditions as above). The isolated products are the 2-acetoxy-quinol derivatives.

There are, however, many different types of electrochemical oxidations of phenol derivatives possible, the results of which largely depend on the methods used as well as the structure of the different phenols. Secondary chemical reactions of factors including the primary or secondary oxidation products can also occur. The various electrochemical methods used are dependent on solvents, p_H values, electrode materials or absorption effects at the electrodes. These all influence the measured potentials. Moreover, the liquid/liquid potentials and the various indicator electrodes can give results, which cannot be safely compared with the general E_0' scala of redox potentials in aqueous solutions. In this review we cannot go into the many details obtained by these methods. For some examples see Ref. [203].

Chemical two electron oxidations of phenols in the presence of nucleophiles have been well reviewed in Ref. [204]. No example of a stable phenoxenium salt of type *2*,

Table 13. Polarographic half wave anodic potentials by a two step oxidation of arylated phenols in $HOAc/H_2O$ (9:1) 5% NaOAc against an aqueous Ag/AgCl electrode

Substance	$E_{1/2}$ (mV)
3,5-dicyano-2,4,6-triphenylphenol	+1061
4-cyano-2,6-diphenylphenol	+ 952
4-(4'-bromophenyl)-2,6-diphenylphenol	+ 931
2,3,4,5,6-pentaphenylphenol	+ 930
3-cyano-2,4,6-triphenylphenol	+ 926
2,4,6-tris-(4'-bromophenyl)phenol	+ 875
3-chloro-2,4,6-triphenylphenol	+ 858
2,3,4,6-tetraphenylphenol	+ 845
4-(4'-cyanophenyl)-2,6-diphenylphenol	+ 845
2,6-diphenyl-4-(phenyl-4'-sulfonic acid)-phenol	+ 803
2,6-bis(4'-bromophenyl)-4-phenylphenol	+ 794
2,4,6-triphenylphenol	+ 786
4-(4'-bromophenyl)-2,6-diphenylphenol	+ 784
4-fluoro-2,6-diphenylphenol	+ 757
2,4,6-tris-(4-phenoxyphenyl)-phenol	+ 747
4-(4'-methoxyphenyl)-2,6-diphenylphenol	+ 710
2,6-bis-(4'-methoxyphenyl)-4-phenylphenol	+ 692
2,4,6-tris-(4'-methoxyphenyl)-phenol	+ 671

however, (or *6*) has been isolated. The first stable phenoxenium salts were isolated by Dimroth et al. [205] who also studied their reactions [206]. It was found that the red 2,4,6-triphenoxyl (*8*) in organic solvents on the addition of strong acids or Lewis acids gave a deep blue color. But it quickly vanished to give a brown solution from which only some 2,4,6-tri-phenylphenol was isolated. We first attributed the blue color to a radical cation such as *9* [9].

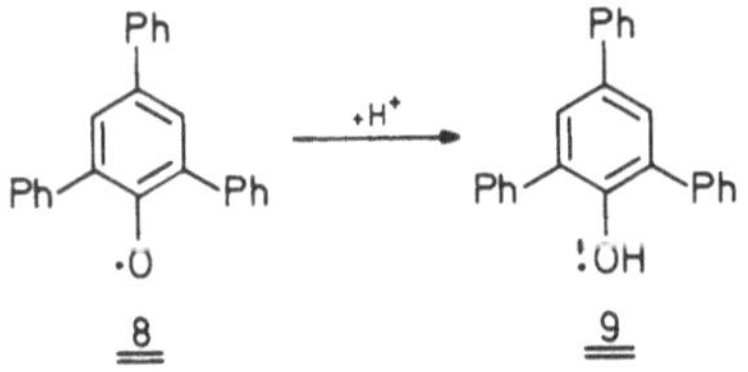

Later on, however, deep blue crystals were isolated, when the 2-acetoxy-quinol-derivative *10* or the dimeric 2,4,6-triphenylquinolether *11a* or the 4-methoxy-quinol derivative *11b* in non-reactive solvents such as benzene or carbondisulfide, were

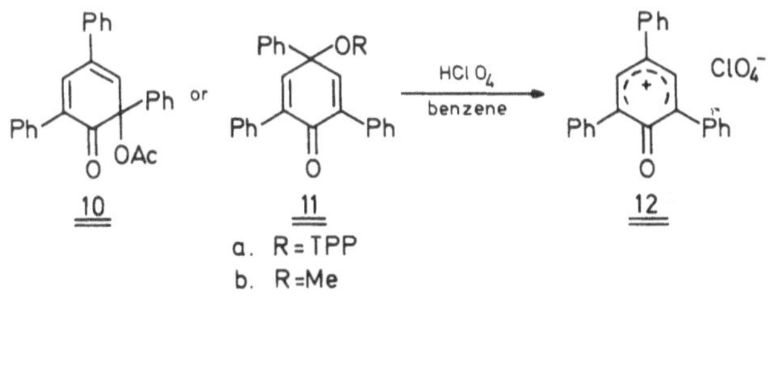

reacted with a few drops of 70% perchloric acid at −60°. Since no ESR spectrum could be observed, a phenoxenium salt *12*, formed via a two electron abstraction must have been produced. The dark blue crystals *12*, however, can only be stored at low temperatures [115].

When methoxyl substituents were introduced into the 4′-positions of the 2,4,6-phenyl rings of 2,4,6-triphenylphenol, stable phenoxenium salts *14a–c*, were isolated as nearly black crystals [205]. In dry dichlormethane they gave deep blue and blue green solutions, respectively, which are stable for hours at 0° and slowly decompose at room temperature. The best starting compounds for isolating these salts *14a–c* (X = BF_4 or $SBCl_6$) are the quinolethers *13a–c*, since the work up is simple. When dimers of the phenoxyls such as *11a* were used, the isolation of *14a–c* was more difficult because of removing the arylated phenols (the secondary products instead of methanol).

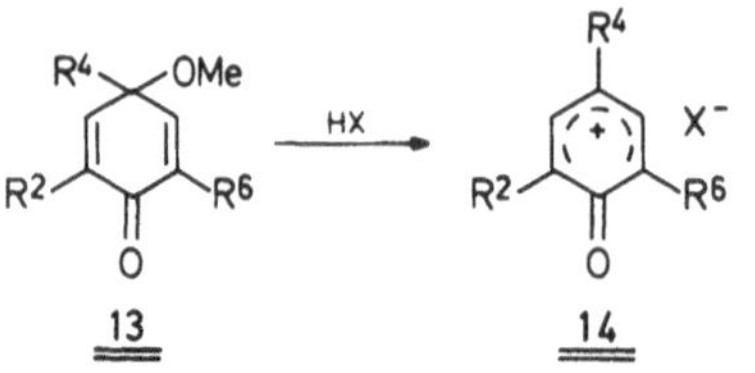

14	R^2	R^4	R^6	X	m, N (°C)	yield (%)
a	Ph	PhOMe(4)	Ph	BF_4	134–135	52
b	PhOMe(4)	Ph	PhOMe(4)	BF_4	138–139	80
c	PhOMe(4)	PhOMe(4)	PhOMe(4)	$SbCl_6$	67–73	95

Table 14 gives the maxima of the visible spectra of the corresponding phenols, 4-methoxy-quinols and the phenoxenium tetrafluoroborates in dichloromethane. The extinction coefficients of the phenoxenium salts are not very exact, since the solutions slowly decompose at 25°. The longer wave absorptions have not been measured. The extinction coefficients of *14a–c* are independent of the concentrations, since, contrary to the phenoxyls, no dimerization takes place. No ESR spectra of the pure phenoxenium salts *14a–c* have been observed.

The IR-spectra (in nujol) of *14a–c* show only absorption bands at 1596 to 1612 cm^{-1}, but not the characteristic C=O bands of the quinols *13a–c* between 1610 to 1670 cm^{-1}. The electron spectra are at a somewhat longer wave length than those of the phenoxyls. The ^1H—NMR spectra of the protons at C-3 and C-5 (in $CDCl_3$) are shifted to a lower field (7.2 ppm) than those of the phenols (6.95 ppm).

All these phenoxenium salts are easily reduced using $LiAlH_4$, $NaBH_4$ or zinc dust in acetic acid to give the arylated phenols in quantitative yield. Sodium hydroxide or sodium methylate, and even water or methanol, immediately decolorize the deep blue-green benzene solutions of the phenoxenium salts to give the 4-quinols, and the 4- (and some 2-)methoxyquinol-derivatives in 70 to 75% yield. The 4-methylquinol

Table 14. Color and absorption maxima in CH_2Cl_2 of the phenols, phenoxyls and phenoxenium-tetrafluoroborates **14a–c**

	Phenols		
	a	b	c
color maxima (nm) (extinction coefficient)	colorless 250 (39250) —	colorless 256 (17600) —	colorless 258 (15000) —
	Phenoxyls		
	a	b	c
color maxima (nm) (extinction coefficient)	blue 588 (—)[a] 378 (—)	blue 568 (—) 412 (—) 350 (—)	bluegreen 599 (—) 414 (—) 379 (—)
	Phenoxeniumtetrafluoroborates *14*		
	a	b	c
color maxima (nm) (extinction coefficient)	dark blue 610 (—) 442 (—)	dark blue 600 (3100) 490 (190) 410 (140)	dark blue green 720 (4450) 480 (2280) 410 (2900) 310 (28100) 263 (54500)

[a]) extinction coefficient dependent on the concentration

ethers are identical with the compounds which have been prepared when the 2,4,6-arylated phenols are oxidized in methanol with lead- or manganesedioxide. Contrary to the arylated quinole derivatives, where the 4-hydroxy or 4-alkoxy derivatives are the thermodynamically most stable compounds, the addition of sodium acetate to the phenoxenium salts *14a–c* gives the 2-acetoxy-quinol-derivatives. In this case, these

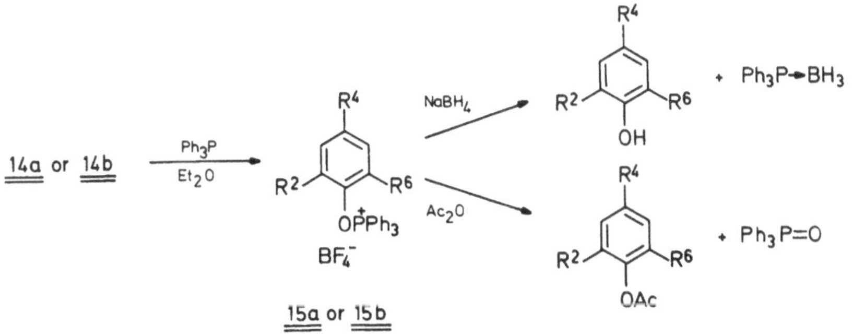

compounds are the thermodynamically stable derivatives. This has been shown by thermal rearrangements in benzene or other solvents. The 2-methoxy-quinol derivatives rearrange to the 4-methoxy-quinols, whereas the 4-acetoxy-quinol derivative rearranges to the 2-acetoxy-quinols. Experiments with *14a* and *b* with triphenylphosphine in dry ether gave colorless salts with no IR-bands in the C=O or OH-groups region. These are probably the quasiphosphonium salts *15* (*a*, *b*) formed by reacting the O-cation with the phosphine. Reduction of *15* with $NaBH_4$, leads to the 2,4,6-triarylphenols and triphenylphosphinborine. Acetanhydride gives the phenol-acetates and tri-phenylphosphinoxide [206].

The interesting intermediates, the areneoxenium ions *17*, are believed to be involved when phenol derivatives which are connected with a good electron accepting leaving group, such as the 4-methoxypyridinium tetrafluoroborate (*16*), are thermolysed. 2-Nitro-dibenzofurane (*18*) (22%) and 4-nitro-2-phenyl-phenol (*19*) (24.8%) have been isolated [207].

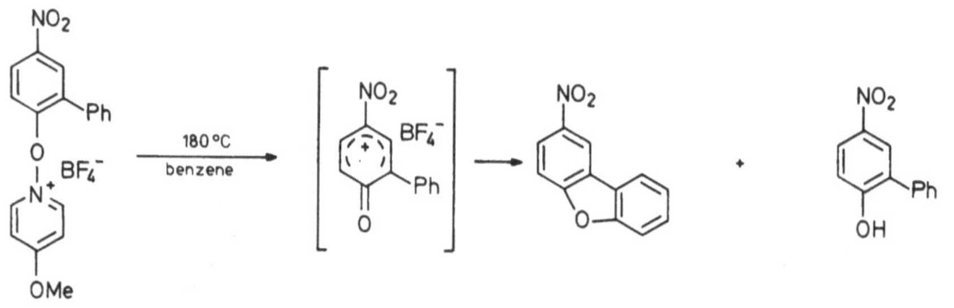

16 17 18 19

Another interesting intermolecular reaction involving the phenoxenium ion *21* as supposed intermediate, was observed by reacting *20* with resorcinol-dimethylether to give *23* and *24* (61% and 13% yield by O—C or C—C-coupling, respectively; some 2-nitro-4-phenoxyphenol (*22*) is also formed [207].

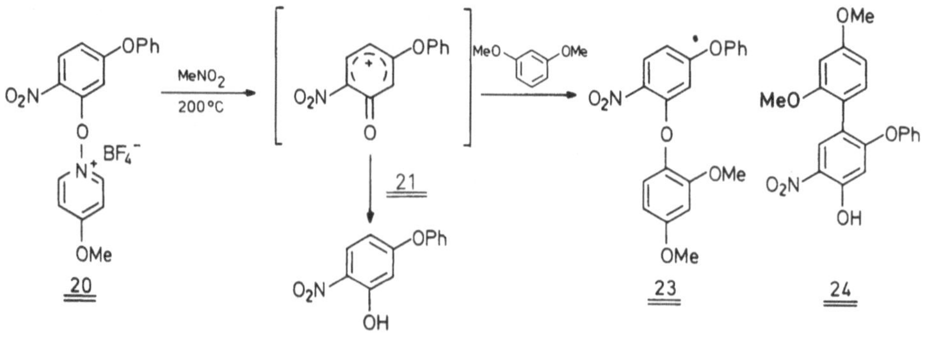

20 22 23 24

After this short digression to non-arylated phenoxenium ions we now return to the reactions of arylated stable phenoxenium salts *14a–c* which are strong oxidants. When the dark blue solution of *14c* in acetonitrile is mixed with the lithium salt of its phenol, the red 2,4,6-aryl triphenylphenoxyls immediately arises.

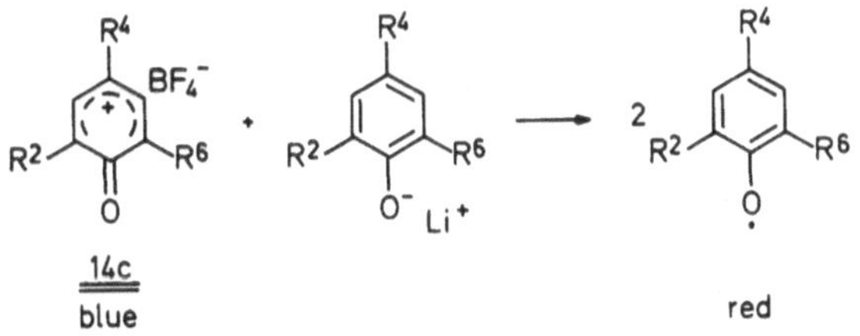

Many other lithium phenolates can be similarly oxidized. Cycloheptatriene and *14c*, suspended in ether, give the tropylium-tetrafluoroborate and the phenol of *14c*.

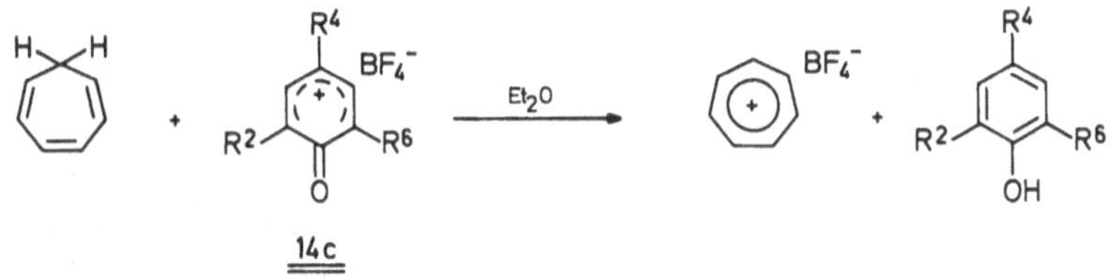

Another hydride transfer has been observed when dioxolane *25* was reacted with *14c* to give the dioxolenium salt *26*. A methoxyl-anion, however, is transferred when *14c* is reacted with 1-methoxy-1-methyldioxolane (*27*) to give 57% of 1-methyldioxolenium-tetrafluoroborate *29* and 4-methoxy-2,4,6-tris-(4′-methoxyphenyl)cyclo-2,5-hexadiene-one-1 (*28*)[205].

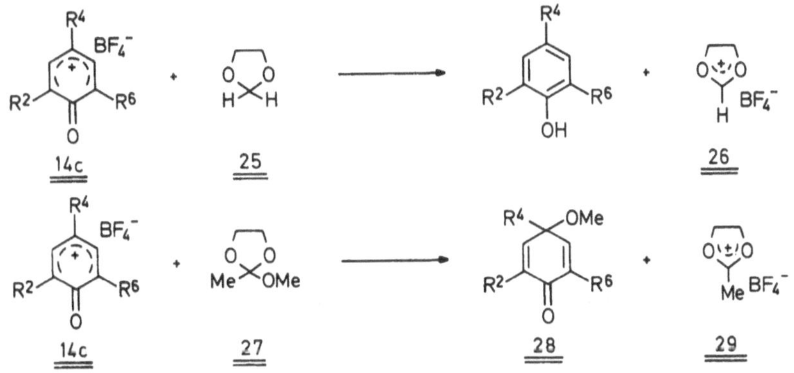

The most important reactions involve addition of nucleophiles to the phenoxenium salts *14a–c* or to the observed deep blue very reactive intermediates *12* via an S_N1 or by an acid supported S_N2 reaction. Reduction with $LiAlH_4$ or $NaBH_4$ may also occur by adding a hydride ion to any of the possible cationic centers of the mesomeric formulas *2* (p. 151) to give the phenol.

4.4.3 o- and p-Quinol Derivatives: Syntheses, Constitution and Reactions

There are four main routes for synthesizing quinol (cyclohexadienolone) derivatives from arylated phenols:

i) Oxidation of the arylated phenols to give phenoxyl radicals, which more or less dimerize to give 4-quinolphenol ethers.

ii) Oxidation of the phenols in the presence of nucleophiles, preferently alcohols with lead dioxide or some other oxidants to give 4- and some 2-alkoxy-quinol derivatives.

iii) Exchange of the alkoxy (or acetoxy-)substituents by water or an alcohol (preferently ethanol but no higher alcohols), or by some carboxylic acids (usually acetic acid). These reactions are catalysed by acids and can either give a phenoxenium ion as intermediate (S_N1) or occur by a S_N2 exchange mechanism with or without an allyl rearrangement. Contrary to the 2,6-di-t-butyl substituted phenol derivatives with varying substituents in position 4, e.g. 4-(or 2)-quinol-derivatives with halogen atoms, nitro- or other non-oxygen bonded groups which are isolable, the corresponding quinol derivatives in the 2,4,6-triphenylphenol series are not known. They may be too instable.

iv) A special route leading only to 2-acetoxy-2,4,6-triphenyl-quinol has been found, using the lead tetraacetate method of Wessely [208]. When 2,4,6-triphenylphenol (*1*) was reacted with lead tetraacetate in acetic acid, the pure 2-acetoxy-2,4,6-triphenylquinol *3* was isolated in 85% yield, the lead phenolate *2* is a possible intermediate:

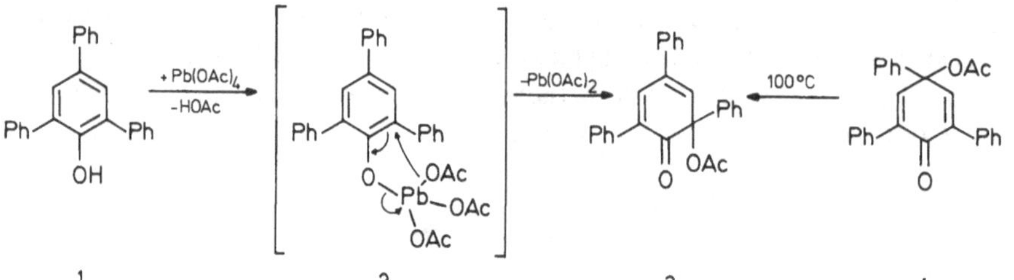

1 2 3 4

This synthesis, by a purely chemical route, confirmed the structure of the thermally stable 2-acetoxy-isomer *3*, since the 4-isomer (*4*), which can be produced from the 4-hydroxyquinol derivative by acetylation with acetanhydride and pyridine, rearranges at 100° in organic solvents to give *3* (see p. 155). On the contrary, the thermodynamic structures of methyl- or ethyl-ethers or the hydroxy quinol derivatives (the last two are only known as the 4-substituted quinol derivatives) are the 4-substituted quinols. Table 15 gives some physical data of the two isomeric 2,4,6-triphenyl-quinol derivatives.

Table 15. Physical data of 2- and 4-quinole derivatives of 2,4,6-triphenol

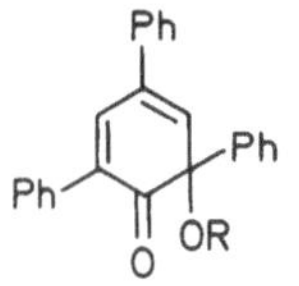

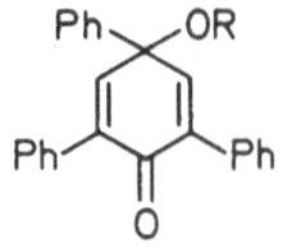

	R=H	R=Me	R=OAc	R=H	R=Me	
m.p. (°C)	—	117–119	129–131	160–161	130–132	151–152
UV (nm, mol. extinction) (in EtOH)	—	242,5 (24200) 350 (3690)	237 (23800) 349 (3850)	225,3 (32200) 283 (4000)	223,5 (34400) 283 (5700)	228 (37200) 283 (7940)
^{1}H—NMR (δ in ppm) ($CDCl_3$)	—	6.75; d $^4J_{H-H}$ = 2.5 Hz	6.65; d $^4J_{H-H}$ = 2.5 Hz	7.14; s	7.00; s	7.15; s

UV- and ^{1}H-NMR, spectra are the most useful in assigning the correct structures. The UV-maxima of the para quinol derivatives are at shorter wave lengths and the compounds are colorless. The two protons in the 3,5-positions are equivalent and give a singlet in the ^{1}H-NMR. The IR-frequencies of the C=O bond have proved useful although the interpretation can be ambiguous since the absorption bands are also influenced by the substituents of the quinol ring in positions 2, 6 and even 4 [209]. Earlier publications [205], therefore, are not in agreement with the corrected data given above.

An interesting reaction has been observed, when the dimeric 2,4,6-triphenylphenoxy-2,4,6-triphenyl-phenyl-4-quinol, its 4-methylether, the 4-acetate-, or the 2-acetoxy-2,4,6-triphenylquinols are treated with phenylpropiolic acid in benzene and anhydrous zinc chloride. Earlier structures (proposed by Sell [210]) need to be corrected. The pure isolated phenylpropiolic acid substance, m.p. 308–310°, (87% yield from the phenylpropiolic acid reaction) has been identified as the 2-oxo-1,3,9-triphenyl-1,2-dihydro-phenenthren-10-carbonic-lacton *7*.

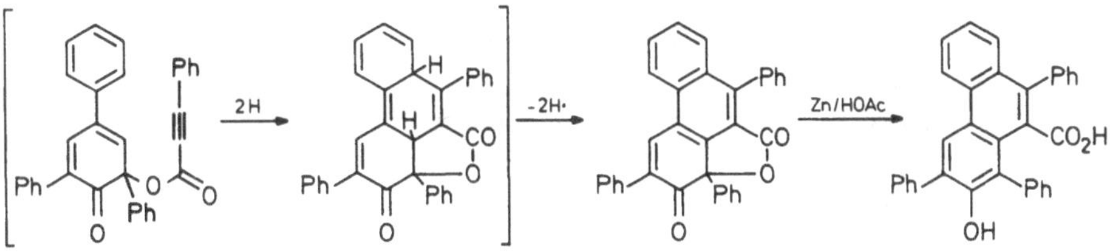

5 6 7 8

Its synthesis has been fully clarified [211]. The quinol-ester *5* is the probable primary product. An intramolecular Diels Alder addition leads to *6* and dehydrogenation gives *7*. Reduction with zinc/acetic affords the phenanthrol carbonic acid *8*. Since these reactions as well as similar reactions using propiolic acid are reported in the literature [211], the experimental details shall not be repeated in this review. The results of the arylated quinol reactions with maleic acid have not been repeated and are surely misinterpreted in Ref. [210].

Another reaction is found when 2-acetoxy-2,4,6-triphenylphenol is hydrolysed with alcoholic potassium hydroxide to give the 2-hydroxy quinol derivative [212]. Since the 4-acetoxyquinol derivative can be easily hydrolysed to give the 4-hydroxy-quinol derivative, the same hydrolysis of the 2-acetoxy quinol derivative *9* was believed to give similar results. However, a high melting thermochromic compound (yield > 60%) was isolated. The reaction mechanism can be understood in terms of an unusual alkaline acyloin rearrangement leading to a ring contraction. The reaction product proved to be a red cyclopentadienone derivative *10* which dimerizes according to a well known route to give the colorless dimer e.g. *11*. Heating of *11* in organic solvents causes dissociation to the red monomer *10*. The following mechanism has been proposed.

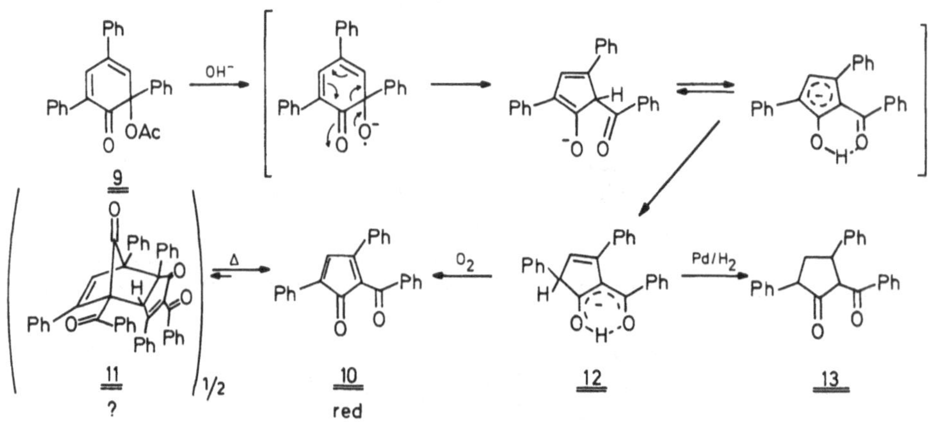

The first isolable compound *12*, which can be catalytically hydrogenated to give *13* is oxidized to the red cyclohexadienone derivative *10*, which reversibly dimerizes to give a colorless dimer of which *11* is one of the possible structures. Similar rearrangements of alkylated o-quinols have been observed by Zbiral et al. [213]. Base induced acyloin rearrangements of 2,4,6-tri-t-butyl-4-hydroxy-quinols give three different compounds, mainly hydroquinone derivatives, by losing or rearranging the t-butyl groups [214].

4-Alkoxy-quinols, when treated with zinc chloride and acetanhydride react to quinones via a [1.4] Thiele addition to give 3-substituted phenol derivatives [168]. 4-Alkoxy-2,4,6-triphenylquinole *14* was easily converted to 2,4,6-triphenylresorcinol

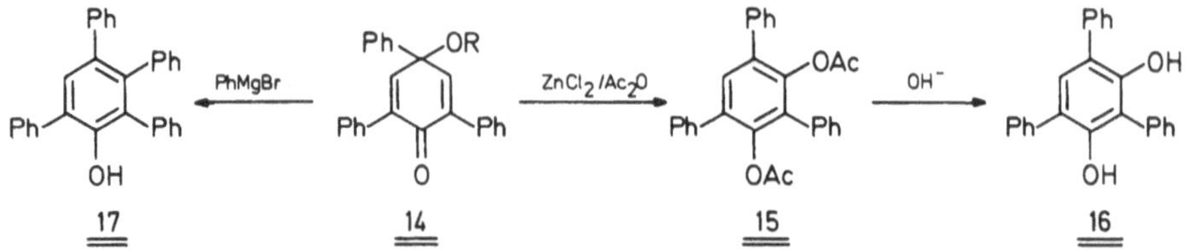

diacetate *15* using this route. It is hydrolysed to the resorcinol derivative *16*. Related to this addition is the substitution in position 3 of *14* by phenyl-magnesium bromides to give 2,3,4,6-tetraphenylphenol *17*.

Similar additions are possible, when 2-acetoxy-2,4,6-triphenylquinol *18* is heated in methanol with potassium cyanide to give 62% of 3-cyano-triphenylphenol *19*. $K_3Fe(CN)_6$-oxidation affords the phenoxyl *20* which has a high oxidation potential. *20* mainly dimerizes to give yellow crystals *21*, m.p. 135–150°. Repetition of the addition reaction, first by preparing the 3-cyano-2-acetoxy-2,4,6-triphenylquinol *22* (m.p. 165°) and then addition of cyanide in methanol, gives the 3,5-dicyano-2,4,6-triphenylphenol *23*, which has such a high oxidation potential that it could not be oxidized to the phenoxyl by either $K_3Fe(CN)_6$/NaOH or using lead dioxide.

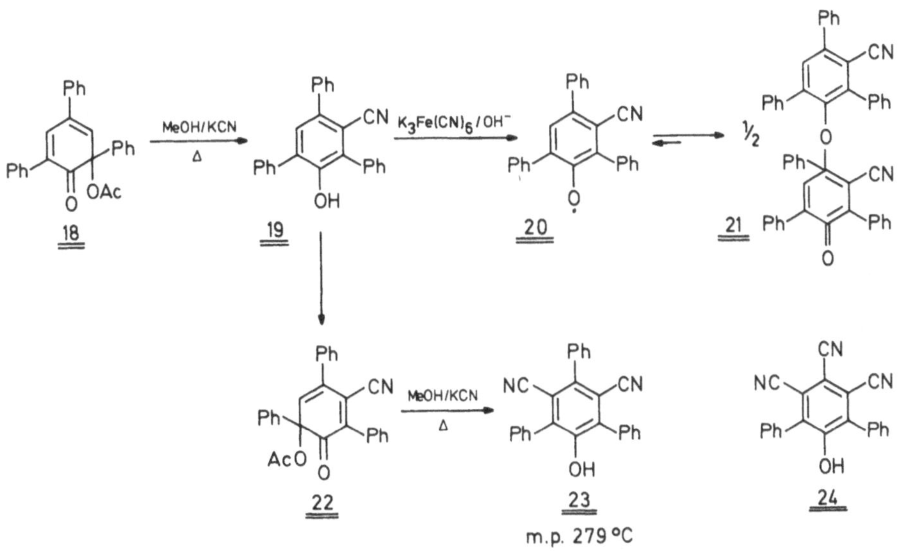

Addition of cyano ion to the 4-methoxy-quinol of 3-chloro-2,4,6-triphenylphenol gives the 3-chloro-5-cyano-2,4,6-triphenylphenol. Similarly 4-cyano-2,6-diphenylphenol leads to 3,4-dicyano-2,6-diphenylphenol and 3,4,5-tricyano-2,6-diphenylphenol *24* [115].

Contrary to these addition reactions at positions 3 and 5 of the quinol derivatives of 2,4,6-triphenylphenol, 4-methoxy-2,3,4,6-tetraphenylquinol (*25*), probably for steric reasons, behaves quite differently. Phenylmagnesiumbromide does not add a phenyl group but gives 2,3,5,6-pentaphenylphenol *26* by reduction. No [1.4] Thiele addition of acetanhydride and zinc chloride takes place. In contrast to the reactions of the 3,4,6-triphenyl derivative *18*, a [1.2] phenyl rearrangement takes place to give the 4-methoxy-2,3,4,6-tetraphenylphenolacetate *27* [115]. Alkaline hydrolysis affords the methoxy-2,3,4,6-tetraphenylphenol *28*. Oxidation with $K_3Fe(CN)_6$ in the usual manner leads to a pale yellow solution, which contains the dimeric phenoxyl *29*. Heating of *29* in organic solvents dissociates partly to the 4-methoxy-2,3,5,6-tetraphenylphenoxyl, a strong oxidant, which immediately oxidizes 2,4,6-tri-t-butylphenoxyl to give its blue air sensitive phenoxyl. The oxidation potential of the phenoxyl of *29* is high enough to dehydrogenate even 2,4,6-triphenylphenol to its red stable phenoxyl. To achieve this, the dimer *29* and 2,4,6-triphenylphenol are heated in o-dichloro-benzene. The equilibrium of these phenoxyls is strongly shifted to the dimer as it is known for other 4-methoxy phenoxyls. The dissociation rate, even at higher temperatures is very slow.

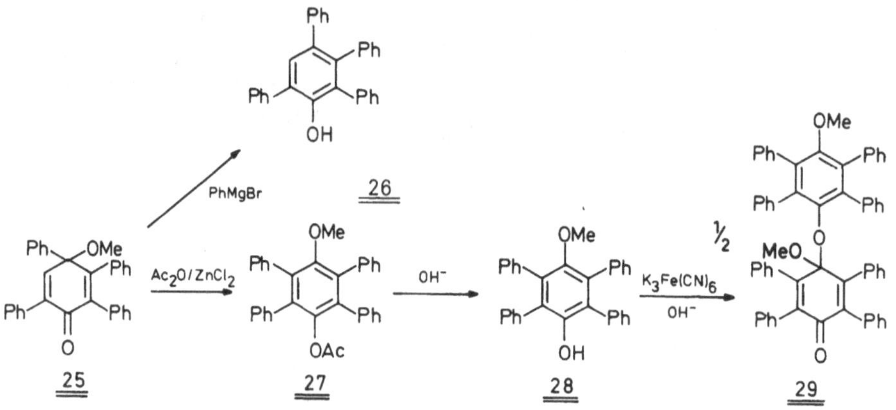

The structure of the rearranged methoxy-phenol *27* was easily determined by demethylation with hydrogen iodide to give the hydroquinol *30*. Oxidation affords the red 2,3,4,6-tetraphenylquinone *31*, identical to the quinone which was first prepared by Kvalnes [116] by arylating quinone with aryldiazonium salts.

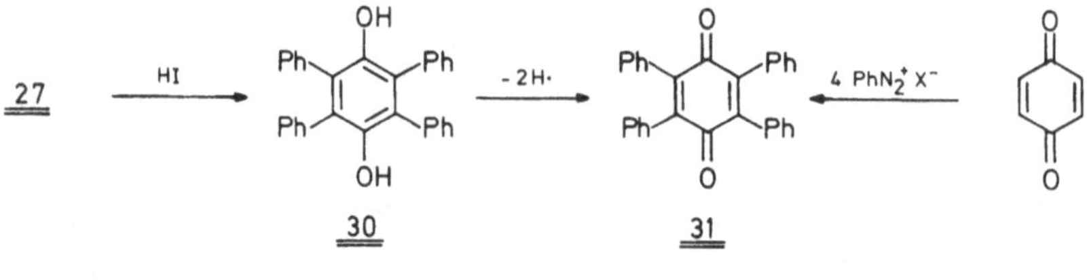

4.4.4 Photoreactions of Arylated o- and p-Quinol Derivatives

The first photoreactions were observed by Perst and Dimroth [215] when 2-acetoxy-2,4,6-triphenylquinol *1* in ethanol or benzene was left for some days in daylight. Only 2% of a new compound *3* was isolated. It was prepared, however, in 90% yield by irradiation of *1* with the short wave light of a mercury lamp. The photoisomer *3* proved to be the bicyclic exo acetate derivative. Heating it in benzene or toluene causes rearrangement back to the 2-acetoxy-quinol *1*. Perst [118,216] later showed by extensive and careful experiments that a ketene *2* is the primary product of this photochemical reaction. It can be trapped with nucleophiles, such as dimethylamine to give *4*. The photochemical rearrangement can be described according to the following scheme: Thermolysis of *2* and *3* leads to the reverse reaction, *2* was an isolable intermediate.

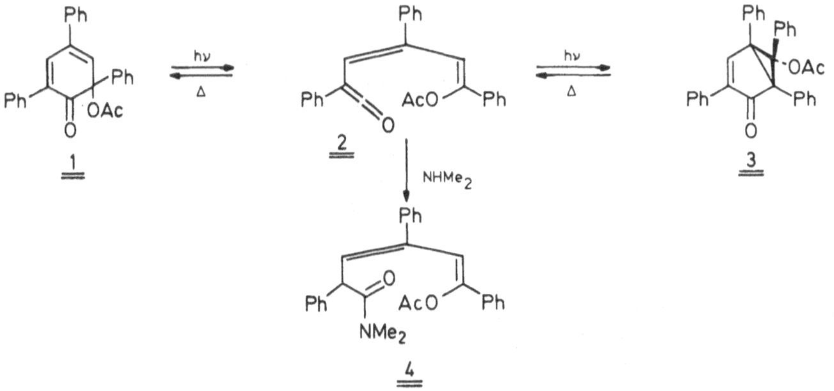

Perst and coworkers [216] have since studied many photochemical and other reactions of aryl substituted quinol derivatives. These are generally quite different from those of alkylated quinol derivatives. The results of these and earler experiments are reviewed in [204].

5 Concluding Remarks

This account describes the syntheses and reactions of the somewhat neglected class of arylated phenols. Some earlier preparative work is mentioned as well as the work carried out by the author's coworkers a part of which has so far only been published in dissertations. It is hoped that the large gap betwen the syntheses and chemical behaviour of alkyl- and aryl-substituted phenols as well as the lack of a comprehensive review of the latter phenols justifies this account. Perhaps some of the unsolved problems will stimulate new research as well as preparative applications of the arylated phenoxyls and phenoxenium ions which belong to the strongest purely organic oxidants. The industrial use of arylated phenols in polymer chemistry as well as the phenoxyls as radical chain inhibitors or antioxidants is hindered by the somewhat difficult accessibility of these compounds.

6 Acknowledgement

The author wishes to thank his many coworkers, cited in this review, who have contributed many of their own ideas and were enthusiastically engaged in this research program. Without the generous financial help of the Deutsche Forschungsgemeinschaft, the Fonds der Chemischen Industrie. Bayer, Leverkusen and Badische Anilin and Sodafabrik, Ludwigshafen, it would not have been possible to accomplish this research. Dr. Edeline Wentrup-Byrne has tried to bring my "home-made" English into a readable form, in itself a difficult enough task. Finally, I wish to thank Dr. h.c. F. Boschke of Springer-Verlag for accepting this account.

7 References

1. Methoden der Organischen Chemie, "Houben-Weyl" 4. edit. (ed. E. Müller), vol. VIc, part 1 and 2 (1976) Stuttgart, G. Thieme Verlag (cited later on only "Houben-Weyl")
2. The Chemistry of the Hydroxyl group, part 1 and 2, ed. S. Patai, Interscience Publ., London, New York, Sidney, Toronto 1971
3. Bartlett, P. D., Roha, M., Stiles, R. M.: J. Am. Chem. Soc. *76*, 2349 (1954); see also Hennion, G. F., Anderson, G. J.: ibid. *68*, 424 (1946); Swanholm, U., Parker, V. D.: J. Chem. Soc. Perkin I, *1973*, 562
4. Müller, E., Ley, K., Kiedaisch, W.: Chem. Ber. *87* 1605 (1954)
5. Schlömer, K.: Dissertation Univ. Marburg 1959
6. Cook, C. D.: J. Org. Chem. *18*, 261 (1953)
7. Müller, E., Ley, K.: Z. Naturforsch. *8b*, 694 (1953)
8. Kalk, F.: Dissertation Univ. Marburg 1959 and unpublished results
9. Dimroth, K., Kalk, F., Neubauer, G.: Chem. Ber. *90*, 2058 (1957)
10. Steuber, F. W., Dimroth, K.: ibid. *99*, 258 (1966); Steuber, F. W.: Dissertation Univ. Marburg 1963. For a similar study with 2,6-di-t-butylated phenols see Šebesta, F., Petránek, J.: Coll. Czech. Chem. Commun. *35*, 2136 (1970)
11. Cook, C. D., et al.: J. Am. Chem. Soc. *75*, 6242 (1953); ibid. *81*, 1176 (1959); J. Org. Chem. *29*, 3716 (1964)
12. Müller, E., et al.: Chem. Ber. *87*, 922 (1954); Liebigs Ann. Chem. *730*, 67 (1969); see also Speiser, B., Rieker, A.: J. Electroanalyt. Chem. *102*, 373 (1979)
13. Pummerer, R. et al.: I. Ber. dtsch. Chem. Ges. *47*, 1472 (1914); XII. Chem. Ber. *86*, 412 (1953)
14. Goldschmidt, St. et al.: Chem. Ber. *55*, 3197 (1922); Liebigs Ann. Chem. *478*, 1 (1930)
15. Coppinger, G. M.: J. Am. Chem. Soc. *79*, 501 (1957); Tetrahedron *18*, 61 (1962)
16. Kharasch, M. S., Joshi, B. S.: J. Org. Chem. *22*, 1435, 1439 (1957), ibid. *27*, 651 (1962)
17. Bartlett, P. D., Funahashi, T.: J. Am. Chem. Soc. *84*, 2596 (1962); ibid. *88*, 3303 (1966)
18. Pokhodenko, W. D., Kalibatschuk, N. N., Khizny, W. A.: Liebigs Ann. Chem. *743*, 192 (1971)
19. Reviews of phenoxyl radicals: a. Altwicker, E. R.: Chem. Rev. *67*, 475 (1967);
    b. Forrester, A. R., Hay, J. M., Thomson, R. H.: Organic Chemistry of Stable Radicals Chapter 7; Academic Press London and New York 1968;
    c. Pokhodenko, V. D., Khizhnyi, V. A., Bidzilya, V. A.: Russ. Chem. Rev. (engl.) *37*, 435 (1968);
    d. Strigun, L. M., Vartanyan, L. S., Emanuel', N. M.: Russ. Chem. Rev. (engl.) *37*, 421 (1968);
    e. Pokhodenko, V. D.: Phenoxyl Radikals (russ.) Dumka Verlag Kiev 1969;
    f. Ershov, V. V., Volod'kin, A. A., Prokof'ev, A. I., Solodovnikov, S. P.: Russ. Chem. Rev. (engl.) *42*, 740 (1973);
    g. Vartanyan, L. S.: Russ. Chem. Rev. (engl.) *44*, 859 (1975)
20. Hopff, H., Haer, A.: Chimia *12*, 105 (1959); $AgBF_4$-synthesis of acetophenone at 70–80° in 1,2-dichloromethane (23% yield): Crist, D. R., Hsieh, Z.-U., Chuicksall, C. O., Sun, M. K.: J. Org. Chem. *49*, 2478 (1984)
21. Vorländer, D., Fischer, E., Wille, H.: Ber. Dtsch. Chem. Ges. *62*, 2836 (1929)
22. Dimroth, K., Bräuninger, G.: Angew. Chem. *68*, 519 (1956); Dimroth, K., Bräuninger, G., Neubauer, G.: Chem. Ber. *90*, 1634 (1957); for electrophilic substitution of 1,3,5-triphenylbenzene see also Ansell, H. V., Clegg, R. B., Taylor, R.: J. Chem. Soc., Perkin II *1972*, 766
23. Dimroth, K., Neubauer, G.: Angew. Chem. *69*, 95 (1957)
24. Kohler, E. P., Blanchard, Jr., L. W.: J. Am. Chem. Soc. *57*, 367 (1935)
25. Saure, M.: Dissertation Univ. Marburg 1959 and unpublished results
26. Lewis, G.: J. Org. chem. *30*, 2798 (1965); ibid. *31*, 749 (1966); Nowack, R.: unpublished results, Univ. Marburg 1962: The sulfonic acid of 1,3,5-triphenyl-

benzene with conc. H_2SO_4 in SO_2 at −30–60 °C is the 4′-sulfonic acid, m.p. 84–85 °, giving the phenol 4′-hydroxy-1,3,5-benzene, m.p. 154–156 °C in molten KOH at 350 °C
27. Dimroth, K.: Angew. Chem. *72*, 331 (1961);
Dimroth, K., Wolf, K. H.: Newer Methods of Preparative Organic Chemistry, Vol. III, 357 (1964) (ed. W. Foerst), Academic Press New York
28. Reviews: Balaban, A. T., Schroth, W., Fischer, G.: Adv. Heterocycl. Chem. *10*, 241 (1969);
Balaban, A. T., Dinculescu, A., Dorofeenko, G., Fischer, G., Koblik, A., Mezheritskii, M., Schroth, W.: ibid. Suppl. *2* (1982)
29. Reviews: Wedemeyer, K. F.: see lit. 1), 1e, 4–84;
Rawlinson, D. J., Sosnovski, G.: Synthesis *1972*, 1–28; ibid. *1973*, 567–603;
for methoxylation of arenes see Konz, E., Pistorius, R.: ibid. *1979*, 603;
Groebel, B. T., Konz, E., Millauer, M., Pistorius, R.: ibid. *1979*, 605
30. see e.g. Williams, G. H.: Homolytic Aromatic Substitution Pergamon Press Oxford 1960
31. Davidson, A. J., Norman, R. O. C.: J. Chem. Soc. (London) *1964*, 5404
32. Eberson, L., Schäfer, H.: Top. Curr. Chem. *21*, 1 (1971);
Eberson, L., Nyberg, K.: Acc. Chem. Res. *6*, 106 (1973);
Eberson, L., Jönsson, L., Wistrand, L. G.: Tetrahedron *38*, 1087 (1982);
Eberson, L.: Adv. Phys. Organ. Chem. *18*, 79 (1982);
see also Enisov, E. T., Metelitsa, D. I.: Russ. Chem. Rev. (engl.) *37*, 656 (1968)
33. Ross, S. D., Finkelstein, M., Petersen, R. C.: J. Am. Chem. Soc. *86*, 4139 (1964);
Leung, M., Herz, J., Salzberg, H. W.: J. Org. Chem. *30*, 310, 2873 (1975);
Adams, R. N.: Acc. Chem. Res. *2*, 175 (1969)
34. Lund, H.: Acta Chem. Scand. *11*, 1323 (1957)
35. Eberson, L., Nyberg, K.: ibid. *18*, 1568 (1966) and lit. 36)
36. Eberson, L., Nyberg, K.: J. Am. Chem. Soc. *88*, 1686 (1966); Eberson, L.: ibid. *89*, 4669 (1967)
37. Blum, Z., Cedheim, L., Nyberg, K.: Acta Chem. Scand. B. *29*, 715 (1975)
38. So, Y. H., Miller, L. L.: Synthesis *1976*, 468
39. Lau, W., Huffman, J. C., Kochi, J. K.: J. Am. Chem. Soc. *104*, 5515 (1982); see also lit. 41)
40. Heiba, E. I., Dessau, R. M., Koehl Jr., W. J.: J. Am. Chem. Soc. *91*, 6830 (1969)
41. Kochi, J. K., Tang, R. T., Bernath, T.: ibid. *95*, 7114 (1973) and lit. cit. therein
42. Nyberg, K., Wistrand, L.-G.: J. Org. Chem. *43*, 2613 (1978);
for catalysis see Eberson, L.: J. Mol. Catalys. *20*, 27 (1983);
Eberson, L., Jönsson, L.: Liebigs Ann. Chem. *1977*, 233
43. Bell, H. C., Kalman, J. R., Pinhey, J. T., Sternhell, S.: Tetrahedron Lett. *1974*, 853;
Kozyrod, R. P., Pinhey, J. T.: ibid. *23*, 5365 (1982) and lit. cit. therein
44. Eberson, L., Gomez-Gonzalez, L.: Acta Chem. Scand. *27*, 1249 (1973);
for similar acetoxylations see Eberson, L., Jönssen, L.: J. Chem. Soc. Chem. Commun. *1974*, 885;
Eberson, L., Jönssen, L.: Acta Chem. Scand. B *28*, 771 (1974); ibid. B *30*, 361 (1976); Liebigs Ann. Chem. *1977*, 233;
Jönssen, L.: Acta Chem. Scand. B *35*, 683 (1981);
Itahara, T.: Chem. and Ind. (London) *1982*, 599
45. Baciocchi, E., Rol, C., Sebastini, G. V.: Gaz. Chim. Ital. *112*, 513 (1982);
Baciocchi, E., Mai, S., Rol, C., Mandolini, L.: J. Org. Chem. *43*, 2919 (1978) and lit. cit. therein
46. Matsuura, T., Omura, K.: Synthesis *1974*, 173
47. Bachmann, W. E., Hoffmann, R. A.: Organic Reactions vol. 2, 224 (1944);
Rüchardt, C., Merz, E.: Tetrahedron Letters *1964*, 2431;
Weinberg, N. L., Weinberg, H. R.: Chem. Rev. *68*, 449 (1968)
48. Hey, D. H., Williams, G. H.: in Rodd's Chemistry of Carbon Compounds, 2nd edit. vol. III, 119 (1971);
Nonhebel, D. C., Walton, J. C.: Free Radical Chemistry *1974*, p. 417, Cambridge University Press;
Minisci, F., Citterio in Adv. in Free Radical Chemistry, vol. 6, 65 (1980) (ed. G. H. Williams), Heyden, London, Philadelphia, Rheine;
Sainsbury, M.: Tetrahedron *36*, 3327 (1980)

49. Caronna, T., Ferrario, F., Servi, S.: Tetrahedron Lett. *1979*, 657
50. Schromm, K.: Dissertation Univ. Marburg 1961;
Dimroth, K., Schromm, K.: unpublished results 1962
51. Kozyrod, R. P., Pinhey, J. T.: Tetrahedron Lett. 23, *1982*, 5365 (1982)
52. Craig, H. C., Pinhey, J. T., Sternhell, S.: Aust. J. Chem. *32*, 1551 (1979);
Bell, H. C., May, G. L., Pinhey, J. T., Sternhell, S.: Tetrahedron Lett. *1976*, 4303
53. Vollbracht, L., Bijkerk, A. H., Iedema, J.: USP 3413356; C. A. *70*, 57402 (1969)
54. Intille, G.: USP 4008266; C. A. *86*, 155349 (1977)
55. Barton, D. H. R., Blazejewski, J-C., Charpiot, B., Lester, D. J., Motherwell, W. B., Papoula, M. T. B.: J. Chem. Soc. Chem. Commun. *1980*, 827
56. Endo, Y., Shudo, K., Okamoto, T.: J. Am. Chem. Soc. *99*, 7721 (1977)
57. Wolf, W., Kharasch, N.: J. Org. Chem. *26*, 283 (1961);
Kharasch, N., Wolf, W., Erpelding, T. J., Naylor, P. G., Tokus, L.: Chem. and Ind. (London) *1962*, 1720;
Wolf, W., Kharasch, N.: J. Org. Chem. *30*, 2493 (1965);
Photoarylation see also Soumillion, J. P., De Wolf, B.: Chem. Commun. *1981*, 436;
for iodophenolether synthesis see Burger, A., Wilson, E. L., Brindley, C. O., Bernheim, F.: J. Am. Chem. Soc. *67*, 1416 (1945)
58. Ugochukwu, E. N., Wain, R. L.: Chem. and Ind. (London) *1965*, 35
59. Schminke, H. D.: Dissertation Univ. Marburg 1961
60. Güsten, H., Kirsch, G., Schulte-Frohlinde, D.: Tetrahedron *24*, 4393 (1968)
61. Hay, A., Clark, R. F.: Makromolecules *3*, 533 (1970)
62. Gerard, W., Lappert, M. F.: Chem. Rev. *58*, 1081 (1958);
Schäfer, W., Franck, K.: Chem. Ber. *99*, 160 (1966);
McOmie, J. F., Watts, M. L., West, D. E.: Tetrahedron *24*, 2289 (1968)
63. Adams, R., Mathieu, J.: J. Am. Chem. Soc. *70*, 2120 (1948)
64. Prey, V.: Ber. Dtsch. Chem. Ges. *74*, 1219 (1941); ibid. *75*, 445 (1942); ibid. *76*, 156 (1943)
65. Hazlet, S. E., Cory, R. A.: J. Org. Chem. *27*, 2671 (1962)
66. Kolka, A. J., Napolitano, J. P., Filbey, A. H., Ecke, G. G.: ibid. *22*, 642 (1957)
67. Smith, W., Sommerfield, E. H.: USP 4001340; C. A. *86*, 120796 (1977);
Britton, E. C.: USP 159283, (1931); C. *1934*, II, 1688; USP 196744 (1932); C. *1935*, II, 1962;
Moose, J. E.: USP 1979116 (1932); C *1935*, I, 3200
68. Lüttringhaus, A., v. Sääf, G.: Angew. Chem. *51*, 915 (1938); Liebigs Ann. Chem. *542*, 241 (1939); *557*, 25 (1947);
Lüttringhaus, A., Ambros, D.: Chem. Ber. *89*, 463 (1956)
69. Gilman, H., Esmay, D. L.: J. Am. Chem. Soc. *75*, 2947 (1953);
Gross, D. E., Fishel, N. A.: USP 4000203; C. A. *86*, 89385 (1977);
Gross, D. E.: USP 3989761; C. A. *86*, 72193 (1977);
Fishel, N. A., Gross, D. E.: USP 4035428; C. A. *87*, 134 589 (1977);
Fishel, N. A., Anvil, S. R., Gross, D. E.: Catalog Org. Synth. *1978*, 119 (7th ed.);
Müller, K. W.: USP 2862035; C. A. *53*, 9157 (1959);
Kagaku, S.: Jap. P. 81,20533, C. A. *95*, 42640 (1981);
La Count, R. B., Friedman, S.: J. Org. Chem. *42*, 2751 (1977);
see also Squires, T. G., Vernier, C. G., Hodgson, B. A., Chung, L. W., Davies, V. A., Panuto, T. W.: J. Org. Chem. *46*, 2373 (1981)
70. Bunnett, J. F., Zahler, R. E.: Chem. Rev. *49*, 273 (1951);
Davies, D. I., Hey, D. H., Williams, G. H.: J. Chem. Soc. (London) *1961*, 562
71. Hawthorne, M. F.: J. Org. Chem. *22*, 1001 (1957);
Kuivila, H. G.: J. Am. Chem. Soc. *76*, 870 (1954);
Yarboro, T. L., Karr, C.: J. Org. Chem. *24*, 1141 (1959)
72. Eberson, L., Jönsson, L., Wistrand, L.-G.: Tetrahedron *38*, 1087 (1982) and lit. cit. therein
73. Eberhardt, M. K.: J. Org. Chem. *42*, 832 (1977); J. Phys. Chem. *81*, 1051 (1977)
74. Zoratti, M., Bunnett, J. M.: J. Org. Chem. *45*, 1769 (1980)
75. Bunnett, J. F.: Acc. Chem. Res. *11*, 413 (1978)
76. Kornblum, N.: Angew. Chem. *87*, 797 (1975); Angew. Chem. engl. *14*, 734 (1975)

77a. Russell, G. A., Danen, W. C.: J. Am. Chem. Soc. *90*, 347 (1968) and later publications
77b. Cerfontain, H.: J. Org. Chem. *47*, 4680 (1982)
78. Suzumura, H.: Bl. chem. Soc. Japan *35*, 108 (1962)
79. Hay, A. S., Clark, R. F.: Makromolecules *3*, 533 (1970);
Hay, A. S.: Fr. P. 1521 799; C. A. *71*, 61006 (1969)
80. Plešek, J.: Chem. Listy *50*, 246 (1956); C.A. *50*, 7732 (1956) Coll. Czech. Chem. Commun. *21*, 275 (1956);
Munk, P., Plešek, J.: Chem. Listy *51*, 771 (1957); C.A. *51*, 11261 (1957);
Plešek, J., Munk, P.: ibid. *51*, 633 (1957);
Munk, P., Plešek, J.: ibid *51*, 771 (1957); C.A. *51*, 11261 (1957);
see also Kunze, H.: Ber. dtsch. Chem. Ges. *59*, 2085 (1926);
de Jonge, C. R. H. I., van Dort, H. M., Vollbracht, L.: Tetrahedron Lett. *1970*, 1881
81. Hay, A. S.: Brit. P. 1207524 (1967); C.A. *77*, 126220 (1967);
Kapner, R. S., Kippax, D. L., Murphy, K. E., Udani, H. H.: Germ. P. 163402; C.A. *83*, 163813 (1975);
see also Brit. P. 1207524; C.A. *77*, 126220 (1972)
82. Hay, A. S.: see lit. 61; Boon, J., Mayer, E. P.: J. Polymer Sci. *7*, 205 (1969)
83. Schwarz, H. H., Metten, J., Koeller, H. Tacke, P., Weissel, O.: Germ. Offen 2.102746 (1971); C.A. *77* 126218 (1972)
84. Naarman, H., Brandstetter, F., Schuster, H. H.: Europ. Appl. P. EP 47824; C.A. *96*, 218455, 218418 (1982)
85a. Hay, A. S., Blanchard, H. S., Endres, G. F., Eustance, J. W.: J. Am. Chem. Soc. *81*, 6335 (1959);
85b. see also Hay, A. S.: J. Org. Chem. *27*, 3320 (1962); the Cu(I) complex is also very valuable for oxidative acetylene coupling;
85c. Hay, A. S.: Tetrahedron Lett. *1965*, 4241
86. Winnacker-Küchler: Chemische Technologie, 4. ed., vol. 6, Carl Hanser Verlag, München 1982;
Hay, A. S.: Advances Polymer Sci. *58*, 581 (1962);
Endres, G. F., Hay, A. S., Eustance, J. W.: J. Org. Chem. *28*, 1300 (1963);
X-ray of two cristalline poly-2,6-diphenyl-4-phenyl-oxydes see Boon, J., Magré, E. P.: Macromol. Chem. *136*, 267 (1970)
87. Davies, G., El Sayed, M. A., Fasano, R. E.: Inorg. Chim. Acta *71*, 95 (1983) and lit. cited therein;
Capdevielle, P., Maumy, M.: Tetrahedron Lett. *24*, 5611 (1983)
88. King, F. D., Walton, D. R. M.: Synthesis *1976*, 40
89a. Pinhey, J. T., Rowe, B. A.: Aust. J. Chem. *33*, 113 (1980)
89b. for other routes of α-arylation of ketons see: Bunnett, J. F., Sundberg J. E.: J. Org. Chem. *41*, 1702 (1976) of potassium cyclohexanenolate in liquid ammonia with PhBr in benzene and light (350 nm) according to a $S_{RN}1$ chain mechanism (72% 2-phenylcyclohexanone);
89c. Meerwein-α-arylation of cyclohexene-enolether or enolacetate with aryldiazonium salt and Cu(I) salts; Allard, M. N., Levisalle, J.: Bull. Soc. chim. France *1972*, 1926;
Adel, Al., Salami, B. A., Levisalles, J., Rudler, H.: ibid. *1976*, 930;
89d. $LiCuPh_2$ and p-tolyazocyclohexene: Sacks, C. E., Fuchs, P. L.: J. Am. Chem. Chem. Soc. *97*, 7372 (1975);
89e. McKillop, A., Rao, D. P.: Syntheses *1977*, 759
90. Krapcho, A. P., Lovey, A. J.: Tetrahedron Lett. *1973*, 957
91. Barton, D. H. R., Papoula, H. T. B., Guilhem, J., Motherwell, W. B., Pascard, C., Huu Dau, E.: J. Chem, Soc. Chem. Commun. *1982*, 732;
Barton, D. H. R., Lester, D. J. L., Motherwell, W., Papoula, M. T.: ibid. *1980*, 246
92. Pinhey, J. T., Rowe, B. A.: Aust. J. Chem. *36*, 789 (1983)
93. Horning, E. C., Horning, M. C.: J. Am. Chem. Soc. *69*, 1359 (1947)
94. Woods, G. F., Tucker, I. W.: ibid *70*, 2174 (1948)
95. Review: Posner, G. H.: Organic Reactions *19*, 1 (1972)
96. Malmberg, H., Nilsson, M., Ullenius, C.: Tetrahedron Lett. *23*, 3823 (1982);
Review: Normant, J. F.: Synthesis *1972*, 63

97. Woods, G. F., Reed, F. T., Arthur, T. E., Ezekiel, H.: J. Am. Chem. Soc. *73*, 3854 (1951)
98. Petrow, A. D.: Ber. dtsch. Chem. Ges. *62*, 642 (1929) and cited lit.; ibid. *63*, 898 (1930)
99. Kenner, J., Shaw, H.: J. Chem. Soc. (London) *1931*, 769
100. Sexsmith, D. R., Rassweiler, J. H.: J. Org. Chem. *25*, 1229 (1960); see also lit. 99)
101. a) Downes, A. M., Gill, N. S., Lions, F.: J. Am. Chem. Soc. *72*, 3464 (1951); for similar syntheses see b) Nasipuri, D., Roy Choudbury, S. R., Blattacharya, A.: J. Chem. Soc. Perkins I, *1973*, 1451
102. Egli, C., Helali, S. E., Hardegger, E.: Helv. Chim. Acta *58*, 104 (1975)
103. Zimmerman, H. E., Schuster, D. I.: J. Am. Chem. Soc. *84*, 4527 (1962);. for the syntheses of the 3,6-diphenyl- and 2,3-6-triphenyl-cyclohex-2-enones see Abdullah, S. M.: J. Ind. Chem. Soc., *12*, 62 (1935); C. *1935* I, 3926
104. Hill, H. B.: Am Chem. J. *22*, 89 (1899); *24*, 5 (1900); Ber. dt. Chem. Ges. *33*, 1241 (1900); see also Borsche, W.: Liebigs Ann. Chem. *312*, 211 (1900)
105. Yates, P., Hyre, J. E.: J. Org. Chem. *27*, 4101 (1962); for the synthesis of 2,3,4,6- and 2,3,5,6-tetraphenyl-phenol see also Eistert, B., Langbein, A.: Liebigs Ann. Chem. *678*, 78 (1964)
106. Eicher, T., v. Angerer, F.: Liebigs Ann. *746*, 120 (1971)
107. Ried, W., König, E.: ibid. *757*, 153 (1972); see also Perst, H., Wesemeyer, I.: Tetrahedron Lett. *1970*, 4189
108. Suter, C. M., Smith, P. G.: J. Am. Chem. Soc. *61*, 166 (1939)
109. Ames, G. R., Davey, W.: J. Chem. Soc. London, *1957*, 3480, *1958*, 1794; Betts, B. E., Davey, W.: ibid, *1961*, 1683, 3440; see also Fichter, F., Grether, E.: Ber. dtsch. chem. Ges. *36*, 1407 (1903)
110. Harrison Jr., E. A.: J. Org. Chem. *44*, 1807 (1979)
111. Ueji, S., Nakatsu, K., Yoshioka, H., Kinoshita, K.: Tetrahedron Lett. *1982*, 1173
112. Jones, E. C. S., Kenner, J.: J. chem. Soc. (London) *1931*, 769; 1842
113. Hay, A. S.: J. Org. Chem. *30*, 3577 (1965)
114. Bordwell, F. G., Wellmann, K. M.: ibid. *29*, 509 (1964)
115. Worschech, K.: Dissertation Univ. Marburg 1960; see also lit. 137)
116. Kvalnes, D. E.: J. Am. Chem. Soc. *56*, 2478 (1934)
117. Zimmerman, H. E.: Angew. Chem. *81*, 45 (1969); engl. *8*, 1; Zimmerman, H. E., Epling, G. A.: J. Am. Chem. Soc. *94*, 7806 (1972)
118. Dannenberg, W., Lemmer, D., Perst, H.: Tetrahedron Lett. 1974, 2133; Dannenberg, W.: Dissertation Univ. Marburg 1975; Rau, W.: Dipl. Arb. Univ. Marburg 1980; the mechanism of these and similar rearrangements has been fully studied by Perst and coworkers, see e.g. Perst, H.: Habil. Arbeit, Univ. Marburg 1972
119. Dannenberg, W., Perst, H.: Liebigs Ann, Chem. *1975*, 1873; Dannenberg, W., Perst, H., Seifert, W. J.: Tetrahedron Lett. *1975*, 3481; Weisskopf, V., Perst, H.: Liebigs Ann. *1978*, 1634
120. Monahan, A. S.: J. Org. Chem. *33*, 1441 (1968)
121. Eistert, B., Langbein, A.: Liebigs Ann. Chem. *678*, 78 (1964); Eistert, B., Thommen, A. J.: Chem. Ber. *104*, 3048 (1971)
122. Dürr, H., Heilkämper, P.: Liebigs Ann. Chem. *716*, 212 (1968)
123. Hart, H. Raggon, J. W.: Tetrahedron Lett. *24*, 4891 (1983)
124. Evans, D. A., Hoffmann, J. M., Truesdale, L. K.: J. Am. Chem. Soc. *95*, 5822 (1973); Evans, D. A., Hart, D. J., Koelsch, P. M., Chain, P. A.: Pure and Appl. Chem. *51*, 1285 (1979) and lit. cit. therein; see also Parker, K. A., Kang, S.-K.: J. Org. Chem. *45*, 1218 (1980)
125. Liotta, D., Saindane, M., Barnum, C.: J. Org. Chem. *46*, 3369 (1981); Fischer, A., Henderson, G. N.: Tetrahedron Lett. *24*, 131 (1983) and lit. cit. therein
126. Rieker, A., Hernes, G.: Tetrahedron Lett. *1968*, 3775; Rieker, A.: Angew. Chem. *81*, 938 (1969); engl. *8*, 918 (1969)
127. Dimroth, K., Berndt, A., Volland, R.: Chem. Ber. *99*, 3040 (1966); Berndt, A.: Dissertation Univ. Marburg 1965
128. Dimroth, K., Vogel, K., Krafft, W.: Chem. Ber. *101*, 2215 (1968)
129. Michel, W.: Dissertation Univ. Marburg 1961

130. Müller, E., Schick, A., Scheffler, K.: Chem. Ber. *92*, 474 (1959);
Müller, E., Schick, A., Mayer, R., Scheffler, K.: ibid. *93*, 2649 (1960)
131. Dimroth, K., Berndt, A., Reichardt, C.: Organic Synthesis, Coll. Vol. *5*, 1128 (1973),
Dimroth, K., Berndt, A., Perst, H., Reichardt, C.: ibid. p. 1130;
Dimroth, K., Reichardt, C., Vogel, K.: ibid. p. 1135
132. Blöcher, K. H.: Dissertation Univ. Marburg 1961
133. Dimroth, K., Kalk, F. Sell, R., Schlömer, K.: Liebigs Ann. Chem. *624*, 51 (1959)
134. Dimroth, K., Wolf, K. H.: Angew. Chem. *72*, 777 (1960)
135. Dimroth, K., Umbach, W., Blöcher, K. H.: Angew. Chem. *75*, 860 (1963); Angew. Chem. engl. *2*, 620 (1963) and unpublished results of Umbach, W.
136. Lendenfeld, H.: Monatsh. *27*, 969 (1906);
Ariyan, Z. S., Mooney, B.: J. Chem. Soc. (London) *1962*, 1519
137. Wache, H.: Dissertation Univ. Marburg 1964;
Dimroth, K., Wache, H.: Chem. Ber. *99*, 399 (1966)
138. Laubert, G.: Dissertation Univ. Marburg 1968
Dimroth, K., Laubert, G., Blöcher, K. H.: Liebigs Ann. Chem. *765*, 133 (1972)
139. Güsten, H., Kirsch, G., Schulte-Frohlinde, D.: Angew. Chem. *79*, 941 (1967); engl. *6*, 948 (1967); Tetrahedron *24*, 4393 (1968)
140. Dimroth, K., Schromm, K.: unpublished results 1962
141. Nelsen, St. F.: in Free Radicals, (ed. J. K. Kochi) vol. II, 527, John Wiley and Sons, New York, London Sidney, Toronto (1973);
Griller, D., Barclay, L. R., Ingold, K. U.: J. Am. Chem. Soc. *97*, 6151 (1975) and lit. cited therein
142. see also Chattaway, F. D., Orton, K. J. P.: J. Chem. Soc. (London) *79*, 461 (1901);
Goldschmidt, St.: Ber. dtsch. chem. Ges. *46*, 2728 (1913)
143. see also Hedayatullah, M., Denivelle, L.: Compt. Rend. Acad. Sci. Paris *258*, 5467 (1964); *259*, 5516 (1965); Bull. Soc. Chim. France *1969*, 4168
144. see also Schultz, G.: Ber. dtsch. chem. Ges. *17*, 461 (1884)
145. Oosterloo, G.: Dissertation Univ. Marburg 1958;
Dimroth, K., Oosterloo, G.: Angew. Chem. *70*, 165 (1958);
2.4.6-Triphenylselenophenol see Kataeva, L. M., Kataev, E. T.: Zh. obć *32*, 2712 (1962)
146. Dimroth, K., Kraft, K. J.: Angew. Chem. *76*, 433 (1964); engl. *3*, 384 (1964)
147. a) Kraft, K. J.: unpublished results, Univ. Marburg 1965;
see also: b) Porowska, N., Polaczkova, W., Kwiatowska, S.: some p_k values in $EtOH/H_2O$, Rocz. Chem. *44*, 375 (1970); C.A. *72*, 131888 (1970);
Bordwell, G. F., Mc Callum, R. J., Olmstead, W. N.: J. Org. Chem. *49*, 1424 (1984)
148. Nakano, F., Sugishita, M.: Jap. P 74 40227; C.A. *82*, 155 777 (1975).
149. Tacke, P.: Germ. Offen. 1930 341; C.A. *74*, 87596 (1971); Smith N. E., Sommerfield, E. H.: USP 4001340; C.A. *86*, 120976 (1977)
150. King, I. R., Hildon, A. M.: USP 4002693; C.A. *86*, 120975 (1977)
151. Raychaudhuri, S. R.: Chem. and Ind. (London) *1980*, 293
152. Kawamoto, A., Uda, H., Harada, N.: Bull. Chem. Soc. Japan *53*, 3279 (1980)
153. Hay, A.: USP 3363008; C.A. *69*, 2702 (1968)
154. Masamune, S., Casellucci, N. T.: J. Chem. Soc. (London), Proc. *1964*, 298
155. Dürr, H.: Tetrahedron Lett. *1966*, 5829; Liebigs Ann. Chem. *711*, 115 (1968)
156. Farnum, D. G., Burr, M.: J. Am. Chem. Soc. *82*, 2651 (1951)
157. Dana, D. E., Hay, A. S.: Synthesis *1982*, 164: Improved synthesis and 4′, 4″ substituted derivatives, see also cited literature
158. Plekhanova, L. G., Cuprova, N. A., Nikiforov, G. A., De Jonge, K., Golubeva, I. A., Ershov, V. V.: Izv. Akad. Nauk. SSSR, Ser Khim. *1980*, 1656; C.A. *94*, 3818 (1981): derivatives from 2,6-diphenylphenol by chloromethylation at C-4: CH_2R: $R = Cl$, OMe, OEt, OH, H, and $CH(CO_2Et)_2$, CH_2CO_2H

159. Dimroth, K., Kaletsch, H.: unpubl. 1982 from (*1*) + CH_2O and NaOH in EtOH
160. De Jonge, C. R. H., Hageman, H. J., Huysmans, W. B. G., Hijs, W. J.: J. Chem. Soc. (London), Perkin II, *1973*, 1276
161. Stillson, G. H., Sawyer, D. W., Hunt, C. K.: J. Am. Chem. Soc. *67*, 303 (1945); Müller, E., Schick, A., Scheffler, K.: Chem. Ber. *92*, 474 (1959)
162. Müller, E., Schick, A., Mayer, R., Scheffler, K.: Chem. Ber. *93*, 2649 (1960)
163. Hammel, D. E.: Dissertation Univ. Marburg 1964
164. Dimroth, K., Tüncher, W., Kaletsch, H.: Chem. Ber. *111*, 264 (1978); Tüncher, W.: Dissertation Univ. Marburg 1976
165. Ristow, K. D.: Dissertation Univ. Marburg 1963; see also Webb, J. L., Hall, L.: US P 3739035; C.A. *79*, 67035 (1973)
166. Umbach, W.: Dissertation Univ. Marburg 1962
167. Umbach, W., Dimroth, K.: 1963 not published
168. Müller, K. H.: Dissertation Univ. Marburg 1963
169. El-Kholy, I., Mishrikey, M. M., Rafla, F. K., Soliman, G.: J. Chem. Soc. (London), *1962* 5153; Ried, W., Kunkel, W., Strätz, A.: Liebigs Ann. Chem. *726*, 69 (1969)
170. Musso, H., v. Grunelius, S.: Chem. Ber. *92*, 3101 (1959); Barker, A. W., Kerlinger H, O., Shulgin, A. T.: Spectrochim. Acta *20*, 1477 (1964)
171. Goddu, R. F.: J. Am. Chem. Soc. *82*, 4533 (1960); Ingold, K. U., Taylor, D. R.: Can. J. Chem. *39*, 471 (1961)
172. Hay, A. S., Boulette, B. M.: J. Org. Chem. *41*, 1710 (1976)
173. Waters, W. A.: J. Chem. Soc. (London) B *1971*, 2026
174. Allmann, R., Hellner, E.: Chem. Ber. *101*, 2522 (1968)
175. For discussion see Rieker, A. in Houben-Weyl vol. VII 3b, part 2, p. 523 (1979)
176. Land, E. J., Porter, G., Strachan, E.: Trans. Faraday Soc. *57*, 1885 (1961); *59*, 2016 (1963)
177. Kuz'min, V. A., Khudjakov, I. V., Levin, Jr., P. P., Emmanuel, N. M., de Jonge, C. R. H.I., Hageman, H. J., Biemond, M. E. F., van der Maeden, F. P. B., Mijs, W. J.: J. Chem. Soc. Perkin II, *1979*, 1540; *1981*, 1234, 1237
Burlatsky, S. F., Levin, P. P., Khudjakov, I. V., Kuzmin, V. A., Ovchinnikov, A. A.: Chem. Phys. Lett. *66*, 565 (1979);
for pulse radiolysis see also Neta, P.: J. Chem. Ed. *58*, 110 (1981) for a short review
178. Ziegler, K., Ewald, L.: Liebigs Ann. Chem. *473*, 163 (1929)
179. Dimroth, K., Berndt, A., Bär, F., Schweig, A., Volland, R.: Angew. Chem. *79*, 69 (1967); engl. *6*, 34 (1967)
180. Hyde, J. S.: J. Phys. Chem. *71*, 68 (1967)
181. Cook, C. D., Depatie, C. B., English, E. S.: J. Org. Chem. *24*, 1356 (1959)
182. Dimroth, K., Kraft, K. J.: Chem. Ber. *99*, 264 (1966); Kraft, K. J.: Dissertation Univ. Marburg 1963
183. Braude, E. A., Brook, A. G., Linstead, R. P.: J. Chem. Soc. (London) *1954*, 3569;
Part 13: Braude, E. A., Jackman, L. M., Linstead, R. P., Lowe, G.: ibid. *1960*, 3133;
Reviews: Walker. D., Hiebert, J. H.: Chem. Rev. *67*, 153 (1967);
Jackman, L. M.: Adv. Organic Chem. vol. 2, 329 (1960); Interscience, New York.
Half wave redox potential of DDCh in acetonitrile using dropping mercury electrode or a rotating platinum electrode and tetramethylammoniumperchlorate as supporting electrolyte have been found to be +0.51 V against a saturated standard calomel electrode: Peover, M. E.: J. Chem. Soc. (London) *1962*, 4540
184. Nickel, B., Mauser, H., Hezel, U.: Z. Phys. Chem. N.F. *54*, 196, 214 (1967); Mauser, H., Nickel, B.: Angew. Chem. *77*, 378 (1965); engl. *4*, 354 (1965).
185. Steenken, S.; Neta, P.: J. Phys. Chem. *86*, 3661 (1982)
186. Gasanov, B. R., Stradyn', Y. P.: J. Gen. Chem. of USSR, engl. *46*, 2469 (1976)
187. Müller, E., Rieker, A., Schick, A.: Liebigs Ann. Chem. *673*, 40 (1964); Rieker, A.: Chem. Ber. *98*, 715 (1965); Becker, H.-D.: J. Org. Chem. *29*, 3068 (1964); 30, 982 (1965)

188. Dimroth, K., Berndt, H.: Chem. Ber. *101*, 2519 (1968)
189. Dimroth, K., Tüncher, W.: Synthesis *1977*, 339;
190. Haaß, K.: Dissertation Univ. Marburg 1962
191. Criegée, R.: Ber. dtsch. Chem. Ges. *69*, 2758 (1936)
192. Reviews: Zollinger, H.: Angew. Chem. *70*, 204 (1958);
Krumbiegl, P.: Isotopieeffekte, Akademie Verlag, Berlin 1970
193. Hageman, H. J., Huysmans, W. G. B.: J. Chem. Soc. Chem. Commun. *1969*, 837.
For similar reactions by heating 2,6,2',6'-tetraphenyldiphenoquinone see Hay, A. S.: J. Org. Chem. *36*, 218 (1971)
194. Sviridov, B. D., Gyzunova, L. P., Kuznetz, V. M., Nikiforov, G. A., De Jonge, K., Hageman, H. J., Ershov, V. V.: Izv. Akad. Nauk SSSR, Ser. Khim. *9*, 2160 (1978); (C.A. *90*, 21981 (1979)
195. Reviews: Rotermund, G. W.: Houben Weyl IV 1b, 761 (1975);
Wedemeyer, K. F.: ibid. VI 1c, (1976);
Ulrich, H., Richter, R.: ibid, VII 3a, 27 (1977);
Boldt, P.: ibid. VII 3b, 196 (1979);
Barton, D. H. R.: Chem. in Britain *3*, 330 (1967) and lit. cit. therein;
Musso, H.: Angew. Chem. *75*, 965 (1963);
see also Bird, C. W., Chauhan, Y.-P.: Tetrahedron Lett. *1978*, 2133
196. Teuber, H. J., Glossauer, G.: Chem. Ber. *98*, 2643 (1965)
197a. De Jonge, C. R. H. I., van Dort, H. M., Vollbracht, L.: Tetrahedron Lett. *1970*, 1981; see cit. ref. for oxidation of phenols in apolar solvents
197b. Finkbeiner, H., Toothacker, A. T.: J. Org. Chem. *33*, 4347 (1965)
198. Dewar, M. J. S., Nakaya, T.: J. Am. Chem. Soc. *90*, 7134 (1968)
199. Vogt Jr., L. H., Wirth, J. G., Finkbeiner, H. L.: J. Org. Chem. *34*, 273 (1969)
200. Bolton, D. A.: ibid. *34*, 2031 (1969)
201a. Abramovitch, R. A., Alvernhe, G., Bartnik, R., Dassanayake, N. L., Inbasekaran, M. N., Kato, S.: J. Am. Chem. Soc. *103*, 4558 (1981) and lit. cited therein;
201b. Martius, C., Eilingsfeld, H.: Liebigs Ann. Chem. *607*, 159 (1957)
201c. Adler, E., Falkehag, I., Smith, B.: Acta Chim. Scand. *16*, 529 (1962)
202. Suttie, A. B.: Tetrahedron Lett. *1969*, 953
203. Rolán, A., Parker, V. D.: J. Chem. Soc. (London) (C) *1971*, 3214;
Rolán, A.: J. Chem. Soc., Chem. Commun. *1971*, 1643;
Nilsson, A., Palmquist, U., Pettersson, T., Rolán, A.: (part 5), J. Chem. Soc. (London), Perkin I, *1978*, 696 and lit. cit. therein;
Rieker, A., Dreher, E.-L., Geisel, H., Khalifa, M. H.: Syntheses *1978*, 851;
Speiser, B., Rieker, A.: J. Elektrol. Chem. *110*, 231 (1980);
Review: Evans, D. H.: Acc. Chem. Res. *10*, 313 (1977)
204. Rieker, A., Bracht, J., Dreher, E. L., Schneider, P.: Houben-Weyl, vol. VII 3b, 529 (1979)
205. Dimroth, K., Umbach, W., Thomas, H.: Chem. Ber. *100*, 132 (1967), see also ref. 166)
206. Dimroth, K., Thomas, H.: Chem. Ber. *102*, 3795 (1969);
Thomas, H.: Dissertation Univ. Marburg 1968
207. Abramovitch, R. A., Bartnik, R., Cooper, M., Dassanayake, N. L., Hwang, H.-Y., Inbasekaran, M. N., Rusek, G.: J. Org. Chem. *47*, 4817 (1982) and ref. cit. therein.
For a related coupling reaction of phenol with hypervalent iodine (III) species see White, J. D., Chong, W. K. M., Thirring, K.: J. Org. Chem. *48*, 2300 (1983)
208. Wessely, F., Sinwel, F.: Monatsh. Chem. *81*, 1055 (1950);
Dimroth, K., Perst, H., Schlömer, K., Worschech, K., Müller, K.-H.: Chem. Ber. *100*, 629 (1967)
209. Rieker, A., Rundel, W., Kessler, H.: Z. Natforsch. *24b*, 547 (1969)
210. Sell, R.: Dissertation Universität Marburg 1960; see also ref. 133)
211. Dimroth, K., Perst, H.: Liebigs Ann. Chem. *708*, 86 (1967),
see also Perst, H.: Dissertation Universität Marburg 1965
212. Dimroth, K., Perst, H., Müller, K.-H.: Chem. Ber. *100*, 1850 (1967); see also ref. 168)
213. Zbiral, E., Wessely, F., Lahrmann, L.: Monatsh. Chem. *91*, 92 (1960)
214. Nishinaga, A., Itahara, T., Matsuura, T., Berger, S., Henes, G., Rieker, A.: Chem. Ber. 109, 1530 (1976)

215. Perst, H., Dimroth, K.: Tetrahedron *24*, 5385 (1968)
216. Perst, H., Sprenger, W.: Tetrahedron Lett. *1970*, 3601;
Perst, H., Wesemeier, I.: ibid. *1970*, 4189;
Lemmer, D., Perst, H.: ibid. 1972, 2735;
Seifert, W. J., Perst, H., Dannenberg, W.: ibid. *1973*, 4999;
Seifert, W., Perst H.: ibid. *1975*, 2419;
Dannenberg, W., Seifert, W. J., Perst, H.: ibid. *1975*, 3481;
Perst, H., Dannenberg, W.: Liebigs Ann. Chem. *1975*, 1873;
Weisskopf, V., Perst, H.: Liebigs Ann. Chem. *1978*, 1634

Natural Polyamines-Linked Cyclophosphazenes. Attempts at the Production of More Selective Antitumorals

Jean-François Labarre

Laboratoire Structure et Vie, Université Paul Sabatier, 118 Route de Narbonne,
31062 Toulouse Cedex, France

Table of Contents

1 Introduction

Any allopathic remedy, even when prescribed carefully, always induces side-effects in treated patients. Drugs are indeed extraneous bodies for the living system who will plus or minus look upon them favourably, depending on the size of the bullet. Thus, posologies in allopathy, even upon sub-acute, chronical and limited schedules, are actually confronted to a dramatical balance between therapeutic benefits and penalizing side-disorders.

This is especially the situation in cancer chemotherapy, antitumor agents having commonly low therapeutic indexes. In other words, anticancer drugs must normally be used close to their toxic doses, thus inducing huge side-effects. In that case, such side-effects are a consequence of the poor selectivity of drugs for the malignant cells, a large amount of the injected dose being spread out over the rest of the body, i.e. over the healthy cells. This poor selectivity constitutes till now a challenge to both chemists who are in charge of designing new drugs and clinicians who are daily in charge of prescribing them. Tremendous multidisciplinary efforts have been made during the last few years for enhancing the selectivity of anticancer drugs, essentially through covalent bindings either to monoclonal antibodies (immunoglobulins) [1] or to natural polyamines (mainly putrescine, cadaverine, spermidine and spermine) [2], antibodies and polyamines playing the role of *tumor finders* and, in few cases, of homing heads. These two new approaches were diversely successful but they are definitely the only way to escape from the uncomfortable dilemmas allopathy in cancer chemotherapy nowadays comes up against.

Anticancer Inorganic Ring Systems [3] do not escape the rule and several attempts at the production of more selective agents have been achieved in our laboratory during the last decade, polyamines being selected as tumor finders.

This contribution tells the story of the covalent binding of anticancer cyclophosphazenes to natural polyamines and of the benefits of such a targeting about selectivity.

2 An up-to-date Survey of "NUDE" (i.e. non vectorized) Inorganic Ring Systems as Anticancer Agents

Interest in Inorganic Ring Systems as anticancer drugs, initially mentioned by Cernov et al. [4], was enhanced in 1978 by the discovery that the aziridinocyclophosphazenes $N_3P_3Az_6$ and $N_4P_4Az_8$ (Az = aziridinyl) were active on a large series of experimental neoplasms [5,6]. Subsequent studies conducted within the framework of the activities of the EORTC Screening and Pharmacology Group [7] and employing a wide range of rodent neoplasms including leukemias and solid tumors of different histological nature, growth rate and chemotherapeutic sensitivity, showed that $N_3P_3Az_6$ (code name: MYKO 63) was the more active of these two chemicals.

In order to make further improvements on MYKO 63 antineoplastic activity and reduce its toxicity, several derivatives were subsequently investigated.

Following the approach of adding methyl groups on the 2 and 2′ C positions of the aziridino ligands of $N_3P_3Az_6$, new drugs were prepared which were shown to display a level of antitumoral activity and of cumulative toxicity essentially comparable to that of MYKO 63 [8]. Further, when the P atoms of the N_3P_3 ring were replaced step-by-step by S atoms, a new series of antitumorally effective compounds was obtained having the general structure $(NPAz_2)_2(NSOX)$ (code names: SOF, when X = F; SOPHi, when X = Ph; SOAz, when X = Az). Interest in these compounds was heightened by the finding that they displayed definite antineoplastic activity together with a reduced cumulative toxicity compared with MYKO 63 [9-11]. Among the members of this series, SOAz was found to be the most antitumorally active [12], being devoid of mutagenicity in bacterial system [13] as well as of significant nephro-, hepato- and cardiotoxicity in preliminary tests in dogs and primates, while inducing in these species controllable hematotoxicity. Phase I clinical trials with SOAz have already been performed and the agent is currently undergoing Phase II clinical evaluation [14].

In concurrent studies it was also found that reduction from 6 to 4 in the number of Az ligands grafted on the N_3P_3 ring, thus giving $N_3P_3Az_5Cl$ and gem-$N_3P_3Az_4Cl_2$ (code name: MYCLAz) produced compounds which possessed a still significant antitumor activity but lower toxicity than MYKO 63 [15,16].

As mentioned above, SOAz underwent clinical trials and, whatever its potentiality as an anticancer agent seems to be, pharmacokinetics proved that 60% of the amount of drug injected to animals and humans are excreted through urins (without any metabolisation) during the next 24 hours after injection. In other words, a large part of the remedy does not reach the tumor and is actually spread out all over the body without any therapeutical efficiency, the main side-effect being thrombocytopenias which can become irreversible.

Thus, in the attempt to increase the selectivity for malignant cells of SOAz and relatives and to decrease their toxicity for normal tissues and consequently obtain compounds possessing a high therapeutic index, it was of interest to investigate the activity of anticancer cyclophosphazenes when linked to natural polyamines. This was the driving idea we obtained early in 1982: we indeed knew that rapidly proliferating cells have a higher capacity for active, carrier-mediated uptake of natural polyamines [2,17,18]. Accordingly, natural polyamines could represent a useful means for targeting to neoplastic tissues compounds possessing cell-inhibitory capacity.

Actually, the covalent linkage of polyamines to persubstituted drugs as MYKO 63 or SOAz is unrealizable: aziridino groups do not react at all with amino functions of polyamines. Thus, our driving idea seemed to be an utopia, just "paper chemistry".

The only drug amongst the series which could be expected to bind polyamines was MYCLAz, linkage occuring through its two labile chlorine atoms. Chemical and physical evidences for such a covalent binding of MYCLAz with putrescine were made conspicuous [19] but the yield of the reaction appears to be very poor, an inextricable medley of final products being obtained.

Thus, we decided to "inverse" the assembly line for the synthesis of the expected vectorized drugs. This reverse line is based on the following set of reactions:

i) reaction of the natural polyamine (P) with $N_3P_3Cl_6$ (as starting material) in order to get suitable P-linked tetrachlorocyclotriphosphazenes;
ii) peraziridinylation of these tetrachlorocyclotriphosphazenes leading to chemicals in which the drug (one N_3P_3 ring bearing 4 Az groups at least) is covalently bound to (P) as the tumor finder.

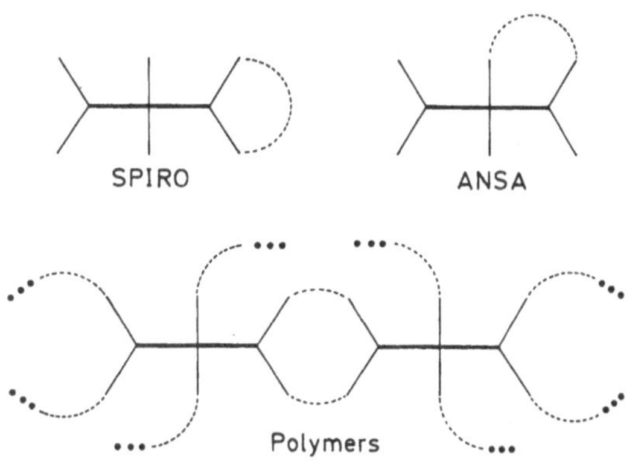

Fig. 1. The three possible routes for the reaction of $N_3P_3Cl_6$ with difunctional reagents

However, this reverse line of synthesis constituted in early 1982 a tricky task. Indeed at that time, although the reactions of $N_3P_3Cl_6$ with monofunctional amines had been extensively investigated [20–25], those of hexachlorocyclotriphosphazene with difunctional reagents were less well documented [20].

There were indeed in the literature three possible routes for the reaction of $N_3P_3Cl_6$ with difunctional reagents (Fig. 1):

i) replacement of two chlorine atoms from the same phosphorus atom to give a SPIRO derivative;
ii) replacement of two chlorine atoms from different phosphorus atoms to give an ANSA compound;
iii) intermolecular condensation to yield cyclolinear and/or cyclomatrix polymers.

The problem as to whether the monomer has a SPIRO or an ANSA structure had been strongly contested, Becke-Goehring and Boppel were resolutely in favour of the ANSA structure [26] whereas Shaw and co-workers support a SPIRO configuration [27]. It must be emphasized, however, that these contradictory conclusions were based on indirect physicochemical data from techniques such as IR, ^{1}H and ^{31}P NMR spectroscopy; unambiguous confirmation of the SPIRO structure had been only provided in 1978 by an X-ray investigation of the molecular structure of $N_3P_3(NMe_2)_4$ [HN$-(CH_2)_2-$NH], the ethylenediamino derivative [28]. Moreover, recent ^{1}H and ^{31}P NMR data for other ethylenediamino and ethanolamino compounds had been found to be consistent with SPIRO structures [29].

In other words, a systematic investigation of the structures of the products from the reactions of $N_3P_3Cl_6$ with difunctional reagents as natural polyamines had to be carried out. The next section deals with the structures of these chlorinated intermediates as *precursors* of the expected vectorized drugs.

3 Syntheses of Chlorinated Precursors upon Reaction of $N_3P_3Cl_6$ with Natural Polyamines

3.1 The Case of the 1,3-Diaminopropane [30]

3.1.1 Synthesis

The synthesis was performed in stoichiometric conditions by both Shaw's method for the synthesis of $N_3P_3Cl_4$[HN$-(CH_2)_2-$NH] in the presence of NEt_3 [27] and that of Becke-Goehring for the synthesis of the so-called (see below) ANSA $N_3P_3Cl_4[HN-(CH_2)_3-NH]$ in the presence of 110% excess of propylenediamine [26]. The resulting two samples were found to be identical from their elemental analyses, melting point (174 °C versus 166 °C from Ref. [26]), IR and ^{31}P NMR spectra. The yield of Becke-Goehring's reaction (70%) was, however, better than that of Shaw's (20–30%). Thin-layer chromatography gives an unique R_f value of 0.69 with CH_3OH as eluant.

3.1.2 Mass Spectrometry

The spectrum was recorded on a R1010 Ribermag quadrupole mass spectrometer using a direct inlet system. The source temperature was 150 °C, electron energy 70 eV. The spectrum was analyzed by means of a DEC PDP 8/M computer and stored on disk. A small sample (about 1 µg) was introduced into the probe, the temperature of which was then gradually increased from ambient temperature to 100 °C, taking care that neither the electron multiplier nor the amplifier were in a saturated condition at any time. The areas under the curves corresponding to the current carried by the various ions were calculated by computer. The 70 eV electron impact mass spectrum of $N_3P_3Cl_4[HN-(CH_2)_3-NH]$ is presented in Fig. 2.

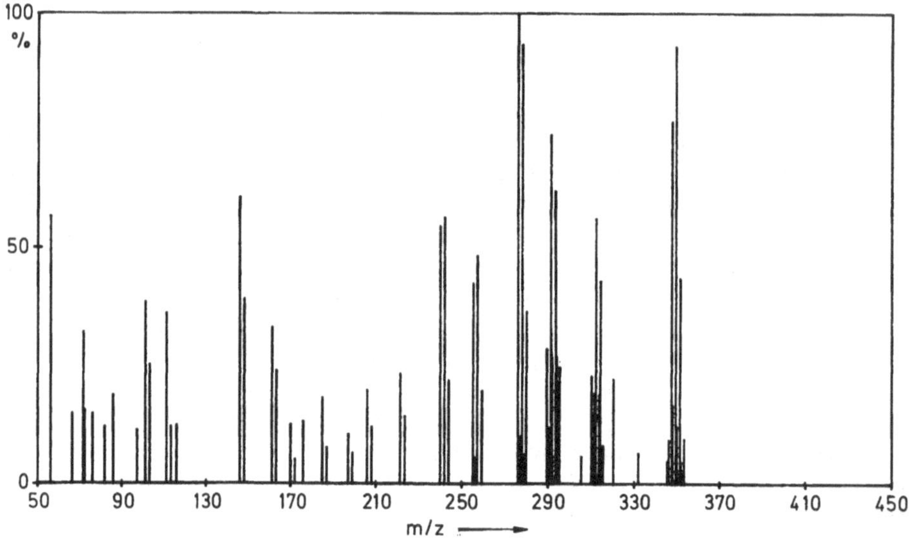

Fig. 2. 70 eV electron impact mass spectrum of $N_3P_3Cl_4[HN-(CH_2)_3-NH]$

The molecular ion is observed at m/z 349 (92.9%) with satellites at m/z 350 (12.1%), 351 (43.6%), 352 (4.7%) and 353 (9.6%). The intensity ratio of these five peaks indicates the presence of four chlorine atoms in the molecule. Four other satellites of the molecular ion are observed at m/z 348 (16.7%), 347 (77.0%), 346 (9.4%) and 345 (4.9%) corresponding to the loss of one to four H atoms. One major fragmentation route is detected involving the successive loss of 1 CH_2, 2 CH_2, 3 CH_2,

3 CH_2 + 1 NH and 3 CH_2 + 2 NH (associated with H transfers) to give maximal peaks at m/z 332 (6.6%), 320 (22.2%), 305 (5.9%), 291 (74.0%) and 276 (100.0%). Further consecutive losses of Cl atoms give peaks at m/z 242 (56.8%), 208 (12.2%) and 170 (12.7%). Another fragmentation route is observed with peaks at m/z 314 (42.9%), 278 (93.3%), 242 (56.8%) and 206 (19.9%) corresponding to the successive loss of one to four chlorine atoms (again associated with H transfers) from the molecular ion.

Due to its simplicity, it was an easy matter to check the spectrum for the presence of any possible contaminant. Thus, in this case, mass spectrometry is an adequate tool for controlling the purity of the compound in question, as was demonstrated previously in the case of other cyclophosphazenes with biological significance [31].

3.1.3 NMR Spectroscopy

The ^{31}P NMR spectrum of $N_3P_3Cl_4[HN—(CH_2)_3—NH]$ was recorded on a Brucker WH 90 instrument. The doublet at 21.39 and 20.09 ppm corresponds to PCl_2 entities and the triplet around 7.58 ppm to a PN_2 moiety (intensity ratio 2:1). However, such an assignment is based upon the assumption of a SPIRO structure and it should be noted that the 7.58 ppm value ascribed to the endocyclic P atom in the spiro loop is not consistent with the NMR data obtained by Shaw and co-workers for SPIRO ethylenediamino derivatives, i.e. $\delta(PCl_2)$ and $\delta(P_{spiro})$ were in the ranges 21.4–22.2 and 22.0–35.5 ppm respectively [29]. These two ^{31}P shifts were found to be identical (22.0 ppm) in $N_3P_3Cl_4[HN—(CH_2)_2—NH]$ when the spectrum was recorded on a Brucker HFX 90 instrument operating at 36.43 MHz [29] but they have recently been resolved (23.5 and 22.8 ppm) by a more sophisticated NMR apparatus [32].

Assuming a SPIRO structure for $N_3P_3Cl_4[HN—(CH_2)_3—NH]$ necessitates an explanation as to why $\delta(^{31}P)$ for the P_{spiro} entity shifts drastically from 22.8 to 7.58 ppm when the size of the loop only increases by one CH_2 group.

No reasonable explanation for such a huge shift can be provided until an X-ray structure of the compound is performed. Crystallography is the only technique which may both conclusively support the SPIRO structure and provide reasons for the anomalous ^{31}P chemical shift in $N_3P_3Cl_4[HN—(CH_2)_3—NH]$.

3.1.4 Infrared Spectroscopy

The infrared spectrum (KBr disks) was recorded on a Perkin-Elmer 683 spectrometer (range 4000–200 cm^{-1}, calibrated with polystyrene lines) and was found to be superimposable on that given in Ref. 26. This spectrum was used by Becke-Goehring and Boppel as evidence for the ANSA structure. It is demonstrated below how careful one has to be when assigning a certain molecular configuration on the bases of indirect structural techniques such as IR spectroscopy.

Incidentally, it was observed that when the compound was left in KBr disk at 37 °C overnight, all traces of water normally present in KBr were absorbed; H_2O-associated N—H bands appeared in the IR spectrum and the broad band due to free water was absent as a consequence of $N_3P_3Cl_4[HN—(CH_2)_3—NH]$ hydration.

3.1.5 X-Ray Crystal and Molecular Structure

Single crystals, suitable for X-ray crystallography, were obtained directly from the synthesis of the compound (solvent = anhydrous methylene chloride). The crystal structure was determined on a CAD-4 ENRAF-NONIUS PDP 8/M computer-controlled single-crystal diffractometer. Atomic positional parameters, bond lengths and bond angles are published elsewhere [33].

A perspective view of the $N_3P_3Cl_4[HN-(CH_2)_3-NH]$ molecule is shown in Fig. 3, together with the numbering of the atoms. This view clearly illustrates the SPIRO structure with the six-membered phosphazene ring perfectly planar. An alternative view of the molecule (Fig. 4) shows it to have a two-fold symmetry axis. A third view (with the N_3P_3 ring in the plane) reveals considerable puckering of the SPIRO loop (Fig. 5); the distances of N_3 and C_1 from the $P_1C_1C_2$ and $P_1N_3C_2$ planes respectively are equal to 0.67 Å. Interestingly, the $N_3P_1N_{3i}$ plane, that which contains the C_2 atom, is not strictly perpendicular to the phosphazenic ring, the angle between these two planes being equal to 99°. In other words, the two exocyclic phosphorus-nitrogen bonds deviate by 9° from the expected σ_v front plane containing atoms P_1 and N_2.

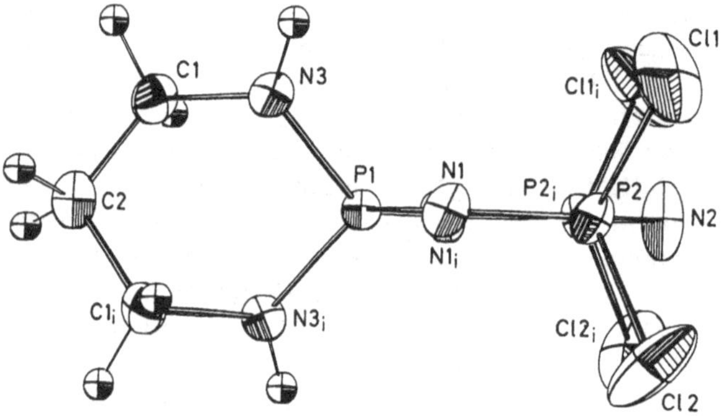

Fig. 3. A perspective view of $N_3P_3Cl_4[HN-(CH_2)_3-NH]$ illustrating the SPIRO structure

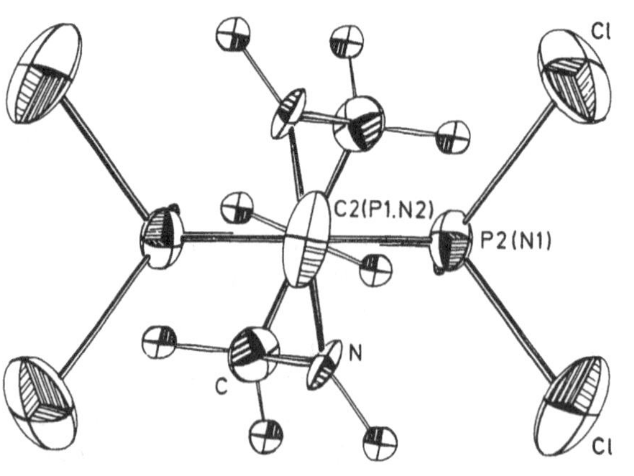

Fig. 4. A perspective view of $N_3P_3Cl_4[HN-(CH_2)_3-NH]$ showing the two-fold axis

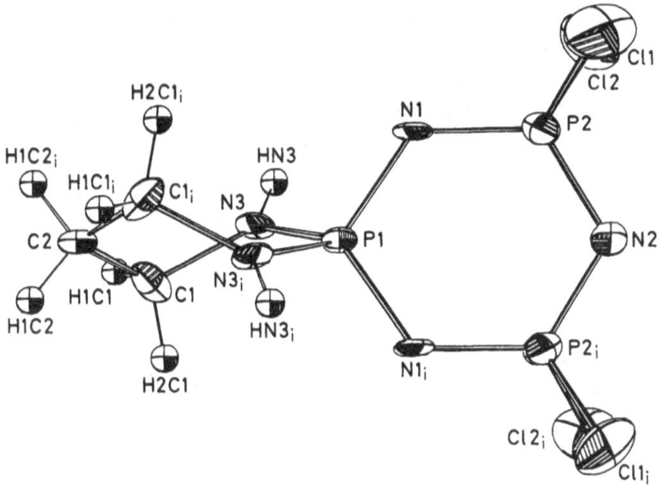

Fig. 5. A perspective view of $N_3P_3Cl_4[HN{-}(CH_2)_3{-}NH]$ revealing the puckering of the SPIRO loop

The angle between the $N_3P_1N_{3i}$ and $C_1C_2C_{1i}$ planes is 32° while that between $P_1C_1C_2$ and the inorganic ring plane is 65.9°.

The main feature of the structure concerns the neighbourhood of atom P_1: the presence of the loop pulls P_1 away from N_2 along the two-fold axis, the endocyclic $P_1{-}N_1$ and $P_1{-}N_{1i}$ bonds being much longer (1.609 Å) than the average value (ca. 1.58 Å) normally expected for a trimeric cyclophosphazenic ring. Consequently, the $N_1P_1N_{1i}$ angle is much smaller (111.5°) than that in other trimeric cyclophosphazenes (ca. 119°) [34].

As a consequence of this internal molecular stretching along the two-fold axis, atom P_1 appears to be in a pseudo-tetrahedral situation, the four associated phosphorus-nitrogen bonds being practically equal. Such a T_d-like environment for a phosphorus atom belonging to a trimeric ring is unique in cyclophosphazenes and may explain the peculiar NMR behaviour manifested by P_1 (see above).

Moreover, there is a very short intramolecular contact distance of 2.66 Å between N_1 of the phosphazene ring and N_3 of the SPIRO loop, which suggests the existence of an intramolecular hydrogen bond. A similar feature was observed in the crystal structures of $N_3P_3(NMe_2)_4[HN{-}(CH_2)_2{-}NH]$ and $N_3P_3Cl_4(NHBu^t)_2$ [28] although with a weaker intensity: the N_1 ... N_3 distance in the former compound is 3.18 Å. Such hydrogen bonds are probably responsible for the peculiar IR spectrum in the (N—H) stretching region, especially since there exist two very strong intermolecular hydrogen bonds in the crystal structure [33].

The hexagonal crystal structure visualized in Ref. [33] provokes the following two remarks.

i) the six loops within the unit cell are coiled up around the six-fold axis in an helicoidal arrangement, the distance between two successive loops being equal to c/6, i.e. 4.349 Å. Such stacking is due to very strong hydrogen bonds between the $N_3{-}H$ bond of one loop and the N_1 atom of the adjacent molecule; the N_1 ... H intermolecular distance is 2.08 Å.

ii) There exist in the unit cell ovoid tunnels which closely resemble the channels characterizing Allcock's clathrates [35]. It is possible that the new clathrate-like structure described here can encage solvent molecules when the single crystal is formed. Thus, the CH_2Cl_2 (b.p. 40 °C) used in the synthesis of $N_3P_3Cl_4[HN-(CH_2)_3-NH]$ is probably the reason why crystal bubbling is observed at 37 °C before collection of the X-ray data. Furthermore, when the organic solvent has left the crystal structure, water molecules may be absorbed through the channels owing to the very high acidic character of the N—H bonds within the loops; such a tendency may explain the gradual increase in the unit cell volume, until complete efflorescence of the crystal with disruption of its structure, which was observed after several days.

The molecular structure of the compound was used to calculate its electronic structure in order to understand better the anomalous ^{31}P NMR shifts described above.

3.1.6 Electronic Structure

Calculations were performed using the SCF-LCAO-MO method within the CNDO/2 approximation [36,37].

Table 1. Electronic Structure of $N_3P_3Cl_4[HN-(CH_2)_3-NH]$

Atom	Charge	Bond	Wiberg Index
P_1	4.60	P_1-N_1	1.486
N_1	5.26	N_1-P_2	1.697
P_2	4.76	P_2-N_2	1.576
N_2	5.20	P_2-Cl_1	1.347
Cl_1	7.06	P_2-Cl_2	1.352
Cl_2	7.06	P_1-N_3	1.364
N_3	5.23	N_3-C_1	0.957
C_1	3.88	C_1-C_2	1.016
C_2	4.03	N_3-H	0.928
$H(N_3)$	0.86	C_1-H_1	0.967
$H_1(C_1)$	1.00	C_1-H_2	0.967
$H_2(C_2)$	1.00	C_2-H_1	0.966
$H_1(C_2)$	0.98		

Charge densities and Wiberg indices [38] are summarized in Table 1. The atom P_1 appears to be considerably discharged ($Q = 4.60$), its acidity in the Lewis sense is the highest ever observed in cyclophosphazenes [39]. Moreover, the four N atoms linked to P_1 appear to be highly basic and, consequently, the H atoms on N_3 are acidic. The Wiberg indices of the P_1-N_1 and P_1-N_3 bonds are similar (1.486 and 1.364 respectively) and smaller than those of other P—N bonds of the molecule. The electron distribution around P_1 reflects its tetragonal geometry and the surrounding isotropic charge density is responsible for the large paramagnetic component of its ^{31}P shift, which contributes to the shift of the P_1 signal towards high field. This phenomenon explains the value of 7.58 ppm observed for P_1 relative to that of 22.0 ppm observed in $N_3P_3Cl_4[HN-(CH_2)_2-NH]$ [29]. Manohar's structure for this molecule illustrates quite different features of its P_1 atom: the P_1—N_3 and P_1—N_1 bonds are quite different (1.67 and 1.60 Å) as are the angles $N_1P_1N_1'$ and $N_3P_1N_3'$ (115.0 and 95.6°). Consequently, the charge distribution around P_1 is predicted to be much more anisotropic than in our case, explaining why the ^{31}P shift for P_1 in Manohar's molecule occurs at a much lower field than here.

3.1.7 Conclusion

Thus, X-ray crystallography shows unambiguously that the final product of the reaction of $N_3P_3Cl_6$ with 1,3-diaminopropane displays a definite SPIRO configuration. No ANSA cousin or cyclo-matrix polymer were ever observed, in stoichiometric conditions at least.

What happens now when increasing by one methylenic group the length of the diamine?

3.2 The case of the 1,4-Diaminobutane (Putrescine) [40]

3.2.1 Synthesis

The synthesis was done following Shaw's method [27] in the presence of NEt_3. A non-polar solvent (70% petroleum ether 60–80 °C, 30% CH_2Cl_2) was used in order to minimize the formation of non-crystalline resins (cyclo-linear and/or cyclo-matrix polymers). It is well-known indeed that polar solvents (e.g. diethyl ether, DMSO) lead to a proton abstraction process from the PNHR groups thus promoting polymerization [23,41]. The yield of the reaction was 65%. Thin-layer chromatography gave two spots: a large one (A) at Rf = 0.51 and a smaller one (B) at Rf = 0.65 with CCl_4—CH_2Cl_2 (3:7) as eluant.

It is noteworthy that the resins are initially soluble in organic solvents but harden and become insoluble on storing for 2–3 days, except in acetone. The IR spectrum of such a resin shows a strong broad absorption band at ca. 1200 cm^{-1} suggesting that the six-membered N_3P_3 ring is retained.

3.2.2 Separation of (A) and (B) by SiO_2 Column Chromatography

The crude final mixture (85% A + 15% B) was submitted to SiO_2 column chromatography using CCl_4—CH_2Cl_2 (3:7) as eluant. A and B were then readily separable in high yield. Their melting points were 179° and 148.5 °C respectively.

3.2.3. Mass Spectrometry of (A)

Spectra was recorded on the R1010 Ribermag system described above.

The 70 eV electron impact mass spectrum of A is shown in Fig. 6. The molecular ion was observed at m/z 363 (100%) with a set of satellites at m/z 364 (14.6%), 365 (55.0%), 366 (6.7%) and 367 (11.2%). The intensity ratio of these five peaks indicates the presence of four chlorine atoms in the molecule. Two other satellites of the molecular ion were observed at m/z 362 (15.7%) and 361 (87.6%), corresponding to the loss of 1 and 2 H atoms respectively. One main fragmentation route involves successive loss of 1 CH_2, 2 CH_2, 3 CH_2, 4 CH_2, 4 CH_2 + 1 NH and 4 CH_2 + 2 NH (associated with H-transfers) to give maximal peaks at m/z 348 (19.1%),

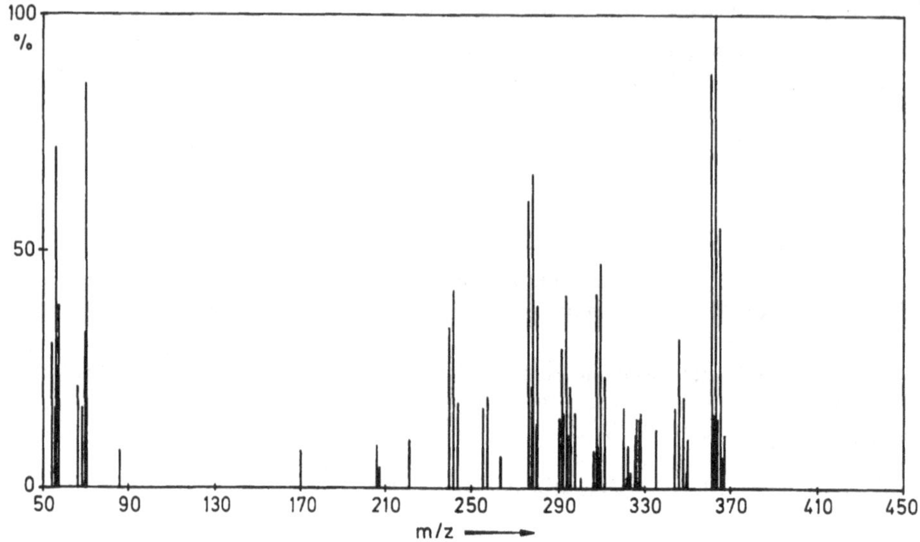

Fig. 6. 70 eV electron impact mass spectrum of $N_3P_3Cl_4[HN-(CH_2)_4-NH]$

335 (12.3%), 320 (16.8%), 307 (40.4%), 292 (15.7%), 277 (21.3%). Further consecutive loss of Cl atoms gives peaks at m/z 242 (41.5%), 206 (8.9%) and 170 (7.8%).

A second fragmentation route is observed: starting from the molecular ion, peaks are observed at m/z 328 (15.7%), 292 (15.7%), 257 (19.1%) and 221 (10.1%) which correspond to the successive loss of one to four chlorine atoms (again associated with H-transfers) from the molecular ion. Each of these four fragments gives a related peak corresponding to the loss of one CH_2 group from the SPIRO loop at m/z 311 (23.5%), 278(66.2%), 243 (17.9%) and 207 (8.9%) respectively.

In the low mass range, several peaks can be attributed to certain fragments: $HN-(CH_2)_4-NH$, m/z 86 (7.8%); $N-(CH_2)_4$, 70 (85.3%); $N-(CH_2)_3-CH$, 69 (32.5%); $N-(CH_2)_3-C$, 68 (16.8%); $N-(CH_2)_2-C-C$, 66 (21.3%); $N-(CH_2)_3$, 56 (71.9%); $N-(CH_2)_2-CH$, 55 (16.8%) and $N-(CH_2)_2-C$, 54 (30.3%).

The intensity of all other peaks is less than 2%. Thus, it is relatively simple matter to check for the presence of any possible contaminant. Mass spectrometry again appears to be an adequate tool for monitoring the purity of a cyclosphazene [31]. Furthermore, since the fragmentation pathways of product A are identical to those previously observed [30] for the SPIRO-$N_3P_3Cl_4[HN-(CH_2)_3-NH]$ derivative, it seemed reasonable to assign a SPIRO structure to A. X-ray crystallography (see below) unambiguously supports this conclusion.

3.2.4 NMR Spectroscopy of (A)

The ^{31}P NMR spectrum of A was recorded on a Brucker WH 90 instrument. The doublet at 21.47 and 20.18 ppm undoubtedly corresponds to PCl_2 entities and the triplet around 13.07 ppm (14.36, 13.07 and 11.78 ppm) to the P_{spiro} moiety (intensity ratio 2:1). These chemical shifts are compared with those previously reported for

Table 2. ^{31}P NMR data (ppm) for some $N_3P_3Cl_4[HN-(CH_2)_n-NH]$ derivatives

$N_3P_3Cl_4[HN-(CH_2)_n-NH]$	Doublet Centered on	Triplet Centered on	Δ (ppm)	Ref.
n = 2	23.5	22.8	0.7	[12]
n = 3	20.74	7.58	13.16	[3]
n = 4	20.82*	13.07	7.75	This work

other $N_3P_3Cl_4[HN-(CH_2)_n-NH]$ homologues (Table 2). It can be seen that the gap, Δ, between the doublet and the triplet does not parallel the variation in n: Δ is ca. 0 for n = 2, reaches its maximum (about 13.2 ppm) for n = 3 and then decreases again to 7.7 ppm for n = 4. Moreover, some preliminary investigations of higher homologues (n = 5 and 6) indicated at that time that Δ falls to zero for such values of n. No reasonable explanation for such a variation in Δ can be proposed until X-ray structures for the whole series have been performed: crystallography is the only technique which may both support the SPIRO assumption and provide the reasons (geometrical and electronic) for the changes in the ^{31}P chemical shifts.

When the NMR spectrum of A is recorded in a non-polar solvent (CCl_4 or n-hexane), a triplet-quadruplet substructure superimposed in a "comb-like" pattern on the primary doublet-triplet main structure (Fig. 7) appears. Such a triplet-quadruplet secondary pattern implies the existence in solution of an intermolecular (P ... P) coupling, probably through hydrogen bonds as it is the case for $N_3P_3Cl_4[HN-(CH_2)_3-NH]$ [33]. This assumption is supported by the fact that the secondary structure disappears in highly polar solvents.

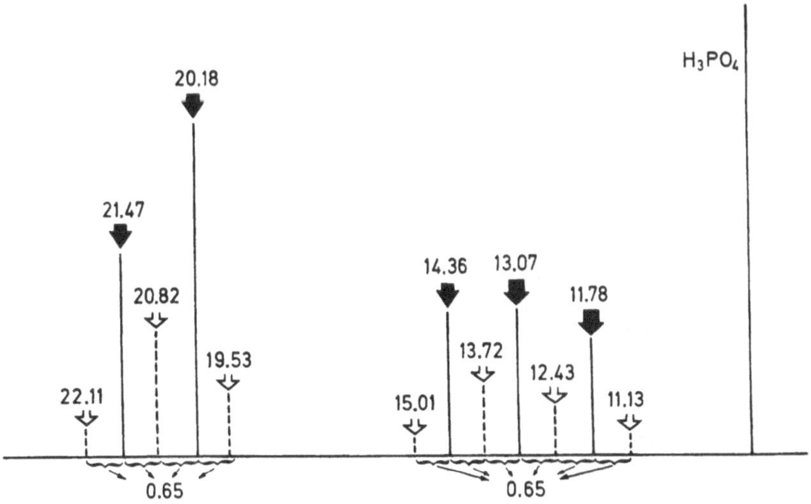

Fig. 7. "Comb-like" triplet-quadruplet substructure of the primary doublet-triplet pattern

3.2.5 Infrared Spetroscopy of (A)

The IR spectrum (KBr disk) was recorded on a Perkin-Elmer 683 spectrometer (range 4000–200 cm^{-1}, calibrated with polystyrene). This spectrum was found to be very similar to that of $N_3P_3Cl_4[HN-(CH_2)_3-NH]$ [30] except in the 3200–3450 cm^{-1} region where A exhibits three (N—H) bands at 3240, 3330 and 3430 cm^{-1} whereas $N_3P_3Cl_4[HN-(CH_2)_3-NH]$ is characterized by a unique band at 3270 cm^{-1}. A thorough investigation of the way in which the intensities and wavelengths of these stretching modes vary with the solvent is now in progress in order

i) to assign the precise nature of the various (N—H) bands and
ii) to measure the acidity (in the Lewis sense) of the corresponding protons.

The latter factor is of real importance in the design of anticancer drugs having a strong affinity for tumor [3]: the higher the acidity, the stronger the affinity of the chemical for tumor and, presumably, the greater its anticancer activity.

3.2.6 X-Ray Crystal and Molecular Structure of (A)

Compound A spontaneously provides suitable single crystals as a result of its synthesis.

The crystal structure of A was determined on an automatic SYNTEX $P2_1$ diffractometer. Data collection required three successive crystals due to the slow decomposition of A on exposure to X-rays. Thus, the R factor was never greater than 0.10. Only approximate bond lengths and bond angles are therefore reported in Tables 3 and 4.

(A) crystallizes in the monoclinic system, space group $P2_1/c$, with cell parameters $a = 10.463(3)$, $b = 17.174(5)$, $c = 16.239(5)$ Å, $\beta = 102.53(3)°$, $V = 2849(2)$ Å^3, $d_x = 1.69$ Mg m^{-3}, $Z = 8$. Two crystallographically independent molecules, A (I)

Table 3. Tentative bond lengths (Å) in the A(I) and A(II) forms of $N_3P_3Cl_4[HN-(CH_2)_4-NH]$

Bond length (Å)	A(I)	A(II)
P_1-N_1	1.61	1.58
P_1-N_{1i}	1.58	1.61
P_1-N_3	1.61	1.62
P_1-N_{3i}	1.62	1.61
P_2-N_1	1.58	1.55
$P_{2i}-N_{1i}$	1.59	1.57
P_2-N_2	1.55	1.55
$P_{2i}-N_2$	1.58	1.58
P_2-Cl_1	2.02	1.99
P_2-Cl_2	1.98	2.00
$P_{2i}-Cl_{1i}$	1.98	2.01
$P_{2i}-Cl_{2i}$	2.00	2.00
C_1-N_3	1.47	1.47
$C_{1i}-N_{3i}$	1.48	1.46
C_1-C_2	1.54	1.51
$C_{1i}-C_{2i}$	1.51	1.54
C_2-C_{2i}	1.52	1.50

Table 4. Tentative bond angles (°) in the A(I) and A(II) forms of $N_3P_3Cl_4[HN-(CH_2)_4-NH]$

Bond Angle (°)	A(I)	A(II)
$N_1-P_1-N_{1i}$	113.5	113.6
$N_3-P_1-N_{3i}$	105.2	104.9
$P_1-N_3-C_1$	120.8	121.9
$P_1-N_{3i}-C_{1i}$	123.7	122.4
$P_1-N_1-P_2$	120.8	121.1
$P_2-N_2-P_{2i}$	118.5	119.4
$N_2-P_{2i}-N_{1i}$	118.0	118.4
$P_1-N_{1i}-P_{2i}$	125.3	123.8
$Cl_1-P_2-Cl_2$	99.2	100.4
$Cl_{1i}-P_{2i}-Cl_{2i}$	99.8	99.1
$N_3-C_1-C_2$	112.7	117.4
$C_1-C_2-C_{2i}$	113.8	117.2
$C_{1i}-C_{2i}-C_2$	116.1	115.5
$C_{1i}-N_{3i}-P_1$	123.7	122.4

and A (II), coexist in the asymmetric unit. A perspective view of the A (I) molecule is shown in Fig. 8 in which the numbering of the atoms is indicated. This view emphasizes the SPIRO structure with the six-membered phosphazene ring strictly non-planar. A second view (with the N_3P_3 ring in the plane) shows significant puckering of the SPIRO loop (Fig. 9) both in A (I) and A (II). The $N_3P_1N_{3i}$ plane is not strictly perpendicular to the phosphazenic ring, the angles between these two planes being equal to 98.0 and 99.2° respectively. In other words, the two exocyclic phosphorus-nitrogen bonds deviate by 8.0 and 9.2° from the expected σ_v forward planes containing the P_1 and N_2 atoms. It is noteworthy, however, that this $N_3P_1N_{3i}$ plane contains atoms C_2 and C_{2i}, carbon atoms C_1 and C_{1i} deviating by ca. 0.70 Å above and below this plane.

The main feature of the structure is observed around P_1: the presence of the loop pulls P_1 away from N_2, endocyclic P_1-N_1 and P_1-N_{1i} bonds being larger (1.595 Å) than the average value (ca. 1.58 Å) of the other phosphorus-nitrogen bonds of the ring. Consequently, the $N_1P_1N_{1i}$ angle is smaller (113.5°) than the value (ca. 119°) observed around P_2 and P_{2i}. In other words, both the $N_1P_1N_{1i}$ and $N_3P_1N_{3i}$ angles are squashed.

As the result of such molecular internal stretch along the P_1-N_2 direction, the P_1 atom appears to be in a pseudo-tetrahedral environmental: $(P_1-N_{exo}) = 1.62(5)$ Å; $(P_1-N_{endo}) = 1.59(5)$ Å; $(N_{exo}-P_1-N_{exo}) = 105.0°$; $(N_{endo}-P_1-N_{endo}) = 113.5°$.

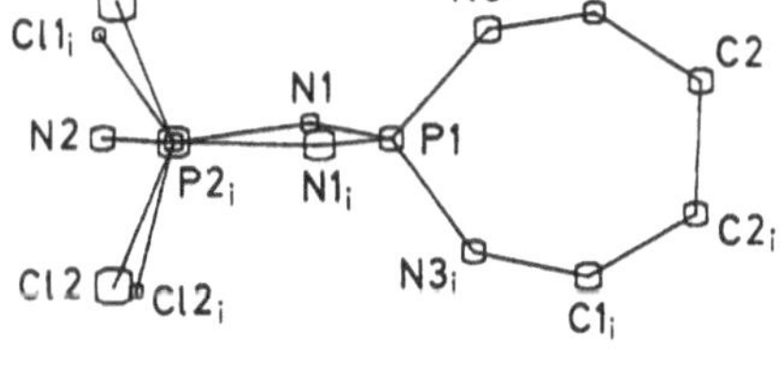

Fig. 8. A perspective view of $N_3P_3Cl_4[HN-(CH_2)_4-NH]$ showing the SPIRO structure

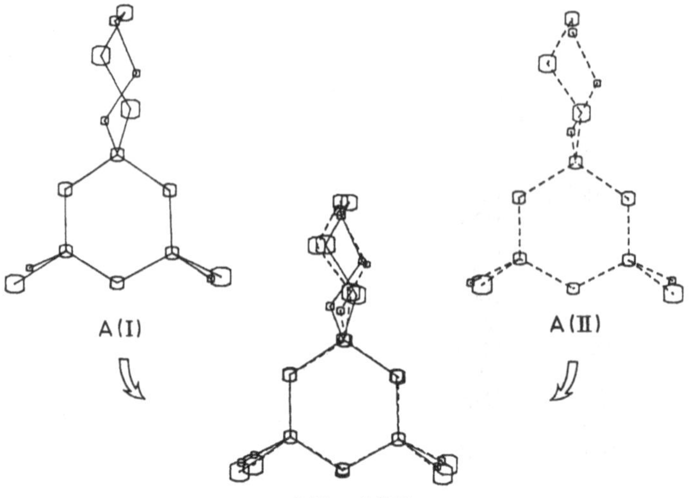

Fig. 9. Perspective views of the A (I) and A (II) forms of $N_3P_3Cl_4[HN-(CH_2)_4-NH]$ showing the puckering of the SPIRO loop

Such a T_d-like environment for a P atom belonging to a trimeric ring is extremely unusual in cyclophosphazenes and may explain the peculiar NMR behaviour manifested by P_1. However, a complete refinement of the X-ray structure of A is necessary to quantify the variations in Δ as a function of n (see above).

The packing of eight molecules of A in the monoclinic unit cell suggests intermolecular hydrogen bonds (as demonstrated by the short distances between some of

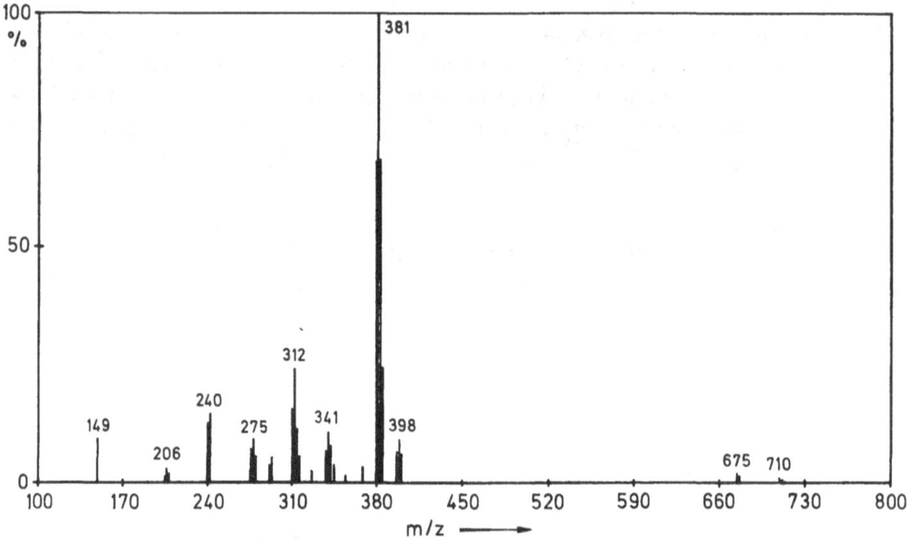

Fig. 10. 70 eV electron impact mass spectrum of compound (B)

the heavy atoms of different molecules). A precise description of these hydrogen bonds awaits a full refinement of the structure, including the location of the hydrogen atoms.

3.2.7 Mass Spectrometry of (B)

The 70 eV electron impact mass spectrum of (B) is shown in Fig. 10. The molecular ion was observed at m/z 710 (1.0%) with a set of very weak satellites at m/z 711 (0.4%), 712 (0.7%), 713 (0.2%) and 714 (0.3%). The intensity ratio of these five peaks indicates the presence of more than four chlorine atoms in the molecule. The base peak at m/z 381 (100%) with satellites at m/z 379 (68.2%), 380 (22.0%), 382 (10.6%), 383 (68.7%), 384 (7.1%) and 385 (24.3%) corresponds to the $N_3P_3Cl_5[HN-(CH_2)_4]$ entity (associated with H-transfers and $^{35}Cl/^{37}Cl$ isotopic satellites) which shows that the molecule is built up from two symmetric halves. Consecutive losses of 1 CH_2, 2 CH_2, 3 CH_2, 4 CH_2 and 4 CH_2 + 1 NH from this fragment give peaks at m/z 369 (3.3%), 355 (1.4%), 341 (10.5%), 327 (2.7%) and 312 (24.0%), also associated with H-transfers. Further consecutive loss of three chlorine atoms from the m/z 312 fragment give peaks at 277 (3.7%), 240 (12.5%) and 206 (2.8%).

A second fragmentation route is observed; starting from the base peak (m/z 381), peaks are observed at m/z 346 (3.8%), 310 (15.9%), 275 (2.3%), 239 (1.3%) and 204 (1.2%) which correspond to the successive loss of one to five chlorine atoms (again associated with H transfers). It is noteworthy that there is some overlap between peaks in the two fragmentation pathways resulting from proximity of masses of two chlorine atoms (70) and of the $HN-(CH_2)_4$ block (71).

Suprisingly, peaks corresponding to the successive loss of one to ten chlorine atoms from the molecular ion M^+ are not observed at all except ā m/z 675 (2.1%) (M^+ − 1 Cl) and 567 (1.2%) (M^+ − 4 Cl).

Peaks at m/z 396 (6.5%), 397 (1.3%), 398 (9.4%), 399 (1.2%) and 400 (6.0%) correspond to the $N_3P_3Cl_5[HN-(CH_2)_4-NH]$ fragment which looses either three (m/z 293, 5.3%) or five (m/z 221, 1.6%) chlorine atoms.

In the low mass range, only one peak is observed at m/z 149 (9.2%) which corresponds to the N_3P_3N fragment. Thus, the N_3P_3 ring remains stable as long as at least one ligand stays linked to it. Indeed the peak at m/z 135 (N_3P_3) is not found here in contrast with what happens in other aminocyclophosphazenes [31].

The intensity of all other peaks is less than 1%. Thus, it is a relatively simple matter to check for the presence of any possible contaminant. Mass spectrometry again appears to be an adequate tool for monitoring the purity of a cyclophosphazene.

3.2.8 Infrared Spectroscopy of (B)

The IR spectrum (KBr disk) was recorded on a Perkin-Elmer 683 spectrometer (range 4000–200 cm^{-1}, calibrated with polystyrene). This spectrum exhibits a unique N—H band at 3278 cm^{-1} as in $N_3P_3Cl_4[HN-(CH_2)_3-NH]$ (3270 cm^{-1}, [30]). According to the electronic structure of the latter, calculated by the SCF-LCAO-MO within the CNDO/2 approximation [36], it may be concluded that protons on N atoms in the bridge are highly acidic in the Lewis sense. This acidity is of real importance in the design of anticancer drugs [3] (see above).

Table 5. Molecular parameters (with e.s.d.'s in parenthesis) for compound (B)

bond lengths		bond angles	
N(4)—P(1)	1.571 (0.008)	N(6)—P(1)—N(4)	119.4 (0.4)
N(6)—P(1)	1.546 (0.007)	CL10—P(1)—N(4)	108.3 (0.3)
CL10—P(1)	1.989 (0.004)	CL10—P(1)—N(6)	108.9 (0.4)
CL11—P(1)	1.986 (0.004)	CL11—P(1)—N(4)	108.0 (0.4)
N(4)—P(2)	1.576 (0.008)	CL11—P(1)—N(6)	109.9 (0.4)
N(5)—P(2)	1.548 (0.007)	CL11—P(1)—CL10	100.8 (0.2)
CL12—P(2)	1.988 (0.003)	N(5)—P(2)—N(4)	119.1 (0.4)
CL13—P(2)	1.983 (0.004)	CL12—P(2)—N(4)	108.9 (0.3)
N(5)—P(3)	1.604 (0.007)	CL12—P(2)—N(5)	109.4 (0.3)
N(6)—P(3)	1.593 (0.007)	CL13—P(2)—N(4)	109.0 (0.3)
N(7)—P(3)	1.605 (0.008)	CL13—P(2)—N(5)	108.6 (0.3)
CL14—P(3)	2.013 (0.004)	CL13—P(2) CL12	100.2 (0.2)
C(8)—N(7)	1.474 (0.011)	N(6)—P(3)—N(5)	117.2 (0.4)
C(9)—C(8)	1.561 (0.016)	N(7)—P(3)—N(5)	108.0 (0.4)
C(9)—C(9)	1.479 (0.021)	N(7)—P(3)—N(6)	110.6 (0.4)
		CL14—P(3)—N(5)	106.9 (0.3)
		CL14—P(3)—N(6)	106.2 (0.4)
		CL14—P(3)—N(7)	107.6 (0.3)
		P(2)—N(4)—P(1)	121.0 (0.5)
		P(3)—N(5)—P(2)	121.4 (0.5)
		P(3)—N(6)—P(1)	121.7 (0.4)
		C(8)—N(7)—P(3)	121.3 (0.6)
		C(9)—C(8)—N(7)	105.6 (0.8)

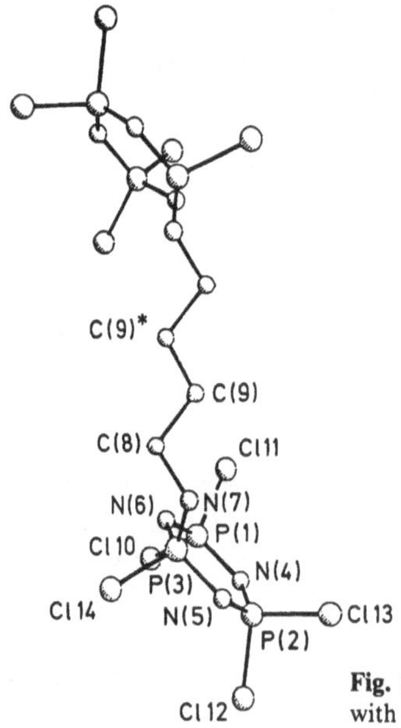

Fig. 11. A perspective view of $N_3P_3Cl_5[HN-(CH_2)_4-NH]Cl_5P_3N_3$ with numbering of atoms (half molecule)

3.2.9 X-Ray Crystal and Molecular structure of (B)

The crystal structure of (B) was measured on an automatic SYNTEX $P2_1$ diffractometer. The R factor was 0.05 and bond lengths and bond angles are reported in Table 5.

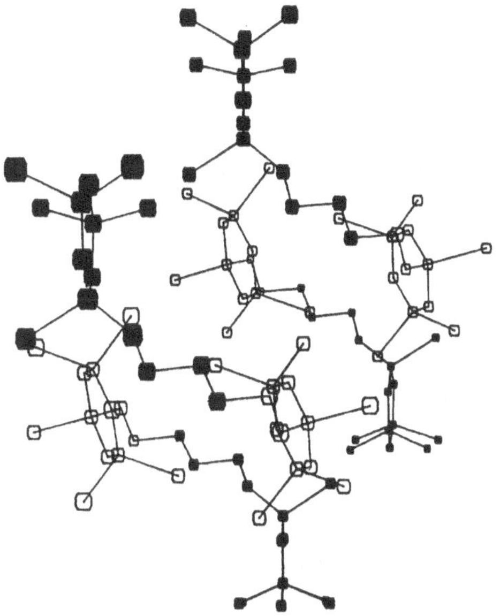

Fig. 12. A perspective view of the four molecules in the unit cell showing the relative situation of N_3P_3 rings

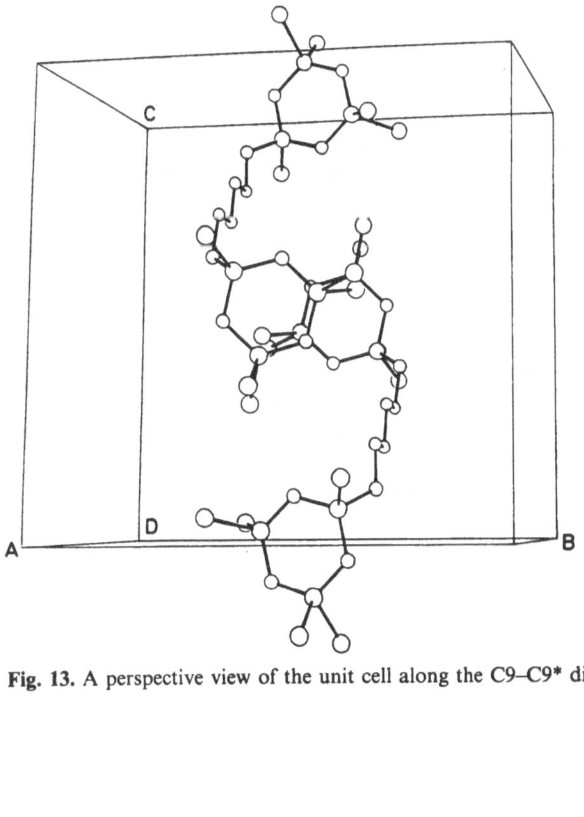

Fig. 13. A perspective view of the unit cell along the C9–C9* direction

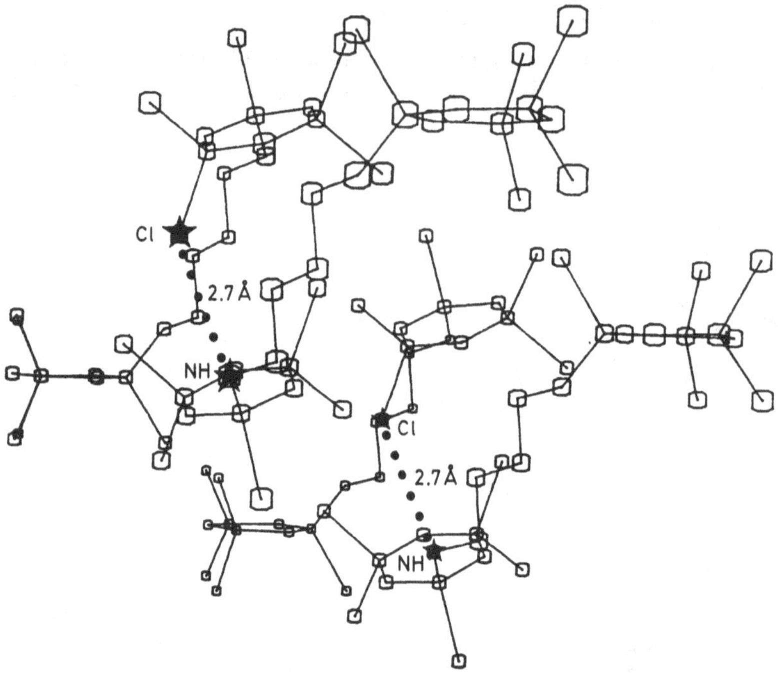

Fig. 14. Visualization of intermolecular hydrogen bonds responsible for crystal stacking

(B) crystallizes in the monoclinic system, space group C2/c, with cell parameters a = 11.746(3), b = 15.144(5), c = 14.708(5) Å, β = 94.64(3)°, V = 2608(1) $Å^3$, d_x = 1.809 Mg m^{-3}, d_m = 1.82 Mg m^{-3}, Z = 4.

A perspective view of the molecule is shown in Fig. 11 in which the numbering of the atoms is indicated. This view emphasizes the two-ring bridged structure with the two six-membered phosphazene rings strictly planar. A second view (with one N_3P_3 ring perpendicular to the plane of the Fig.) shows that the two N_3P_3 planes within the molecule are not parallel, the angle between them being equal to 23.6(1)° (Fig. 12). A third view along the C_9—C_9^* direction visualizes the relative spatial arrangement of N_3P_3 rings both in every molecule and in the unit cell (Fig. 13).

A thorough investigation of intermolecular distances shows that the stacking in the crystal is due to significant hydrogen bonds between some chlorine atom of molecule 1 and exocyclic N(H) atoms of molecule 2 (Fig. 14). Such an (N ... Cl) distance is indeed equal to 2.7 Å, to be compared with the value of 3.144 Å observed in the $N_3P_3(NC_2H_4)_6 \cdot 3\ CCl_4$ anticlathrate structure [42].

Evidently, the hydrogen bonds present in the solid break up in solution, as demonstrated by IR spectroscopy (a unique N—H band, see above).

3.2.10 Conclusion

The reaction of $N_3P_3Cl_6$ with 1,4-diaminobutane (1:1) leads to a mixture of the SPIRO $N_3P_3Cl_4[HN-(CH_2)_4-NH]$ compound (A) as the major product and of a minor product (B) which has a two-ring assembly structure in which the $N_3P_3Cl_5$ moieties are bridged through a [HN—$(CH_2)_4$—NH] entity. This new structure will be coded as the BINO configuration. Thus, 1,4-diaminobutane does not behave like 1,3-diaminopropane and 1,2-diaminoethane where the SPIRO product was obtained in a pure state. We shall see in the next paragraph that the yield of (B) increases markedly with the size of the diamine at the expense of the SPIRO product.

3.3 The Case of the 1,5-Diaminopentane (Cadaverine) and of its higher Homologues [43]

3.3.1 Synthesis

Reactions of several diamines, H_2N—$(CH_2)_n$—NH_2 with n varying from 5 to 10, with $N_3P_3Cl_6$ (1 : 1) were achieved following Shaw's method [27] in the presence of NEt_3. Contrarily to what we mentioned above about the reaction of 1,4-diaminobutane with $N_3P_3Cl_6$, the use of polar solvents like diethyl ether does not favour here the formation of non-crystalline resins. Cadaverine and higher cousins then lead to final products whose PNHR groups seem to be poorly sensitive to the proton abstraction process responsible for polymerization. The yield of reactions was about 60% for n = 5 to 10. Thin layer chromatography gave a unique spot for n = 6 to 10, Rf = 0.78 with Et_2O—CCl_4 (1:9) as eluant. In contrast, two spots are observed for n = 5, the major one at Rf = 0.78 and the second one, hardly imperceiveable, at Rf = 0.65 with CCl_4—CH_2Cl_2 (3:7) as eluant. In that case, SiO_2 column chromatography with CCl_4—CH_2Cl_2 (3:7) as eluant allowed to get the major product (Rf = 0.78) in a very pure state but we could not recover enough of its by-product (Rf = 0.65) to assign its structure. It may be assumed reasonably however that the Rf = 0.65 by-product for n = 5 is the SPIRO $N_3P_3Cl_4$[HN—$(CH_2)_5$—NH] derivative, owing to the similarity of its Rf value with the one obtained above for the SPIRO—$N_3P_3Cl_4$ [HN—$(CH_2)_4$—NH].

In other words, the relative yield in SPIRO and BINO final products passes bluntely from 90%–10% for n = 4 to traces-100% for n = 5 and to 100% for n = 6 to 10. The n = 5 value then corresponds to a boarder between SPIRO and BINO areas.

Melting points of the whole series of BINO derivatives are gathered in Table 6. BINO *n* means $N_3P_3Cl_5$[HN—$(CH_2)_n$—NH]$Cl_5P_3N_3$. It is noteworthy that melting point varies monotonously within the two odd and even sub-series but in a serrated manner in the whole series.

Table 6. Melting point of BINO compounds

Compound	Melting Point (°C)
BINO 4	148
BINO 5	108
BINO 6	98
BINO 7	64
BINO 8	80
BINO 9	65
BINO 10	74

3.3.2 Mass Spectrometry of BINO Compounds

Mass spectra of BINO 4 to BINO 10 are gathered in Figs. 15 to 21. They are remarkably straightforward and every peak can be assigned in a very facile way.

In the high mass range, i.e. between m/z 500 and M^+ peaks, very few fragments (with very low intensity except in BINO 6) are observed which correspond to the successive loss of 1 to 5 Cl atoms from M^+. In every case, the $M^+ - 1$ Cl peak is the highest within the series.

In the low mass range, i.e. between m/z 100 and m/z 350, 6 features are common to the whole spectra: m/z 135 (N_3P_3), m/z 242 ($N_3P_3Cl_3$), m/z 277 ($N_3P_3Cl_4$), m/z 293 ($N_3P_3Cl_4{-}NH$), m/z 312 ($N_3P_3Cl_5$) and m/z 341 ($N_3P_3Cl_5{-}NH{-}CH_2$).

It is noteworthy that the intensity of the m/z 312 peak grows up from BINO 4 (24%) to BINO 5 (56%) and to BINOs 6 to 10 in which this peak becomes the base peak (I = 100%).

Intensities of the m/z 341 peak within the series are: 10% for BINO 4, 53% for BINO 5, 24% for BINO 6, 41% for BINO 7, 60% for BINO 8, 21% for BINO 9

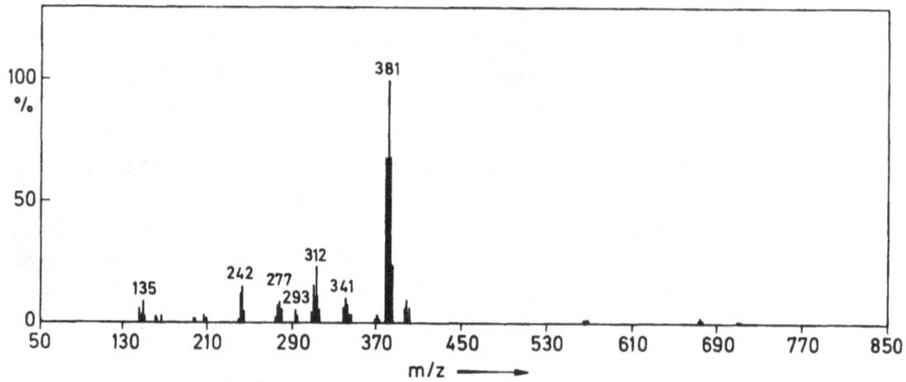

Fig. 15. 70 eV electron impact mass spectrum of BINO4

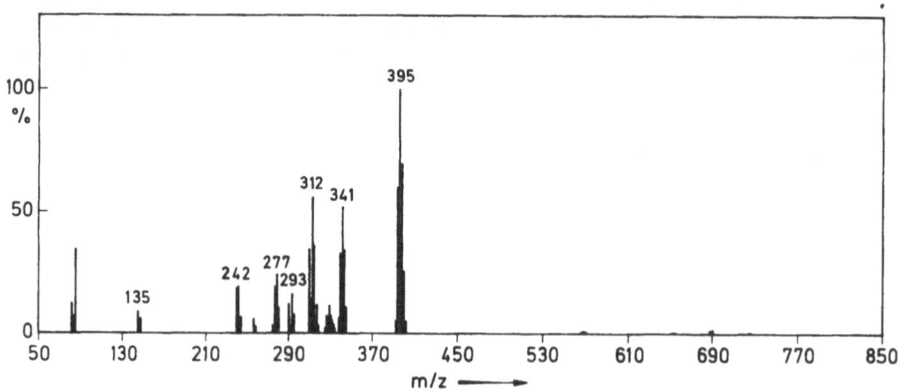

Fig. 16. 70 eV electron impact mass spectrum of BINO5

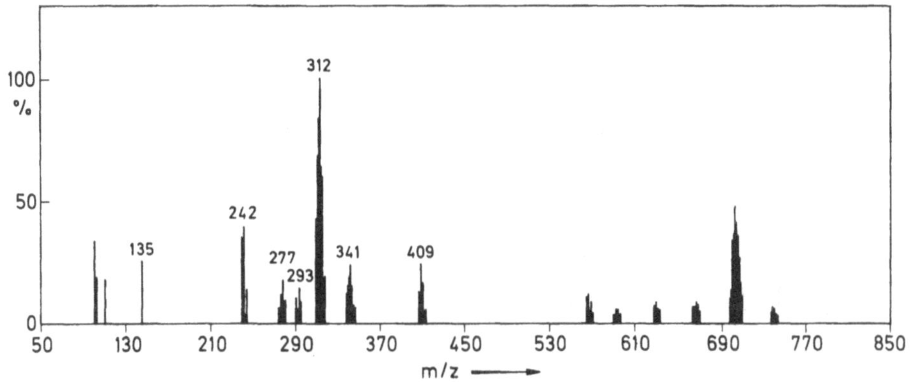

Fig. 17. 70 eV electron impact mass spectrum of BINO6

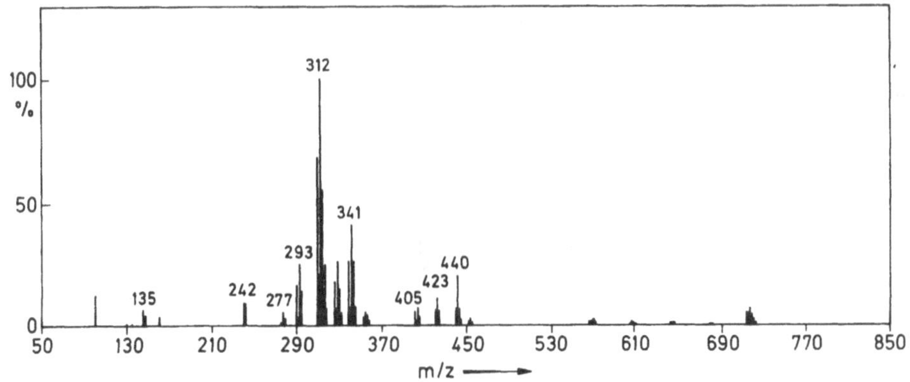

Fig. 18. 70 eV electron impact mass spectrum of BINO7

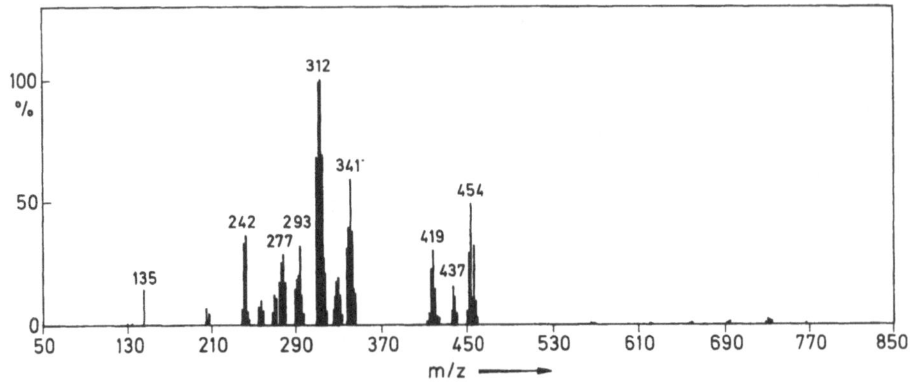

Fig. 19. 70 eV electron impact mass spectrum of BINO8

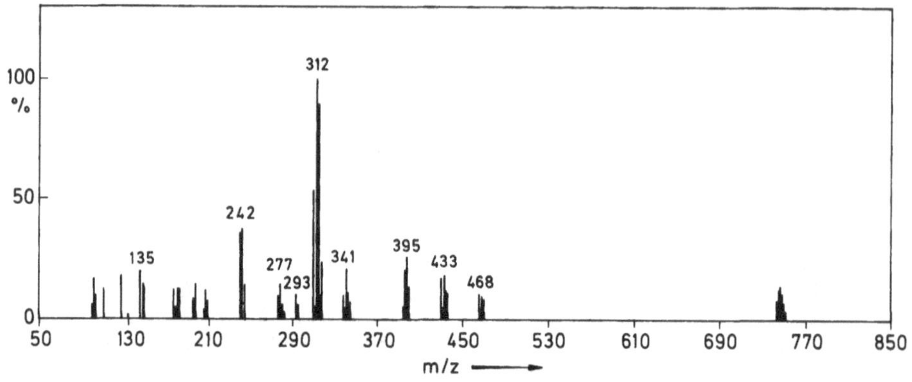

Fig. 20. 70 eV electron impact mass spectrum of BINO9

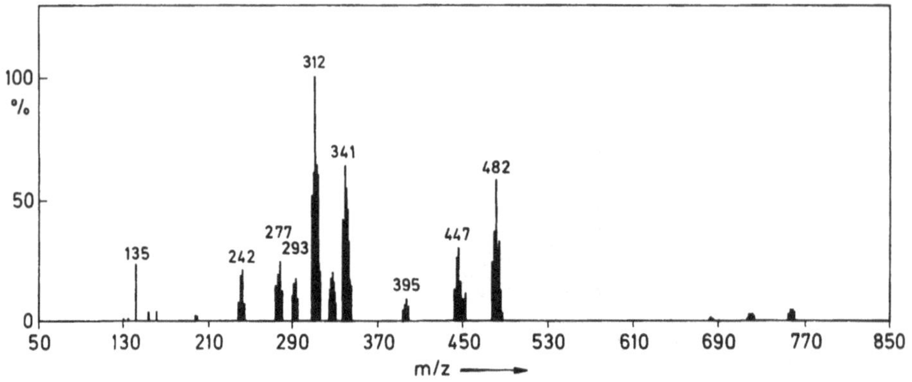

Fig. 21. 70 eV electron impact mass spectrum of BINO10

and 64% for BINO 10. Thus, the intensity of the m/z 341 peak increases with *n* within the odd sub-series and decreases within the even one.

The four other peaks are of minor importance and their intensities within the series fluctuate in a rather random manner. However, the m/z 135 signal, which characterizes the N_3P_3 cyclophosphazene ring itself, is ever observed as the outcome of the fragmentation: this observation, quite trivial in cyclophosphazenes, supports once more the non-aromatic Dewar islands electronic structure for phosphorus nitrogen rings [37].

In the medium mass range, i.e. between m/z 350 and m/z 500, 3 main peaks are observed:

1. the one, coded as F1, corresponding to the $N_3P_3Cl_5$—NH—$(CH_2)_n$ which is the base peak (m/z 381 and m/z 395) for BINO 4 and BINO 5; its intensity decreases sharply from BINO 6 (m/z 409, 25%) to BINO 8 (m/z 437, 16%) and becomes rather negligible in BINO 9 (m/z 451, 3%) and BINO 10 (m/z 465, 5%);
2. the one, coded as F2, corresponding to F1—1 Cl, which appears only for BINOs 7 to 10 (m/z 405 (8%), 419 (30%), 433 (18%) and 447 (30%) respectively);

3. the one, coded as F3, corresponding to F1 + 1 NH, which appears too only for BINOs 7 to 10 (m/z 440 (20%), 454 (50%), 468 (10%) and 482 (59%) respectively).

Owing to the remarkable simplicity of their mass spectra and to the trends we just detailed about the magnitude of main peaks, some conclusions may be reasonably drawn about the relative fragility of chemical bonds in BINOs structures when the diamino bridge is lengthening.

a) The rupture of the carbon-carbon bond in β position of NH groups occurs in the whole series (considering the m/z 341 fragment) with a propensity to increase with n in odd terms and to decrease with n in even terms;
b) the likelihood of the rupture of the nitrogen-carbon bond in α position of NH groups is maximal for BINO 4 and BINO 5 (considering the F1 fragment) and drops down dramatically till BINO 10;
c) the likelihood of the rupture of the exocyclic phosphorus-nitrogen bonds (considering the m/z 312 and the F3 fragments) increases with n, prevailing definitely on any other bond breaking from BINO 7;
d) phosphorus-chlorine bonds are intrinsically more stable than exocyclic phosphorus-nitrogen bonds upon electron impact, in contrast with what exists in SPIRO [30,40] derivatives.

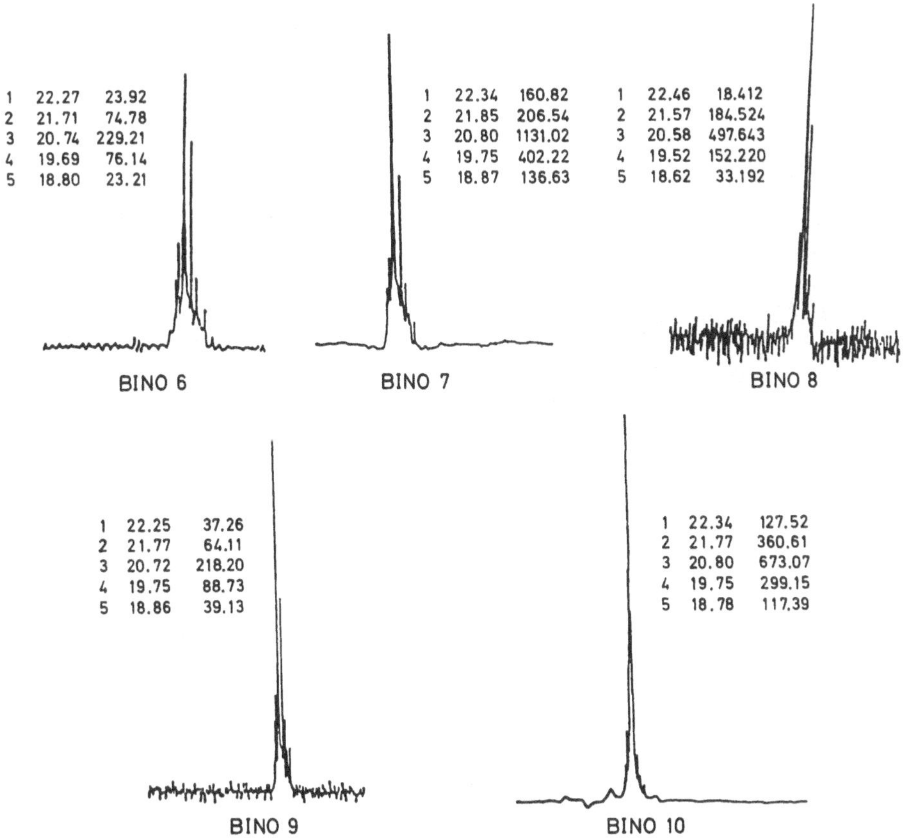

Fig. 22. ^{31}P NMR spectra (WH 90) of BINO compounds

3.3.3 NMR Spectroscopy of BINO Compounds

^{31}P data

The ^{31}P NMR spectra of any BINO term, as recorded on a Brucker WH 90 instrument (36.43 MHz) is a complex multiplet ca. 20–21 ppm in CD_2Cl_2 with H_3PO_4 85% as a standard. Such multiplets give almost first-order A_2B spectra when recorded on a Brucker WH 250 instrument (101.27 MHz) and pure A_2B spectra from a Brucker WH 400 instrument (162.08 MHz).

Spectra for BINO 6 to 10 are gathered in Figs. 22 (36.43 MHz) and 23 (101.27 MHz). A comparison of spectra of BINO 6 recorded at 36.43, 101.27, 162.08 MHz, together with the simulated spectrum at 162.08 MHz, is visualized in Fig. 24.

Chemical shifts at 101.27 MHz (*external standard*) are higher by 1.36 ± 0.07 ppm than those recorded at 36.43 MHz (*internal standard*). Table 7 gathers real chemical shifts values in internal standard conditions together with J(P—P) coupling constants. Spectra are nearly superimposable, J(P—P) being constant along the series, 47.27 ± 0.3 Hz.

^{13}C data

^{13}C spectra from Brucker WH 250 instrument (62.90 MHz) for BINOs 6 to 10 are visualized in Fig. 25.

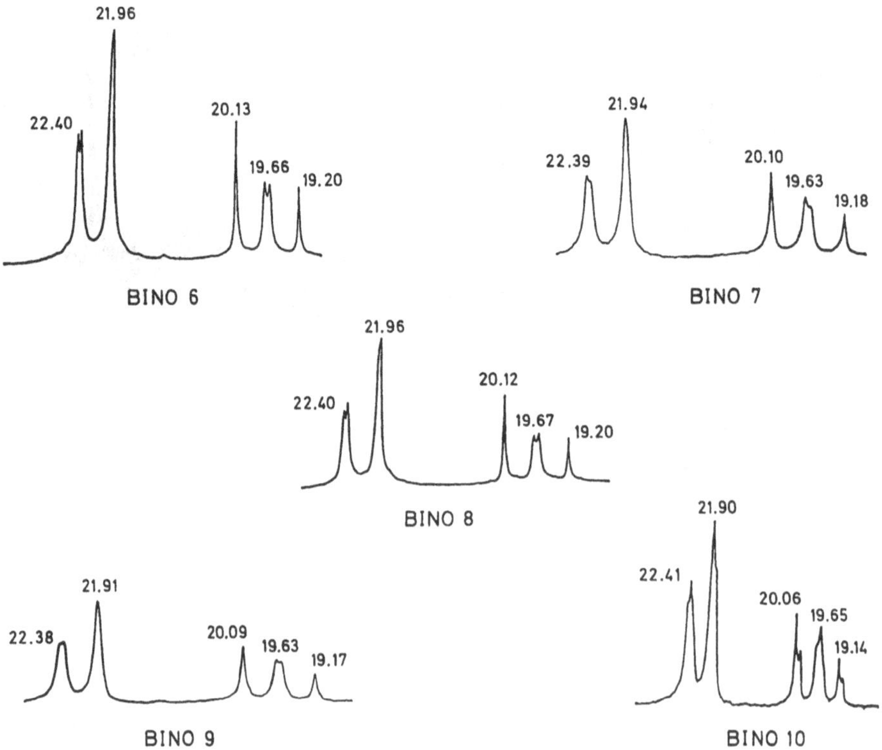

Fig. 23. ^{31}P NMR spectra (WH 250) of BINO compounds

Table 7. ^{31}P chemical shifts and J(P—P) coupling constants of BINO compounds

Compound	WH 90 δ (ppm)	WH 250 δ (ppm)	J(P—P) (Hz)
BINO 6	22.27	22.40	47.12
	21.71	21.96	
	20.74	20.13	
	19.69	19.66	
	18.80	19.20	
BINO 7	22.34	22.39	47.61
	21.85	21.94	
	20.80	20.10	
	19.75	19.63	
	18.85	19.18	
BINO 8	22.46	22.40	47.12
	21.57	21.96	
	20.58	20.12	
	19.52	19.67	
	18.62	19.20	
BINO 9	22.25	22.38	47.18
	21.77	21.91	
	20.72	20.09	
	19.75	19.63	
	18.86	19.17	
BINO 10	22.34	22.41	47.35
	21.77	21.90	
	20.80	20.06	
	19.75	19.65	
	18.78	19.14	

Table 8. ^{13}C chemical shifts (WH 250) of BINO compounds

	δC_{11}	$\delta C_{22'}$	$\delta C_{33'}$	$\delta C_{44'}$	$\delta C_{55'}$
BINO 6	41.28	30.71	26.20		
		30.61			
BINO 7	41.34	30.75	26.55	28.81	
		30.63			
BINO 8	41.34	30.79	26.55	29.14	
		30.64			
BINO 9	41.40	30.81	26.61	29.19	29.52
		30.69			
BINO 10	41.43	30.84	26.64	29.25	29.58
		30.70			

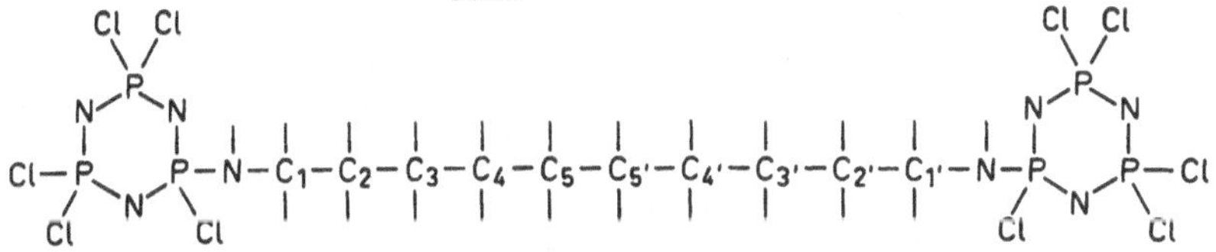

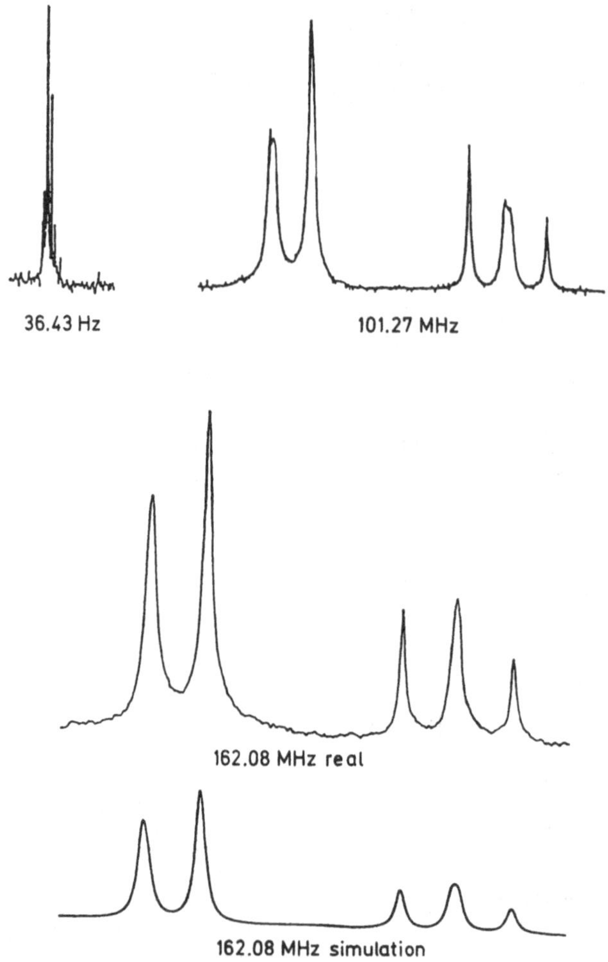

Fig. 24. ^{31}P NMR spectra of BINO6 recorded at 36.43, 101.27, 162.08 MHz and simulated spectrum at 162.08 MHz

A step by step analysis of these spectra provides the assignement of Table 8:

i) for every carbon atom of the bridge, the corresponding chemical shift slightly increases with n: $\delta C_{11'}$, as an example, varies from 41.28 ppm for BINO 6 to 41.43 ppm for BINO 10;

ii) C_1 and $C_{1'}$ atoms, in α position of NH groups, are down-field shifted versus other carbon atoms of the bridge;

iii) $\delta C_{nn'}$ steadily decreases from $C_{11'}$ to $C_{33'}$, increasing again from $C_{33'}$ to $C_{55'}$. This would suggest a lenghtening of C_4—$C_{4'}$ (in BINO 8) and C_5—$C_{5'}$ (in BINO 10) versus other carbon-carbon bonds of the chain. This assumption will have to be supported by X-ray structures but it is verified yet from X-ray structure of $N_3P_3Cl_4[HN-(CH_2)_3-N](CH_2)_4[N-(CH_2)_3-NH]Cl_4P_3N_3$ (see be-

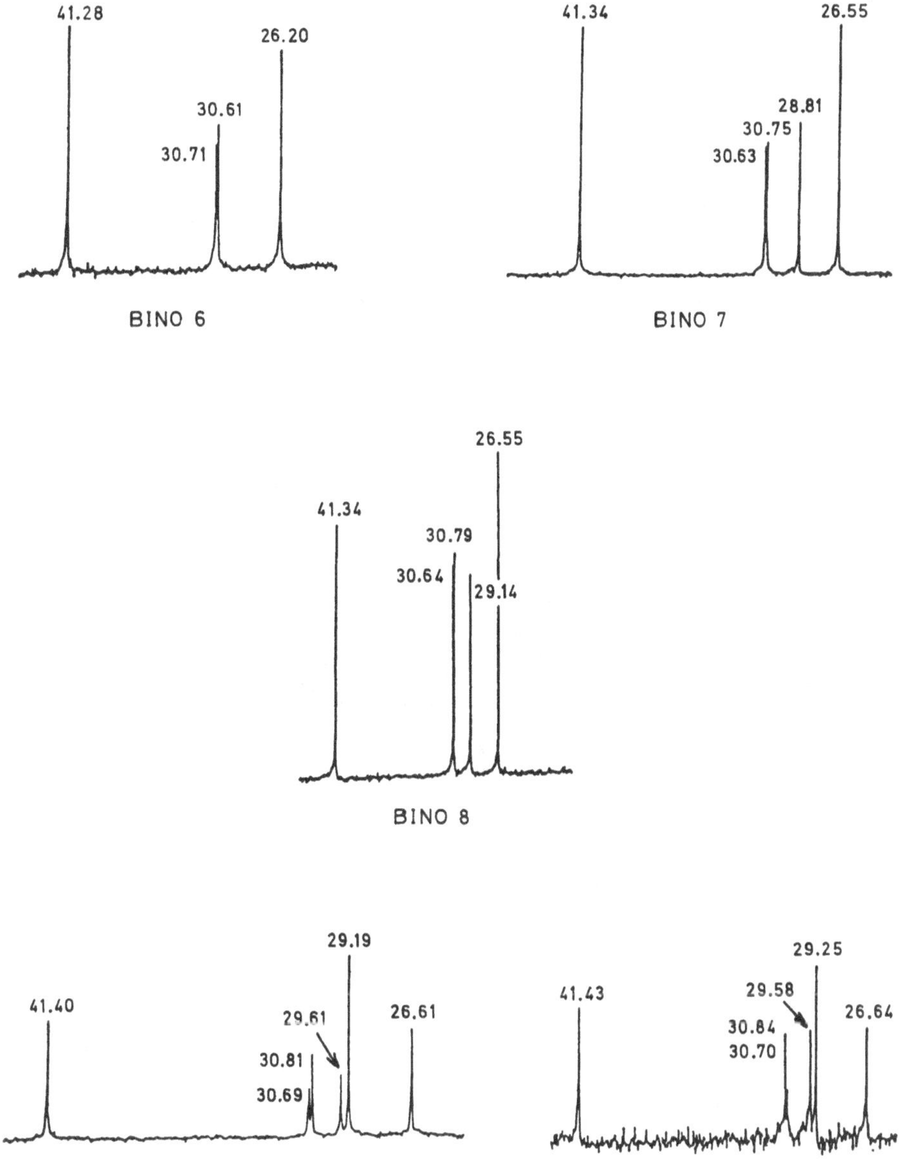

Fig. 25. ^{13}C NMR spectra (WH 250) of BINO compounds

low) where the central "C_5—$C_{5'}$-like" bond is noticeably longer (1.581 Å) than other C—C bonds in the $(CH_2)_4$ bridge (1.529 Å).

^1H data

^1H data (250 MHz) are gathered in Fig. 26 and Table 9. $\delta H_{nn'}$ is quite constant for a given nn′ couple and decreases gently when *n* varies from 1 to 3 to keep a

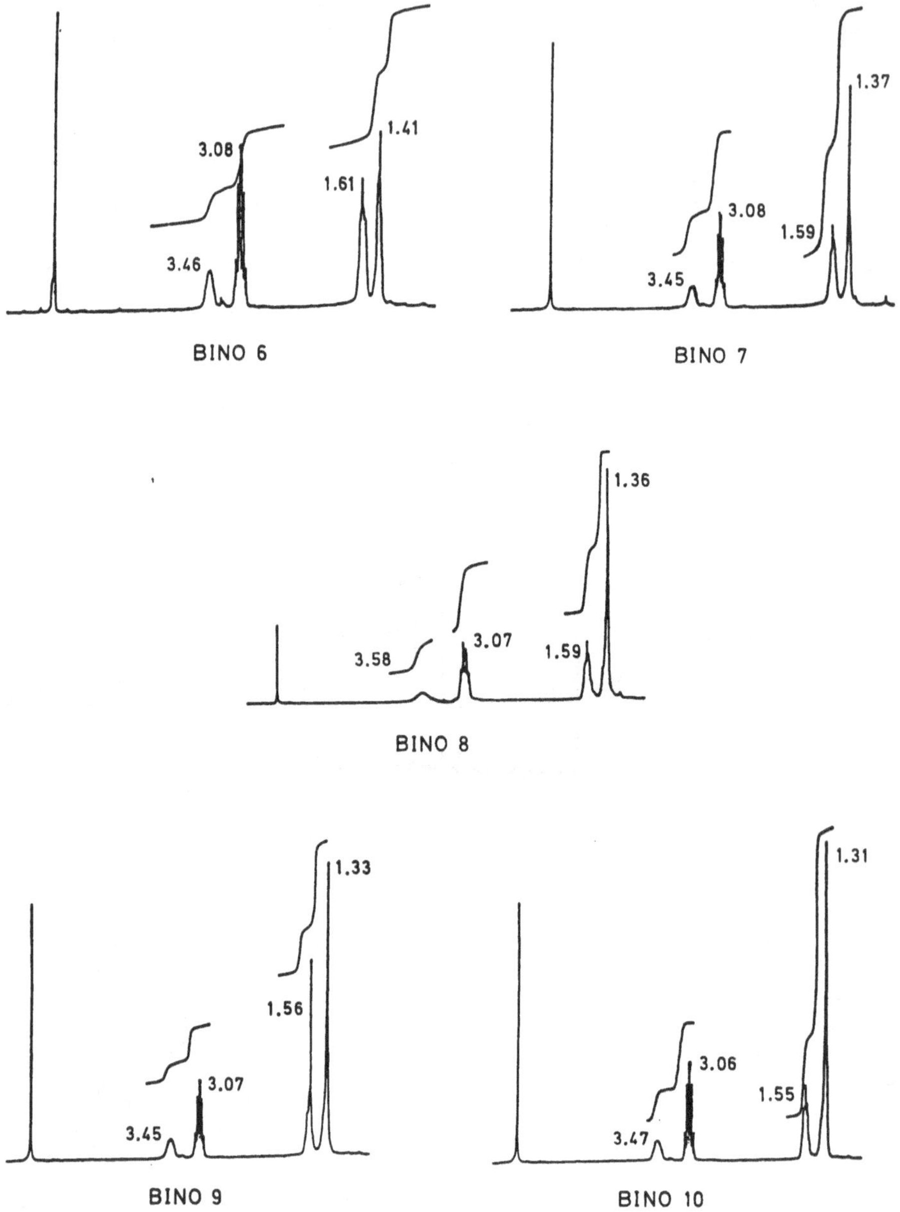

Fig. 26. ^{1}H NMR spectra (WH 250) of BINO compounds

constant value for *n* larger than 3. That means that the shortest C—H bonds in the bridge are linked to the most central carbon atoms (like C_4 and $C_{4'}$ for BINO 8, C_5 and $C_{5'}$ for BINO 10) of the chain. This conclusion will have again to be backed up by X-ray analyses now in progress [44].

Table 9. ^1H chemical shifts (WH 250) of BINO compounds

	δ(N)H	$\delta H_{11'}$	$\delta H_{22'}$	$\delta H_{33'}$	$\delta H_{44'}$	$\delta H_{55'}$
BINO 6	3.46	3.08	1.61	1.41		
BINO 7	3.45	3.08	1.59	1.37	1.37	
BINO 8	3.58	3.07	1.59	1.36	1.36	
BINO 9	3.45	3.07	1.56	1.33	1.33	1.33
BINO 10	3.47	3.06	1.55	1.31	1.31	1.31

3.3.4 Conclusion

The reaction of $N_3P_3Cl_6$ with 1,5-diaminopentane and higher cousins (1:1) leads to pure BINO $N_3P_3Cl_5[HN-(CH_2)_n-NH]Cl_5P_3N_3$ ($n \geqq 5$) compounds, the corresponding SPIRO $N_3P_3Cl_4[HN-(CH_2)_n-NH]$ relatives being never observed except as traces for n = 5. Thus, this n = 5 value constitutes a clear boarder between "preferred" SPIRO ($n \leqq 4$) and "preferred" BINO ($n \geqq 5$) areas.

In conclusion, reaction of $N_3P_3Cl_6$ with natural diamines appears highly stereoselective (SPIRO or BINO configuration) and we shall see below that such a stereoselectivity is maintained when $N_3P_3Cl_6$ reacts with natural polyamines.

3.4 The Case of Spermidine [45)]

3.4.1 Synthesis

Experimental conditions for the reaction of $N_3P_3Cl_6$ with spermidine (1:1) are detailed here, in order to point out some specific difficulties which arose in the synthetic process.

31.5 mmole of spermidine in solution in 50 ml of a 3:5 mixture of Et_2O and n-hexane were added dropwise to a solution of 28.73 mmole of $N_3P_3Cl_6$ and 94.7 mmole of NEt_3 in 800 ml of the same mixture of solvents. The medium was stirred under argon pressure at room temperature for 4 days. Hydrochlorides were filtered off and solvents removed in vacuo at 30 °C to give 19.25 g (91.6%) of a colourless oil.

This residue was submitted to SiO_2 column chromatography using CH_2Cl_2 as eluant. This process allowed elimination of the remaining traces of chlorhydrates and yielded 12 g (60%) of a colourless oil giving one t.l.c. spot at Rf = 0.49 with CH_2Cl_2 as eluant. This oil when dissolved in n-hexane gave upon slow evaporation prismatic single crystals which were quite convenient for X-ray study: m.p. 102.5 °C.

Analytical data: C, 11.40%; H, 2.20%; N, 17.16%; P, 25.50% and Cl, 43.45%. They are consistent with the crude structure $N_3P_3Cl_4$ (spermidine minus 3 H) $Cl_5P_3N_3$ (mol. wt. = 731.186). In other words, elemental analysis would confer to the final product a SPIROBINO configuration in which the (NH_2, NH) couple in $H_2N-(CH_2)_3-NH-(CH_2)_4-NH_2$ would link one $N_3P_3Cl_4$ moiety in a SPIRO configuration when the last NH_2 group would link one $N_3P_3Cl_5$ moiety.

Incidentally, the oily aspect of the final product (coded: SPD) from solution in CH_2Cl_2 is due to the existence in the liquid state of a clathrate with CH_2Cl_2. Thus, it was shown on this occasion that cyclophosphazenes, which so often give clathrates[46,47] and anticlathrates[42] in the solid state, can also generate interactions with solvents in the liquid state which are strong enough to drastically modify the actual physical state of the compounds in question.

3.4.2 Mass Spectrometry of SPD

The mass spectrum of the final crystallized product was recorded on the R1010 mass spectrometer.

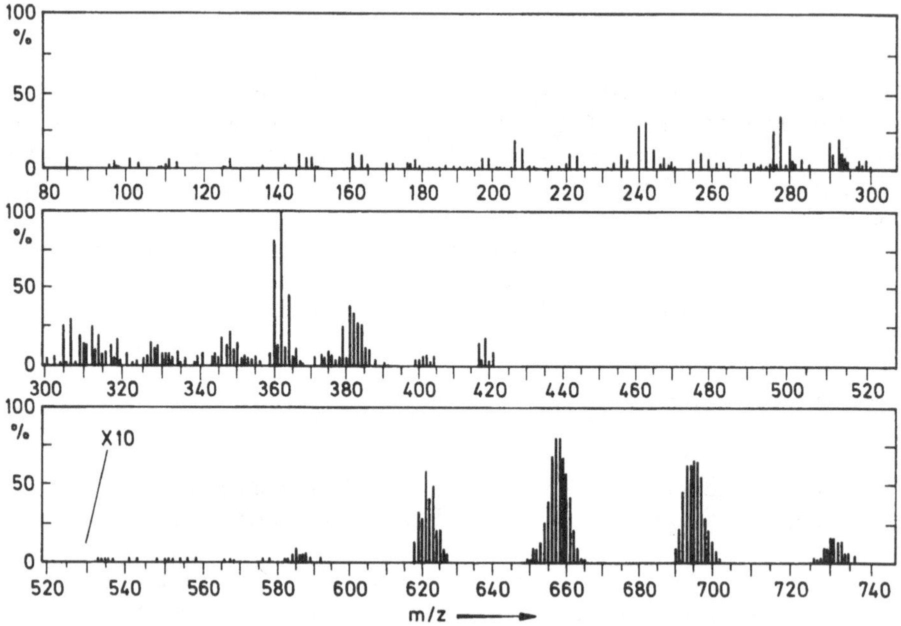

Fig. 27. 70 eV electron impact mass spectrum of SPD

The 70 eV electron impact mass spectrum is shown in Fig. 27. The molecular ion M^+ was observed, as expected, at m/z 731 (1.6%) with 5 satellites at m/z 732 (1.3%), 733 (1.3%); 734 (0.6%), 735 (0.6%) and 736 (0.3%).

The base peak at m/z 362 (100%) corresponds to the $N_3P_3Cl_4-HN-(CH_2)_3-N-CH_2$ fragment (F1). The complementary fragment F2, m/z 731–362 = 369, gives a peak of very weak intensity (1.6%).

One main fragmentation route for F1 involves successive loss of (i) 1 to 4 Cl and (ii) then of 1 CH_2, 2 CH_2, 3 CH_2, 4 CH_2, 4 CH_2 + 1 NH and 4 CH_2 + 2 NH (associated with H-transfers) to give maximal peaks at m/z 327 (14.0%), 291 (11.1%), 256 (10.4%), 220 (9.8%), 206 (18.3%), 192 (1.9%), 178 (5.5%), 164 (2.9%), 149 (7.5%) and 135 (2.6%).

A similar fragmentation route is observed for the (F1−1 CH_2) ion, m/z 348 (21.5%). This ion may lose indeed either (i) from 1 to 4 Cl to give peaks at m/z 313 (25.1%), 277 (34.6%), 242 (29.7%) and 206 (18.3%), or (ii) firstly 1 CH_2,

2 CH_2, 3 CH_2, 3 CH_2 + 1 NH, 3 CH_2 + 2 NH and then from 1 to 4 Cl to give peaks at 332 (9.1%), 318 (16.9%), 304 (25.8%), 289 (2.6%), 275 (24.1%), 240 (28.7%), 204 (1.6%), 169 (0.9%) and 135 (2.6%).

Similarly, the complementary fragment of (F1—1 CH_2), i.e. F3 (m/z 383, 2.77%), loses successively 1 CH_2, 2 CH_2, 3 CH_2, 4 CH_2, 4 CH_2 + 1 NH and then from 1 to 5 Cl to give peaks at m/z 369 (1.6%), 355 (5.5%), 341 (8.1%), 327 (14.0%), 312 (25.1%), 277 (34.6%), 241 (29.7%), 206 (18.3%), 170 (3.2%) and 135 (2.6%).

It may be noticed that a second fragmentation route for M^+ gives a couple of sub-fragments, (i) F4 = $N_3P_3Cl_5$, m/z 312 (25.1%) and (ii) F5, corresponding to the rest of the molecule, m/z 419 (16.6%). The successive loss of 1 NH, 1 NH + 1 CH_2 and 1 NH + 2 CH_2 from F5 gives peaks at m/z 404 (4.5%), 390 (2.6%) and 376 (7.1%) before leading to the F1 fragment.

All the peaks just assigned are actually associated with an exceptionally high number of H-transfers, much larger than in any other cyclophosphazene. These numerous H-transfers and the presence of nine chlorine atoms in the molecule, which induces many isotopic satellites, explain the gaussian patterns of peaks distribution, mainly around the m/z 696, 657 and 621 values.

No mass peak at m/z 767, corresponding to the purely $N_3P_3Cl_5$ (spermidine minus 2 H) $Cl_5P_3N_3$ BINO structure, could ever be detected, whatever the sensitivity of the instrument.

Thus, mass spectrometry provides reasonable proof of a SPIROBINO configuration for the final product of the reaction of spermidine on $N_3P_3Cl_6$. This new SPIROBINO structure is visualized in Fig. 28.

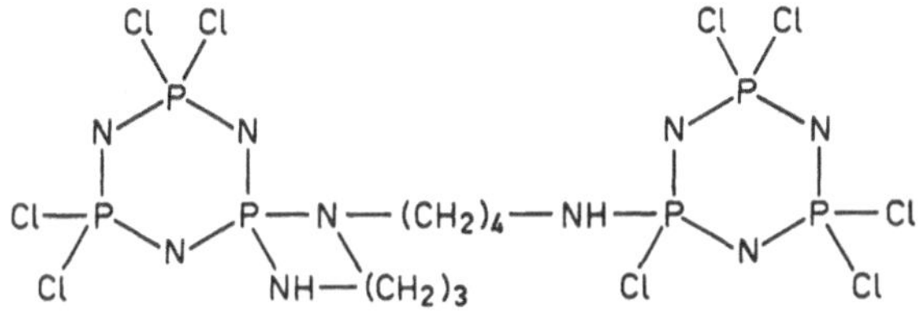

Fig. 28. Assumed SPIROBINO structure for SPD

3.4.3 NMR Spectroscopy of SPD

Let us draw a priori the pattern of the ^{31}P NMR spectrum from data previously obtained both for pure SPIRO and for pure BINO configurations. The ^{31}P NMR spectrum (WH 90) of SPIRO $N_3P_3Cl_4[HN-(CH_2)_3-NH]$ displays one doublet at 21.39 and 20.09 ppm and a triplet centered on 7.58 ppm [30] (Fig. 29) whereas the ^{31}P NMR spectrum (WH 90) of BINO $N_3P_3Cl_5[HN-(CH_2)_4-NH]Cl_5P_3N_3$ exhibits a multiplet centered on 20.66 ppm [48] (Fig. 29). Thus, by roughly summing these two spectra, we may predict that the ^{31}P theoretical spectrum (WH 90) of the SIROBINO product would exhibit a multiplet around 20.6 ppm and a triplet at higher field.

The spectra as recorded on the Brucker WH 90 instrument supports this prediction (Fig. 30): there indeed exist a multiplet centered on 20.72 ppm and a triplet at 11.68, 10.55 and 9.50 ppm (intensity ratio of the two main signals, 3.5). The fact

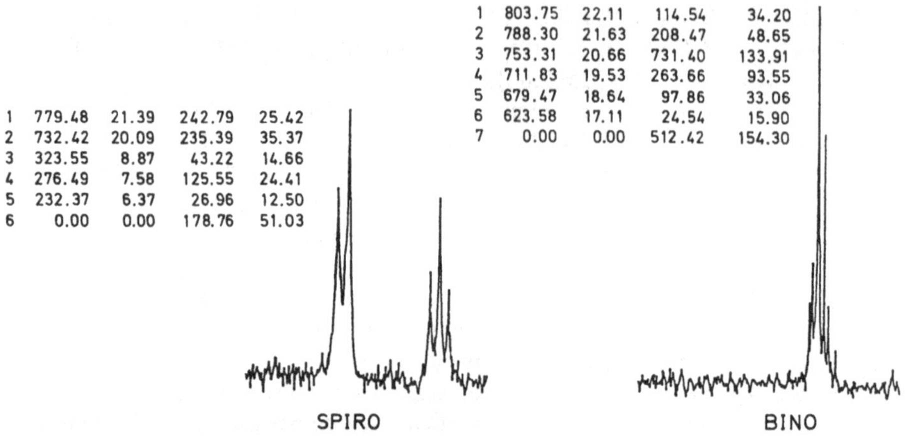

Fig. 29. ^{31}P NMR patterns for pure SPIRO and BINO derivatives

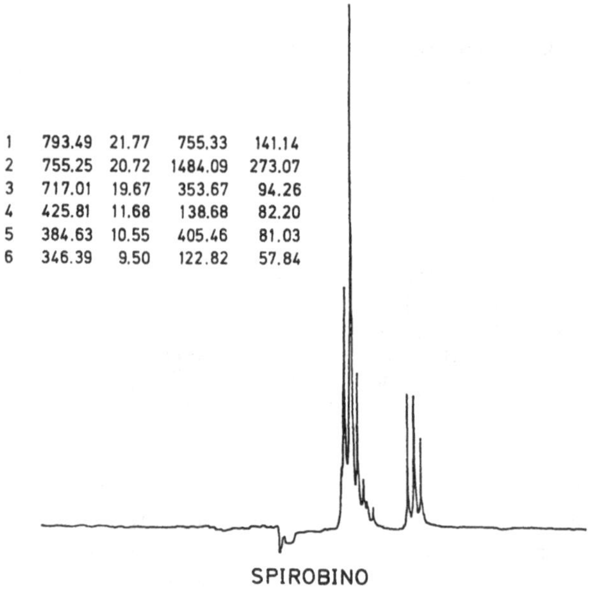

Fig. 30. ^{31}P NMR spectrum (WH 90) of SPD

that the triplet, corresponding to the P_{spiro} atom, has shifted from 7.58 to 10.55 ppm may indicate the influence of the $(CH_2)_4$ BINO chain grafted on the SPIRO loop (Fig. 28).

The spectrum of Fig. 30 was resolved using the Brucker WH 250 instrument. Figure 31 displays the expected one doublet plus two triplets, in agreement with previous spectra obtained for the purely SPIRO and the purely BINO structures [30,48].

3.4.4 X-Ray Crystal and Molecular Structure of SPD [50)]

Single crystals were obtained through a low evaporation of a solution of SPD in n-hexane (colourless cubic pieces with hexagonal cuts).

Preliminary investigations used the precession method and allowed the assignment

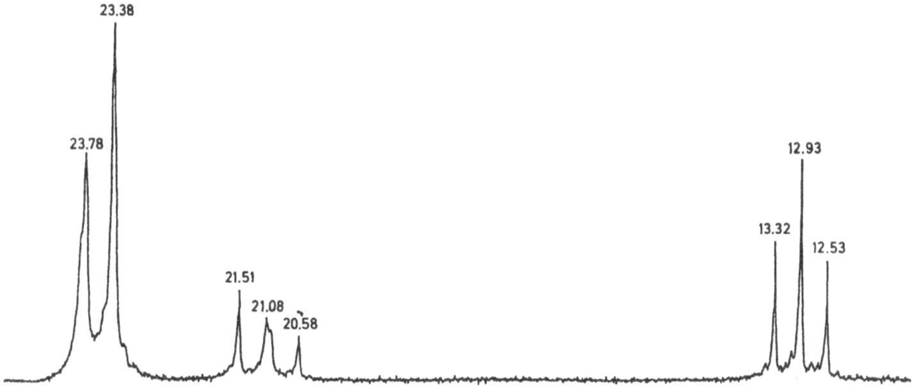

Fig. 31. ^{31}P NMR spectrum (WH 250) of SPD

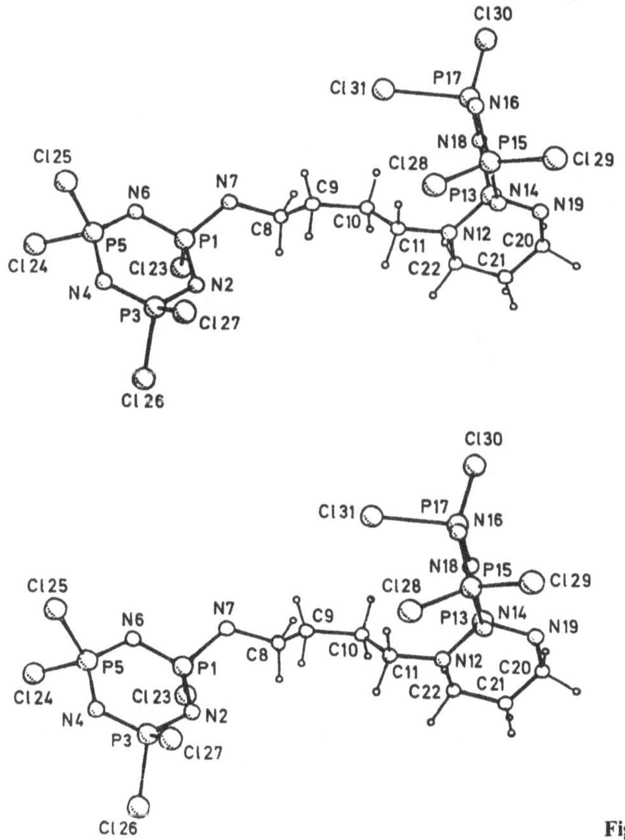

Fig. 32. Stereoscopic view of SPD

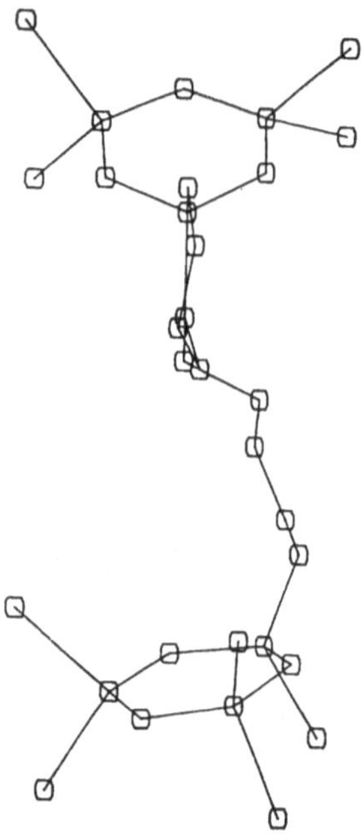

Fig. 33. Molecular pattern of spermidine derivative with the best plane of the SPIRO loop perpendicular to the plane of the figure

of a monoclinic cell, actually $P2_1/n$. The crystal was transferred to a SYNTEX $P2_1$ computer-controlled diffractometer. 25 reflections were used in order to orient the crystal and to refine the cell dimensions. Cell parameters are a = 11.674(8), b = 27.833(12), c = 8.910(4) Å, β = 102.2(4)°, V = 2829(2) $Å^3$, Z = 4. The final R factor was 0.050.

A stereoscopic view of the molecule is shown in Fig. 32 in which the numbering of the atoms is indicated. This view emphasizes the SPIROBINO structure which could be expected to exist on the basis of the 731 value for molecular weight as provided by mass spectrometry. The two six-membered phosphazene rings are not strictly planar and the two N_3P_3 planes within the mollecule are not parallel, the angle between them being equal to 57.6(5)° (Fig. 33). A stereoscopic view of the four molecules in the unit cell is visualized in Fig. 34.

3.4.5 Conclusion

The reaction of $N_3P_3Cl_6$ with spermidine (2:1) leads to a unique final product in which the configuration of the polyamino ligand does not belong either to the SPIRO or to the BINO type. Elemental analysis, mass spectrometry, high resolution NMR and X-ray crystallography prove that the final product has a two-ring assembly structure in which one SPIRO $N_3P_3Cl_4[HN-(CH_2)_3-N]$ moiety and one $N_3P_3Cl_5$ entity are bridged through a $(CH_2)_4$ chain in a BINO configuration. We propose to call this new configuration *SPIROBINO*.

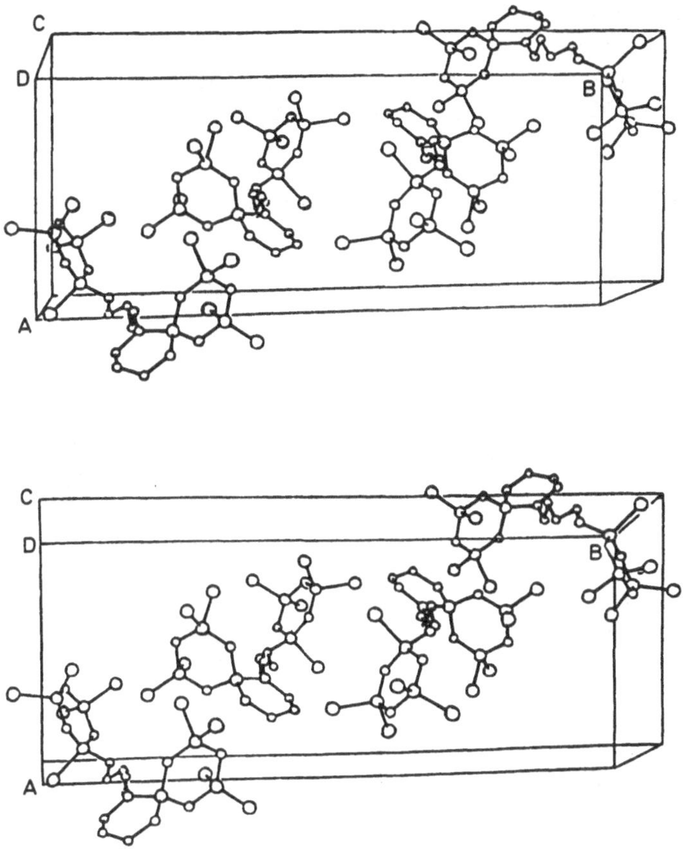

Fig. 34. Stereoscopic view of the unit cell for SPD

3.5 The Case of Spermine[45]

3.5.1 Synthesis

42.9 mmole of spermine in 400 ml of Et_2O were added dropwise in 2 hours to a mixture of 43.1 mmole of $N_3P_3Cl_6$ and 47.7 mmole of NEt_3 in 250 ml of Et_2O. The medium was stirred under argon pressure at room temperature for 5 days. Hydrochloride was then filtered off and solvent removed in vacuo at 30 °C. The residue was washed three times with 200 ml of n-hexane each to eliminate unreacted $N_3P_3Cl_6$ ($N_3P_3Cl_6$ is soluble in n-hexane whereas the reaction product is not). The insoluble residue in n-hexane was recrystallized in 300 ml of Et_2O to give 9 g (28%) of a sample, m.p. 222 °C. Thin layer chromatography revealed one spot at Rf = 0.14 with CH_2Cl_2 as eluant. Analytical data C, 15.88%; H, 2.95%; N, 18.50%; P, 24.58% and Cl, 37.50%. Firstly, this elemental analysis does not fit any SPIRO or ANSA configuration. A structure as $N_3P_3Cl_4$ (spermine minus 2 H) (mol. wt. = 477.083) would have led to the following figures: C, 25.17%; H, 5.07%; N, 20.55%; P, 19.47% and Cl, 29.72%. In contrast, the elemental analysis of final product was closer to the figures calculated for a BINO configuration, i.e., for $N_3P_3Cl_5$ (spermine minus 2 H)

$Cl_5P_3N_3$ (mol. wt. = 826.818): C, 14.52%; H, 2.92%; N, 16.94%; P, 22.47% and Cl, 42.87%. However, the gap between experimental and calculated values for all elements except hydrogen is unacceptable and indicates that the actual structure of the final product (SPM) contains some peculiar features to be elucidated.

3.5.2 Mass Spectrometry of SPM

The 70 eV electron impact mass spectrum of SPM is presented in Fig. 35.

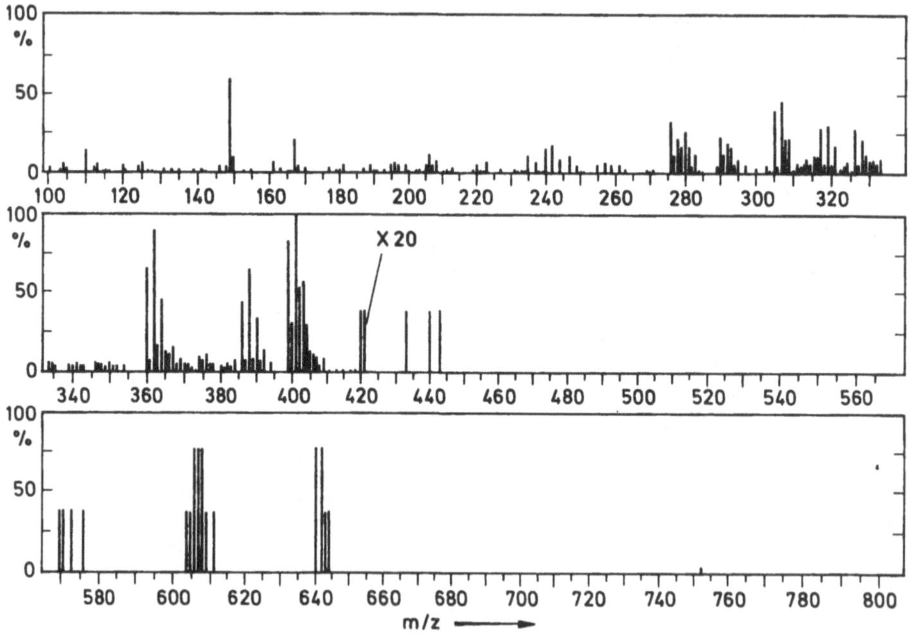

Fig. 35. 70 eV electron impact mass spectrum of SPM

The molecular ion M^+ is observed at m/z 753 (0.3%). The extremely weak intensity of its satellites does not allow prediction of the actual number of Cl atoms in the molecule. Mass spectrometry indicates that the structure of the final product would correspond to the formula $(N_3P_3Cl_4)_2$ (spermine minus 4 H). Taking into account the fact that reaction of 1,3-diaminopropane on $N_3P_3Cl_6$ leads to the purely SPIRO final product in contrast with 1,4-diaminobutane, and looking to the distribution of amino groups in spermine, a molecular weight of 753 would correspond nicely to a *DISPIROBINO* configuration in which each of the two (NH_2, NH) couples in $H_2N{-}(CH_2)_3{-}NH{-}(CH_2)_4{-}NH{-}(CH_2)_3{-}NH_2$ would link one $N_3P_3Cl_4$ moiety in a SPIRO configuration, the two resulting $N_3P_3Cl_4[HN{-}(CH_2)_3{-}N]$ entities being linked by the $(CH_2)_4$ methylenic chain in a BINO configuration. In other words, a possible structure for the final product as suggested by mass spectrometry would be: $N_3P_3Cl_4[HN{-}(CH_2)_3{-}N](CH_2)_4[N{-}(CH_2)_3{-}NH]Cl_4P_3N_3$. We shall see below that X-ray crystallography definitely supports this hypothesis.

Incidentally, calculated figures for this new DISPIROBINO configuration fit experimental analytical data well (C, 15.93%; H, 2.94%; N, 18.58%; P, 24.65% and Cl, 15.93%). Furthermore, when recording the mass spectrum of the final product up to m/z 900, no peak at m/z 826 (corresponding to the purely $N_3P_3Cl_5$ (spermine minus 2 H) $Cl_5P_3N_3$ BINO structure) was ever detected, whatever the sensitivity of the instrument.

We now return to the assignment of the various peaks of Fig. 35. The base peak, corresponding to the $N_3P_3Cl_4[HN-(CH_2)_3-N](CH_2)_4$ fragment (hereinafter coded M1) is centered on m/z 401 (100%) with satellites at m/z 402 (52.9%), 403 (54.9%), 404 (29.4%), 405 (13.7%), 406 (11.7%), 407 (9.8%) and 408 (3.9%). One major fragmentation route for M1 is detected involving the successive loss of 1 CH_2, 2 CH_2, 3 CH_2, 4 CH_2, 5 CH_2, 6 CH_2, 7 CH_2, 7 CH_2 + NH, 7 CH_2 + NH + N (associated with H-transfers) to give maximal peaks at m/z 388 (64.7%), 376 (11.7%), 362 (90.1%), 348 (5.8%), 333 (7.8%), 319 (29.4%), 305 (45.0%), 290 (21.5%) and 276 (31.3%). Further consecutive losses of Cl atoms give peaks at m/z 242 (15.6%), 206 (9.8%), 170 (3.9%) and 135 (1.9%). Another minor fragmentation route for M1 is observed with peaks at m/z 369 (7.8%), 333 (7.8%), 298 (3.9%) and 262 (3.9%) corresponding to the successive loss of one to four chlorine atoms (again associated with numerous H-transfers). Further consecutive losses of one to four CH_2 give peaks at m/z 249 (3.9%), 235 (9.8%), 221 (3.9%) and 207 (9.8%). It is noteworthy that there is some overlap between peaks in the various fragmentation pathways resulting from the similarity of mass of two chlorine atoms (70) and of the two $HN-(CH_2)_3-N$ and $HN-(CH_2)_4$ blocks (71).

In the low mass range, two main peaks are observed at m/z 167 (19.6%) and m/z 149 (58.8%) which correspond to the $N_3P_3(NH_2)_2$ and N_3 P_3N fragments.

The intensity of all other peaks is less than 1%. Thus, it is once more a relatively simple matter to check for the presence of any possible contaminant.

3.5.3 NMR Spectroscopy of SPM

The ^{31}P NMR spectrum was recorded on a Brucker WH 90 instrument. The doublet at 21.77 and 20.69 ppm corresponds to PCl_2 entities and the triplet at 11.36, 10.23 and 9.18 ppm to the P_{spiro} atoms (intensity ratio 2:1, in CD_2Cl_2 with 85% H_3PO_4 as a standard). The J(P—P) coupling constant is 41.18 Hz.

The general pattern of this spectrum, however, makes its first order character questionable. Thus, we recorded the ^{31}P, ^{13}C and ^{1}H spectra on a Brucker WH 250 instrument. The ^{31}P spectrum then appeared as a real first order one with the following figures: doublet at 23.64 and 23.24 ppm, triplet at 12.95, 12.55 and 12.16 ppm, J(P—P) = 40.28 Hz. These spectra clearly support the assumed structure for the final product as revealed by mass spectrometry.

3.5.4 IR Spectroscopy of SPM

The IR spectrum (KBr disks) was recorded on a Perkin-Elmer 683 spectrometer (range 4000–200 cm^{-1}, calibrated with polystyrene lines). This spectrum displayed two NH bands at 3278 and 3345 cm^{-1}. The 3278 cm^{-1} band may be related to the 3270 cm^{-1} band in SPIRO $N_3P_3Cl_4[HN-(CH_2)_3-NH]$ [30].

3.5.5 X-Ray Crystal and Molecular Structure of SPM

Single crystals were obtained through slow evaporation of a solution of the final product in CH_2Cl_2 (colourless cubic pieces with nice hexagonal cuts).

Preliminary investigations used the precession method and allowed the assignment of a monoclinic cell, actually $P2_1/c$.

The crystal was transferred to a CAD-4 ENRAF-NONIUS PDP 8/M computer-controlled diffractometer. 25 reflections were used in order to orient the crystal and to refine the cell dimensions. The conditions for the data collection are reported in Table 10.

The structure was determined using direct methods included in the MULTAN 80 program [49]. The refinement was performed taking into account anisotropic vibration for all atoms, except hydrogen. Difference Fourier syntheses phased by the P, Cl, N and C atoms revealed the H atoms. In the last refinement cycles, the H atoms were positioned geometrically with a bond length of 0.97 Å and isotropic thermal parameters were assigned with a $U = 0.05$ Å^2. The final R values are indicated in Table 10. Atomic coordinates are given in Table 11, anisotropic thermal parameters in Table 12, hydrogen atomic parameters in Table 13, and important bond lengths and angles in Table 14.

In addition to local programs, the following were used on the CICT CII-HB DPS 8 Multics System: Main and co-workers' Multan; Busing, Martin and Levy's Orffe; Johnson's Ortep.

A perspective view of SPM is shown in Fig. 36 on which the numbering of the

Table 10. Crystallographic data and conditions for data collection and refinement

Formula: $[N_3P_3Cl_4HN(CH_2)_3N(CH_2)_2]_2$

Monoclinic, space group $P2_1/c$
$a = 10.397(2)$, $b = 7.869(2)$, $c = 17.898(3)$ Å, $\beta = 102.1(1)°$,
$V = 1431.6(4)$ Å^3, $Z = 2$ (considering the dimer).
$D_m = 1.75(5)$ Mg · m^{-3}
$D_c = 1.748$ Mg · m^{-3}
Radiation Mo, $\lambda K\bar{\alpha} = 0.71069$ Å
Crystal size: 0.50 mm × 0.25 mm × 0.20 mm
Linear absorption coefficient: $\mu = 1.053$ mm^{-1}
No absorption correction
$F(000) = 756$
$T = 298$ K
θ range of reflections used for measuring lattice parameters 8–22°
Controls of intensity: reflections 0 0 14, 4 1 $\bar{3}$, 4 2 $\bar{3}$, each 7200 s
2968 measured reflections, 2814 unique reflections
2764 utilized reflections with $I > 3\sigma(I)$
Use of F magnitudes in least-squares refinement
Parameter refined: 154
Reliability factors:

$$R = \Sigma(||F_0| - |F_c||)/\Sigma|F_0| = 0.042$$

$$R_w = [\Sigma w(|F_0| - |F_c|)^2/\Sigma w F_0^2]^{1/2} = 0.046$$

Maximum and minimum height in final difference Fourier synthesis:
0.6 eÅ^{-3}, −0.6 eÅ^{-3}

Table 11. Final least-squares atomic coordinates with e.s.d.'s for SPM

Atom	x/a	y/b	z/c
P(1)	0.3036(1)	0.3771(1)	0.18216(5)
P(2)	0.1638(1)	0.1113(1)	0.10865(5)
P(3)	0.23856(9)	0.3781(1)	0.02265(5)
Cl(1)	0.4891(1)	0.3591(2)	0.24041(7)
Cl(2)	0.2309(1)	0.5372(2)	0.24996(6)
Cl(3)	0.2331(1)	−0.1262(1)	0.10782(7)
Cl(4)	−0.0211(1)	0.0605(2)	0.11721(8)
N(1)	0.2359(4)	0.1991(5)	0.1856(2)
N(2)	0.1644(3)	0.2010(4)	0.0314(2)
N(3)	0.3012(3)	0.4653(4)	0.1037(2)
N(4)	0.1339(3)	0.5105(4)	−0.0278(2)
N(5)	0.3527(3)	0.3543(4)	−0.0266(2)
C(1)	0.1056(5)	0.4904(6)	−0.1117(2)
C(2)	0.2300(5)	0.5107(6)	−0.1400(2)
C(3)	0.3282(5)	0.3729(6)	−0.1098(2)
C(4)	0.4631(4)	0.2411(5)	0.0051(2)
C(5)	0.4446(4)	0.0583(5)	−0.0242(2)

Table 12. Final anisotropic thermal parameters[a] (A × 100) with e.s.d.'s for SPM

Atom	U_{11}	U_{22}	U_{33}	U_{12}	U_{13}	U_{23}
P(1)	4.60(5)	3.14(5)	2.22(4)	−0.36(4)	0.92(4)	−0.11(4)
P(2)	4.81(6)	2.94(5)	3.07(5)	−0.65(4)	0.95(4)	0.09(4)
P(3)	3.68(5)	2.61(4)	2.31(4)	0.50(4)	0.97(3)	0.11(3)
Cl(1)	5.02(6)	7.03(8)	4.83(6)	0.06(6)	0.03(5)	0.83(6)
Cl(2)	8.06(8)	5.59(7)	3.64(5)	0.76(6)	2.61(5)	−0.84(5)
Cl(3)	8.71(9)	3.25(5)	6.01(7)	0.69(6)	0.97(6)	0.55(5)
Cl(4)	6.23(7)	6.62(8)	8.05(9)	−2.20(6)	3.22(7)	−0.58(7)
N(1)	8.5(3)	4.1(2)	2.5(2)	−2.2(2)	1.0(2)	0.3(1)
N(2)	5.4(2)	3.5(2)	2.7(2)	−1.0(2)	0.7(1)	−0.1(1)
N(3)	5.6(2)	2.8(2)	2.4(1)	−0.5(1)	1.2(1)	−0.1(1)
N(4)	4.6(2)	4.1(2)	3.5(2)	1.8(2)	1.0(1)	0.5(1)
N(5)	4.0(2)	3.5(2)	2.6(1)	1.1(1)	1.2(1)	0.2(1)
C(1)	6.3(3)	4.9(3)	3.8(2)	1.7(2)	−0.7(2)	0.2(2)
C(2)	8.9(4)	5.4(3)	2.3(2)	2.0(3)	1.5(2)	0.7(2)
C(3)	7.7(3)	5.6(3)	2.9(2)	2.2(3)	2.3(2)	0.4(2)
C(4)	3.8(2)	4.0(2)	4.3(2)	1.1(2)	0.8(2)	−0.3(2)
C(5)	4.7(2)	3.6(2)	4.0(2)	1.5(2)	1.3(2)	−0.1(2)

[a] Anisotropic thermal paramèters are of the form: $\exp(-2\pi^2(U_{11}h^2a^{*2} + \dots + 2U_{23}klb^*c^*))$

atoms is indicated; Fig. 37 shows the packing. As all intermolecular distances are greater than 3.5 Å, one may say that the dimers are well individualized and without noticeable interaction.

The cyclophosphazene six-membered rings are quasi planar, as usual in this type of moiety. In fact, the distance from atoms to the best plane is lower than 0.056(3) Å. In contrast, the SPIRO rings present a slight "chair" conformation

Table 13. Hydrogen atomic positional and thermal parameters for SPM

Atom	x/a	y/b	z/c	U(A × 100)
H(N4)	0.092	0.599	−0.004	5.0
H1(C1)	0.069	0.378	−0.125	5.0
H2(C1)	0.043	0.576	−0.135	5.0
H1(C2)	0.210	0.506	−0.195	5.0
H2(C2)	0.269	0.620	−0.124	5.0
H1(C3)	0.411	0.400	−0.124	5.0
H2(C3)	0.295	0.266	−0.133	5.0
H1(C4)	0.542	0.287	−0.008	5.0
H2(C4)	0.473	0.239	0.060	5.0
H1(C5)	0.486	0.013	−0.064	5.0
H2(C5)	0.417	−0.034	0.005	5.0

Table 14. Bond lengths (Å) and angles (degrees) in SPM

Parameter	Value	Parameter[a]	Value
P1—Cl1	1.996(2)	N4—C1	1.478(5)
P1—Cl2	2.004(1)	N5—C3	1.464(4)
P2—Cl3	2.004(2)	N5—C4	1.470(5)
P2—Cl4	2.001(2)		
		C1—C2	1.493(6)
P1—N1	1.575(3)	C2—C3	1.510(6)
P1—N3	1.562(3)	C4—C5	1.529(5)
P2—N1	1.582(3)	C5—C5i	1.581(7)
P2—N2	1.554(3)		
P3—N2	1.616(3)		
P3—N3	1.613(3)		
P3—N4	1.634(3)		
P3—N5	1.631(3)		
N1—P1—N3	120.5(2)	N4—C1—C2	109.3(3)
N1—P2—N2	119.1(2)	N5—C3—C2	113.4(4)
N2—P3—N3	113.0(1)	C1—C2—C3	111.7(4)
P1—N1—P2	119.3(2)	P3—N5—C4	117.3(3)
P2—N2—P3	124.7(2)	N5—C4—C5	113.9(3)
P1—N3—P3	123.2(2)	C4—C5—C5i	109.5(4)
N2—P3—N5	111.6(2)	C1—N4—P3	117.5(3)
N3—P3—N4	108.8(2)	C3—N5—P3	123.2(3)

[a] Symmetry code: none, X, Y, Z; i, 1 — X; —Y; —Z.

(Fig. 38). This to be compared with the folding of the SPIRO loop in the $N_3P_3Cl_4$ [HN—$(CH_2)_3$—NH] molecule [30], which can be described as an "open book" conformation and with the quite planar SPIRO loop appearing in the spermidine derivative (see above) [50]. Actually, the versatility of the six-membered SPIRO loop PNCCCN is remarkable considering these three different molecules. Quantum calculations are now in progress to try to explain such a conformational variation [51].

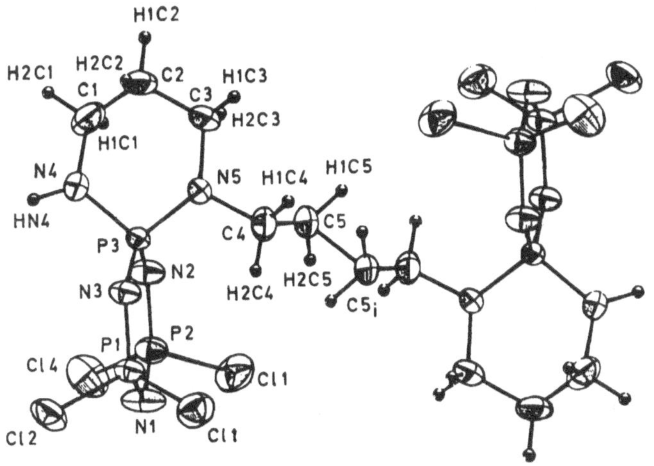

Fig. 36. A perspective view of SPM with numbering of atoms

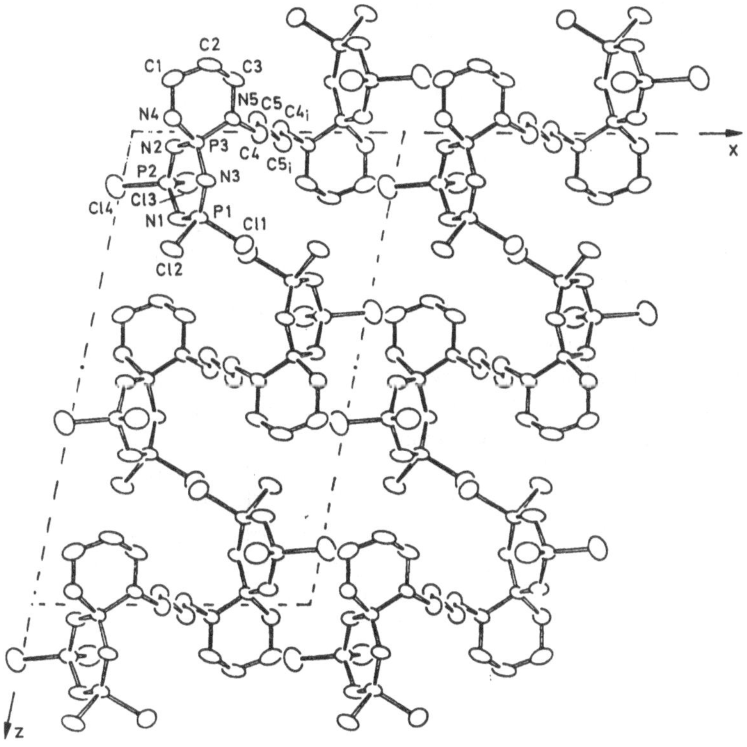

Fig. 37. Projection of the structure onto the plane (010)

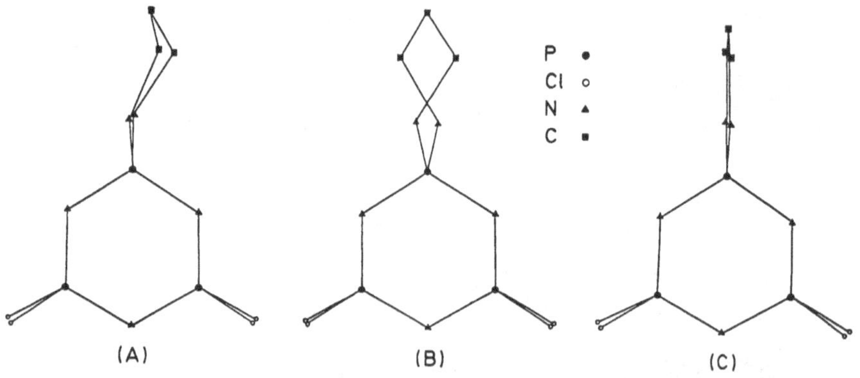

Fig. 38. Comparison of SPIRO loops conformations in (A) SPM, (B) $N_3P_3Cl_4[HN-(CH_2)_3-NH]$ and (C) SPD

On the other hand, the C4—C5—C5i—C4i bridge shows a very nice "stair" conformation, with the center of symmetry between C5 and C5i. This particular location explains the unusually long C5—C5i bond and the short C4—C5 and C4i—C5i bond lengths.

Lastly, the cyclophosphazene ring conformation is almost identical to that found in other similar structures. The NPN angles range between 113.0(1) and 120.5(2)°, while the P—N bonds range from 1.554(3) to 1.634(3) Å. This is to be compared to the crystal structure of SPIRO $N_3P_3Cl_4[HN-(CH_2)_3-NH]$ [30] for example.

3.5.6 Conclusion

The reaction of $N_3P_3Cl_6$ with spermine (2:1) leads to a unique final product which has a two-ring assembly structure in which two SPIRO $N_3P_3Cl_4[HN-(CH_2)_3-N]$ moieties are bridged through a $(CH_2)_4$ chain in a BINO configuration. We propose to call the new configuration observed here *DISPIROBINO*.

These *SPIROBINO* (in SPD) and *DISPIROBINO* (in SPM) new configurations are perfectly consistent with what we have observed previously about the SPIRO/BINO respective areas: when looking to the chemical formulas of spermidine, $H_2N-(CH_2)_3-NH-(CH_2)_4-NH_2$ and of spermine, $H_2N-(CH_2)_3-NH-(CH_2)_4-NH-(CH_2)_3-NH_2$, we could predict that the couple(s) (NH_2, NH) beside the $(CH_2)_3$ group(s) would link $N_3P_3Cl_6$ in a SPIRO configuration preferentially to the (NH, NH_2) or (NH, NH) couples beside the $(CH_2)_4$ group. Remember indeed the behaviour of 1,3-diaminopropane and 1,4-diaminobutane respectively versus $N_3P_3Cl_6$. Thus, the *SPIROBINO* and *DISPIROBINO* configurations could be expected to be the ones of the major final product upon reaction of $N_3P_3Cl_6$ with spermidine and spermine. But it is really amazing to note that these conformations are the only ones which are crystallographically observed. The stereo-selectivity of the reaction of $N_3P_3Cl_6$ with diamines we mentioned above occurs also straightforward for polyamines.

We shall see below how this conclusion may be enlarged to other molecules and how it is nowadays possible to graft on demand several SPIRO loops and/or BINO bridges on cyclophosphazenes in an univocal way.

Anyhow, no ANSA structure (as defined by Becke-Goehring [26]) was ever observed in our works and we could conclude that the SPIRO and BINO configurations do exist in final products of the reaction of $N_3P_3Cl_6$ with diamines and polyamines.

However, Harris and Williams recently reported the designed synthesis and spectral characteristics of the *first ANSA cyclotriphosphazene* [52,53], by the reaction of 3-amino-1-propanol with monomethyl-pentachlorocyclotriphosphazene [54]. The announcement of this discovery was made at the International Conference on Phosphorus Chemistry held in Nice, September, 1983, and it made some outstanding colleagues sceptic . . .

We had the privilege to solve the X-ray structure of the first ANSA derivative and we shall report in the next section the crystal and molecular structure of a cyclophosphazene linked to a difunctional tumor finder-like in a new configuration, i.e. the ANSA one [55].

4 Synthesis of the First Chlorinated Ansa Precursor upon Reaction of $N_3P_3Cl_5(CH_3)$ with 3-Amino-1-Propanol

4.1 Synthesis and Characterization [52,53]

The synthetic route to this ANSA derivative (III Fig. 39) is as follows: treatment of methylpentachlorocyclotriphosphazene with two equivalents of 3-amino-1-propanol in dry dichloromethane (freshly distilled from P_4O_{10}) led to the isolation of the gem-alkylamino compound (II) as a colourless oil in 65% yield. This compound was characterized from the following data [53]: mass spectra: found m/z 364 ($^{35}Cl_4$) calculated m/z 364 ($^{35}Cl_4$); ^{31}P NMR (proton decoupled, $CDCl_3$ solution): $P(CH_3)(NHR)$ 25.0 ppm (a triplet, $J_{PNP} = 19.5$ Hz, this peak broadened significantly upon proton coupling), $P(Cl)_2$ 19.8 ppm (a doublet, $J_{PNP} = 19.5$ Hz, this peak remained virtually unchanged upon proton coupling); ^1H NMR ($CDCl_3$ solution, after treatment with D_2O to remove NH and OH protons): $P{-}CH_3$ 1.67 δ (a doublet of triplets, $J_{PCH} = 15.9$ Hz, $J_{PNPCH} = 2.1$ Hz, $-NHC\mathit{H}_2CH_2CH_2OH$ 3.10 δ (a complex multiplet, couplings were unresolved), $NHCH_2C\mathit{H}_2CH_2OH$ 1.80 δ (a complex multiplet, couplings were unresolved), $NHCH_2CH_2C\mathit{H}_2OH$ 3.75 δ (a triplet, $J_{HCCH} = 7.0$ Hz); infrared spectrum (liquid film): 3320 cm^{-1} (m, br. $\nu_{NH,OH}$), 2940 cm^{-1} (m), 2890 cm^{-1} (m, ν_{CH}), 1230 cm^{-1} (vs), 1180 cm^{-1} (vs. ν_{PN}). Correct microanalytical data were also obtained [56].

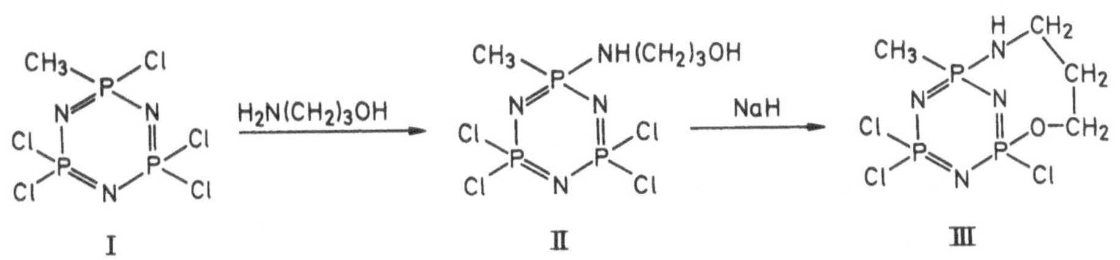

Fig. 39. Synthetic route to the (III) ANSA derivative

Formation of the ANSA derivative (III) was accomplished by treatment of a highly dilute solution of compound II in dry tetrahydrofuran (freshly distilled from a sodium-benzophenon ketyl) with an excess of sodium hydride for 50 hours at room temperature. After filtration of the mixture to remove sodium chloride and any unreacted sodium hydride, compound III was isolated in 40–60% yield as white crystals, m.p. 144–5 °C, from *n*-hexane. Compound III was identified as the ANSA derivative from the following data: mass spectra; found m/z 328 ($^{35}Cl_3$), calculated 328 ($^{35}Cl_3$); ^{31}P NMR (proton decoupled, $CDCl_3$ solution): $P(CH_3)$ (NHR) 31.2 ppm (a doublet of doublets, J_{PNP} = 9.8 Hz, J_{PNP} = 4.0 Hz, this peak broadened to an unresolved multiplet upon proton coupling), P(Cl) (OR) 29.3 ppm (a doublet of doublets, J_{PNP} = 4.0 Hz, J_{PNP} = 48.8 Hz, this peak became a triplet of doublet of doublets upon proton coupling, J_{POCH} ~ 20 Hz), PCl_2 24.5 ppm (a doublet of doublets J_{PNP} = 48.8 Hz, J_{PNP} = 9.8 Hz, this peak remained virtually unchanged upon proton coupling); ^1H NMR ($CDCl_3$ solution): P—CH_3 1.71 δ (a doublet of triplets, J_{PNP} = 16.9 Hz, J_{PNPCH} = 3.3 Hz), —N*H*$CH_2CH_2CH_2O$— 3.04 δ (a complex multiplet, couplings were unresolved), —NHC$H_2$$CH_2CH_2O$— 3.30 δ (a complex multiplet, couplings were unresolved), —NH—CH_2—CH_2—CH_2O— 1.85 δ (a complex multiplet, couplings were unresolved), —NH—CH_2—CH_2—CH_2O— 4.45 δ (a complex multiplet, couplings were unresolved); infrared spectrum (KBr disk): 3310 cm^{-1} (m, ν_{NH}), 2970 cm^{-1} (m), 2940 cm^{-1} (m), 2890 cm^{-1} (w, ν_{CH}), 1190 cm^{-1} (vs, ν_{PN}). Correct microanalytical data were also obtained [56].

4.2 X-ray Crystal and Molecular Structure of III [55]

4.2.1 Crystal Data

The compound crystallized in the orthorhombic $P2_12_12_1$ space group; unit cell parameters at 293 K were as follows: a = 8.033(2), b = 11.534(7) and c = 13.450(4) Å; V = 1246(1) $Å^3$, ϱ_x = 1.76 Mg · m^{-3}, ϱ_{exp} = 1.74 Mg · m^3, Z = 4.

Due to the slow decomposition of the compound at room temperature, the single crystal chosen for this study was rapidly mounted on a CAD4 NONIUS diffractometer and cooled to 123 K using a thermostatically controlled stream of cold nitrogen gas. All the data concerning the new cell parameters and the conditions used for data collection at 123 K are listed in Table 15.

4.2.2 Structure Determination and Refinement

Direct methods were used to determine the structure; groups of atoms whose geometry was known were introduced to help the procedures, i.e., the $N_3P_3Cl_2$ moiety from gem-$N_3P_3Az_4Cl_2$ [57]. The best map yielded the positions of all the non-hydrogen atoms. Isotropic refinement in the space group $P2_12_12_1$ gave a reliability index of 0.045. Introduction of anisotropic temperature parameters reduced the R index to 0.030.

The difference Fourier synthesis, phased by the P, Cl, N, C and O atoms, revealed the hydrogen atoms with their expected locations. Thus, the final refinement could be performed on the entire set of atoms including hydrogens with fixed isotropic thermal parameter factor, B_H = 4 $Å^2$. Final R and S values are 0.022 and 0.948, respectively. The last difference Fourier map showed no values to be greater than ± 0.3 $eÅ^{-3}$ (Table 15). Atomic scattering factors were corrected for anomalous dispersion from Cromer and Waber [58].

Table 15. Crystallographically important data collection and data processing information

Formula: $N_3P_3Cl_3CH_3[HN{-}(CH_2)_3{-}O]$

Data collection:
Unit cell: Orthorhombic, space group $P2_12_12_1$, Z = 4
a = 8.033(2), b = 11.534(7), c = 13.450(4)A, V = 1246(1)A^3 at 298 K
a = 7.984(3), b = 11.406(3), c = 13.380(3)A, V = 1218.5(6)A^3 at 123 K
$\varrho_{exp} = 1.74\ g \cdot cm^{-3}$
$\varrho_x = 1.76\ g \cdot cm^{-3}$
Graphite monochromated MoKα,λ = 0.71069A
Crystal size: 0.25 mm × 0.40 mm × 0.60 mm
Linear absorption coefficient: $\mu = 1.030\ mm^{-1}$
No absorption correction
F(000) = 664
θ range of reflections: 1.5–26°
θ/2θ scan technique
Controls of intensity: reflections 2 1 $\bar{4}$, 1 2 7, 2 $\bar{2}$ 0, each 3600 s
CAD4 Nonius diffractometer
Take-off angle: 2.5°
25 reflections with $5° < \theta < 13°$ used for measuring lattice parameters
Space group (identified by precession method) verified by rapid measurement of
h01, 0k1, hk0 reflections implying $P2_12_12_1$ space group
θ — 2θ scan with $\Delta\theta$ scan = 1.0 + 0.35 tan θ, prescan speed = $10°\ min^{-1}$
σ(I)/I for final scan = 0.018
Maximum time for final scan: 80 s
No significant variation during the whole data collection

Structure determination and refinement:

1413 measured reflections, 1246 unique reflections, 1226 utilized reflections with $I > 3\sigma(I)$
Use of F magnitudes in least-square refinement
Parameters refined:
Reliability factor: $R = \Sigma(||F_0| - |F_c||)/\Sigma|F_0| = 0.0218$
$Rw = |\Sigma w(|F_0| - |F_c|)^2/\Sigma wF_0^2]^{1/2} = 0.0250$ w = 1
$S = [\Sigma w(|F_0| - |F_c|)^2/(NO - NV)]^{1/2} = 0.948.$

In addition to local programs, the following programs were used on the CICT-CII-HB DPS8 Multics systems: Zalkin's Fourier; Main & Germain's Multan; Busing, Martin and Levy's Orffe; Sheldrick's Shelx; Johnson's Ortep.

Final atomic coordinates and anisotropic temperature factors are listed in Table 16. Selected bond lengths and bond angles are listed in Table 17, along with the main intermolecular interatomic distances.

4.2.3 Results and Discussion

The X-ray structure determination of the title compound clearly indicated the ANSA structure. The amino end of the bridge is bound to the methylated phosphorus atom while the oxygen end of the 3-amino-1-propanol group is bound to an adjacent phosphorus atom. The general structure of the molecule, the numbering system used for all non-hydrogen atoms, and selected bond angles and distances are shown in Fig. 40. A perspective view of the molecule, clearly displaying the ANSA structure, is shown in Fig. 41. No obvious intermolecular hydrogen bonds were observed, in

Table 16. Fractional atomic coordinates, equivalent temperature factors and final anisotropic chemical parameters (with e.s.d.'s in parentheses) for compound (III)

Atom	x/a	y/b	z/c	Ueq
P1	0.8795(1)	0.55665(8)	0.23743(6)	1.15(4)
P2	0.6588(1)	0.38767(9)	0.29586(7)	1.27(4)
P3	0.9114(1)	0.46432(8)	0.42392(7)	1.31(4)
Cl1	1.0145(1)	0.56922(9)	0.11168(7)	1.99(5)
Cl2	0.4137(1)	0.42316(9)	0.31996(7)	2.08(5)
Cl3	0.6397(1)	0.22683(8)	0.23422(7)	1.97(5)
N1	0.7222(4)	0.4747(3)	0.2110(2)	1.6(2)
N2	0.7530(4)	0.3798(3)	0.3976(2)	1.8(2)
N3	0.9992(4)	0.5159(3)	0.3245(2)	1.4(2)
N4	0.8586(4)	0.5752(3)	0.4968(2)	1.6(2)
O	0.8115(3)	0.6855(2)	0.2524(2)	1.8(1)
C1	0.8852(5)	0.7612(3)	0.3297(3)	1.7(2)
C2	0.7716(5)	0.7690(3)	0.4198(3)	1.8(2)
C3	0.7188(5)	0.6515(4)	0.4651(3)	1.9(2)
C4	1.0588(6)	0.3822(4)	0.4958(3)	2.2(2)
HN4	0.848(7)	0.560(5)	0.557(4)	4.0
H1C1	0.995(7)	0.734(5)	0.347(4)	4.0
H2C1	0.903(7)	0.829(5)	0.303(4)	4.0
H1C2	0.825(7)	0.813(5)	0.477(4)	4.0
H2C2	0.671(7)	0.810(4)	0.405(4)	4.0
H1C3	0.650(7)	0.613(4)	0.417(4)	4.0
H2C3	0.651(7)	0.666(5)	0.529(4)	4.0
H1C4	1.143(7)	0.426(5)	0.510(4)	4.0
H2C4	1.017(7)	0.363(5)	0.554(4)	4.0
H3C4	1.094(8)	0.327(5)	0.462(4)	4.0

$$Ueq = 1/3 \sum_i \sum_j U_{ij} a_i^* a_j^* \vec{a}_i \cdot \vec{a}_j$$

Atom	U11	U22	U33	U12	U13	U23
P1	1.17(5)	1.19(4)	1.08(4)	−0.15(4)	0.03t4)	0.01(4)
P2	1.24(5)	1.27(4)	1.28(4)	−0.28(4)	0.02(4)	−0.07(4)
P3	1.30(4)	1.46(5)	1.17(4)	−0.01(4)	−0.18(4)	0.05(4)
Cl1	2.16(5)	2.27(5)	1.51(4)	−0.32(4)	0.64(4)	0.14(4)
Cl2	1.26(4)	2.21(5)	2.77(5)	−0.16(4)	0.23t4)	−0.25(4)
Cl3	2.21(5)	1.34(4)	2.33(5)	−0.29(4)	−0.06t4)	−0.42(4)
N1	1.7(2)	1.6(2)	1.4(2)	−0.5(1)	−0.4(1)	0.1(1)
N2	1.9(2)	1.6(2)	1.7(2)	−0.6(2)	−0.1(2)	0.2(1)
N3	1.0(2)	1.6(2)	1.4(2)	0.0(1)	−0.1(1)	−0.1(1)
N4	1.9t2)	1.8(2)	1.1(1)	−0.1(2)	−0.1(1)	−0.2(1)
O	1.9(1)	1.4(1)	1.9(1)	0.3(1)	−0.2(1)	−0.3(1)
C1	1.7(2)	1.2(2)	2.0(2)	−0.2(2)	0.0(2)	−0.1(2)
C2	1.8(2)	1.4(2)	1.9(2)	0.1(2)	0.1(2)	−0.4(2)
C3	1.4(2)	2.3(2)	1.9(2)	−0.2(2)	0.3(2)	−0.2(2)
C4	2.2(2)	2.5(2)	1.8(2)	0.3(2)	−0.8(2)	0.8(2)

Table 17. Selected intramolecular bond lengths and angles

P1—N1	1.606(4) Å	P1—N3	1.577(3) Å
P1—Cl1	2.003(2)	P1—O	1.579(3)
P2—N1	1.591(3)	P2—N2	1.557(4)
P3—N2	1.629(4)	P3—N3	1.615(3)
P3—C4	1.785(5)	P2—Cl2	2.024(2)
P2—Cl3	2.017(2)		
P3—N4	1.651(4)	N4—C3	1.478(6)
C3—C2	1.529(6)	C2—C1	1.512(6)
C1—O	1.470(5)		
C1—H1C1	0.96(5)	C1—H2C1	0.87(5)
C2—H1C2	1.02(5)	C2—H2C2	0.95(5)
C3—H1C3	0.95(5)	C3—H2C3	1.02(6)
C4—H1C4	0.86(6)	C4—H2C4	0.88(5)
C4—H3C4	0.83(5)		
Cl1 ... O	2.816(3)	Cl2 ... Cl3	3.564(2)
P1 ... P2	2.726(2)	P1 ... P3	2.720(2)
P2 ... P3	2.787(2)		
N1—P1—N3	117.7(2)°	P1—N3—P3	116.9(2)°
N3—P3—N2	112.0(2)	P3—N2—P2	122.0(2)
N2—P2—N1	120.4(2)	P2—N1—P1	117.1(2)
O—P1—Cl1	103.0(1)	Cl1—P2—Cl3	100.02(6)
P3—N4—C3	118.3(3)	N4—C3—C2	114.9(3)
C3—C2—C1	115.5(3)	C2—C1—O	110.8(3)
C1—O—P1	119.8(2)	O—P1—N3	112.9(2)
N3—P3—N4	108.5(2)	N4—P3—C4	104.6(2)

H—C—H angles are in the range 104(5)–116(5)°

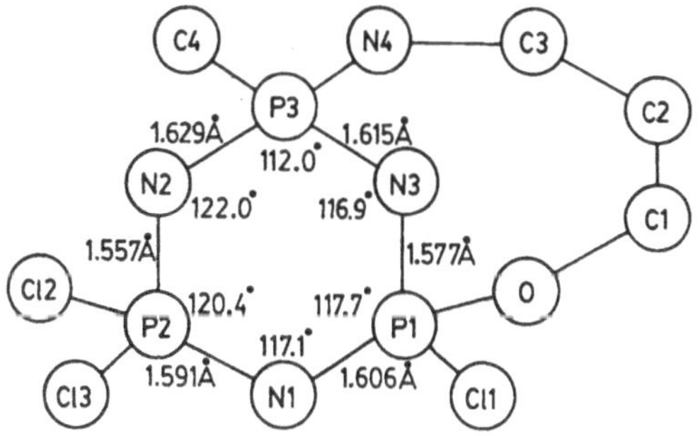

Fig. 40. Numbering scheme and selected bond angles and distances

sharp contrast to the SPIRO relative $N_3P_3Cl_4[HN—(CH_2)_3—NH]$ where two hydrogen bonds (2.08 Å) per molecule are considered to be responsible for the crystal packing arrangement [33].

Several interesting features of the molecule were revealed by the structure determination. These are discussed in the next sections.

4.2.4 The Spatial Conformation of the ANSA Arch

The conformation of the 3-amino-1-propanol group is shown in Figs. 42 and 43. The central carbon atom in the methylene chain (C2) belongs both to the σ_v-like N3—Cl2—P2—Cl3 plane, and to the N4—P3—P1—O plane. Carbon atoms C1

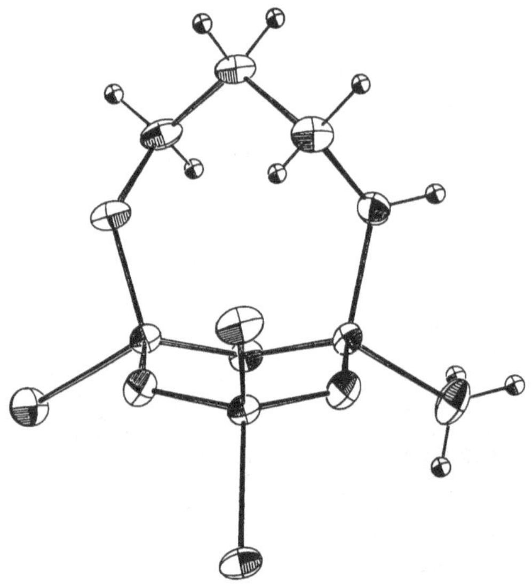

Fig. 41. Perspective view of (III)

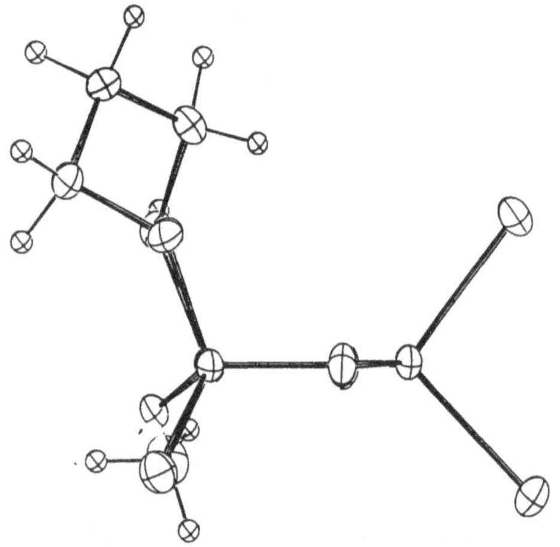

Fig. 42. Configuration of the ANSA arch in (III)

and C3 are ± 0.8 Å distant from the latter plane and the carbon atom C3 is situated over the mean plane of the phosphazene ring.

4.2.5 The Conformation of the Phosphazene Ring

The phosphazene ring in the compound is clearly non-planar, as illustrated in Fig. 43. The nitrogen atom, N3, which is a part of both the ANSA ring and the phosphazene ring is 0.435(3) Å below the mean P1—N1—P2—N2—P3 plane. The dihedral angle between the P1—P2—P3 and the P1—N3—P3 planes is 34.6°. However, despite the lack of planarity within the phosphazene ring, the transannular (P ... P) distances are all approximately 2.7 Å, while the exocyclic P3—N4 and P1—O

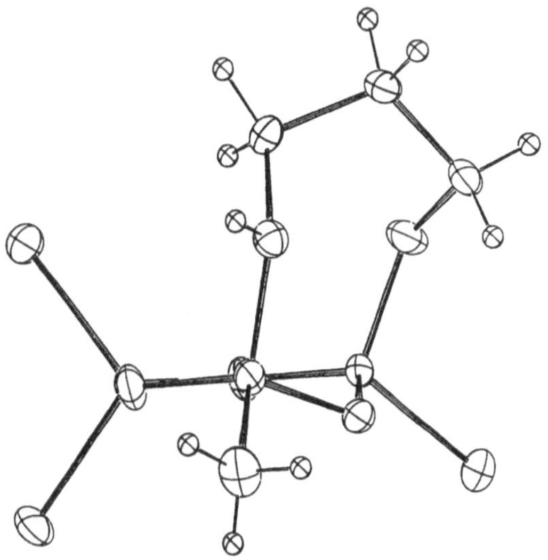

Fig. 43. Puckering of the PN ring in (III)

bond lengths are 1.651(4) and 1.579(3) Å, respectively, all typical of cyclotriphosphazene bond lengths. The reason the phosphazene ring is puckered, therefore, appears to be as follows:

In order for the 3-amino-1-propanol group to span phosphorus P1—P3 without affecting the distance between these two atoms, a rotation about bonds P1—N1 and P3—N2 must occur. This rotation is clearly indicated in Fig. 41. The rotation brings atoms N4 and O to within 3.52 Å of each other, and forces Cl1 and C4 apart. The Cl1—C4 distance in this ANSA compound is 5.58 Å while the Cl—C distances in the unpuckered $N_3P_3Cl_4$i—PrH compound are 4.67 Å and 4.88 Å [59]. Thus, the rotation about P1—N1 and P3—N2, which is necessary to allow the ANSA arch to bridge these two atoms, must force nitrogen N3 down below the plane of the phosphazene ring if the pseudo tetrahedral geometry around phosphorus P1 and P3 is to be maintained. This, indeed, appears to be the case.

4.2.6 Bonding Within the Phosphazene Ring

The variations observed in the phosphorus nitrogen bond distances within the phosphazene ring are very interesting. As has been observed in other cases [59], the lengths of the P—N bonds vary with the electronegativity of the substituents bound to the phosphorus atoms at the ends of a P—N—P island. Consider the island P2—N2—P3. Phosphorus atom P2 clearly contains the most electronegative substituents in the ring, while phosphorus atom P3 contains the least electronegative groups. The P2—N2 length of 1.557 Å, the shortest in the molecule, and the P3—N2 length of 1.629 Å, the longest in the compound, clearly reflect these electronegativity differences. Similar changes can be observed in the islands P1—N1—P2 and P1—N3—P3. The P1—N1—P2 island has the smallest difference in electronegativity between the phosphorus substituents and also has the smallest variation in PN bond lengths. However, the shorter PN bond in this island lies between the slightly more electronegative phosphorus P2—N1.

5 The Basic System

The main conclusion for the first paragraphs of this contribution is that the reaction of $N_3P_3Cl_6$ on polyamines in suitable stoichiometric conditions, leads stereoselectively to SPIRO, BINO, SPIROBINO and DISPIROBINO configurations depending on the nature of the polyamine.

Moreover, the reaction of $N_3P_3Cl_5Me$ with difunctional reagents such as 3-amino-1-propanol leads stereoselectively to an ANSA configuration. Thus, such a stereoselectivity invited us to see if it would be possible to design new molecules in which several BINO (B), ANSA (A) and SPIRO (S) configurations would be simultaneously present in the molecular structure.

We called humorously this chemical game BASIC challenge, BASIC initials being for *B*INO/*A*NSA/*S*PIRO *I*N *C*YCLOPHOSPHAZENES.

This challenge was found to be highly successful and we shall report now three peculiar examples.

5.1 The DISPIRO and TRISPIRO Cyclotriphosphazenes [60]

We mentioned that the synthesis of SPIRO derivatives had to be achieved in non-polar solvents in order to minimize the formation of non-crystalline resins. It is well-known indeed that polar solvents lead to a proton-abstraction process from the PNHR groups of the loop, thus promoting polymerization. Such polymerization was even claimed to be responsible for failure to prepare *monomeric* DISPIRO and TRISPIRO derivatives of $N_3P_3Cl_6$ [29] containing PNHR groups, in contrast with reaction of $N_3P_3Cl_6$ with secondary diamines, HRN—$(CH_2)_n$—NRH, where DISPIRO chemicals could be obtained [61] (no PNHR groups in the loops; no major polymerization). Thus, it seemed interesting to reinvestigate the reaction of $N_3P_3Cl_6$ with primary diamines in *non-polar solvents* with the aim of preparing PNHR-containing DISPIRO and TRISPIRO chemicals.

The next paragraphs report on the synthesis and physico- chemical identity of two DISPIRO derivatives, $N_3P_3Cl_2[HN{-}(CH_2)_3{-}NH]_2$ and $N_3P_3Cl_2[HN{-}(CH_2)_4{-}NH]_2$ (hereafter called DISPIRO 3 and DISPIRO 4, respectively) and of the TRISPIRO $N_3P_3[HN{-}(CH_2)_3{-}NH]_3$ coded as TRISPIRO 3.

5.1.1 The DISPIRO 3 and DISPIRO 4 Derivatives

Synthesis

The DISPIRO products were synthesized by following two methods with a (7:3) mixture of n-hexane and CH_2Cl_2 as the solvent:

(1) the one-step route in which a mixture of $N_3P_3Cl_6$ and the suitable diamine (1:4) is allowed to react at room temperature for 4 hours (Chivers' technique [61]);

(2) the two-step route in which the SPIRO derivative is obtained and purified in the first step and then allowed to react with a suitable excess of diamine to lead to the expected DISPIRO chemical. (1:2) ratios of $N_3P_3Cl_6$ and diamine were used in each of these two steps, the diamine grafting the SPIRO loops onto the N_3P_3 ring and trapping HCl.

The resulting samples from these two techniques were found to be identical from their elemental analyses, melting points (DISPIRO 3 and DISPIRO 4 were higher than 280 °C (decomposed)) and IR spectra. The yields were about 60%. Thin-layer chromatography gives Rf values of 0.68 and 0.66 with CH_3OH as eluant and iodine vapour as developer.

Mass Spectrometry

The 70 eV electron impact mass spectrum of DISPIRO 3 (M = 349.95) is shown in Fig. 44. The molecular ion M^+ was observed at m/z 348 (9.1%) with a set of satellites at m/z 349 (80.4%), 350 (14.6%) and 351 (51.1%). The intensity ratio of these peaks indicates the presence of two chlorine atoms in the molecule.

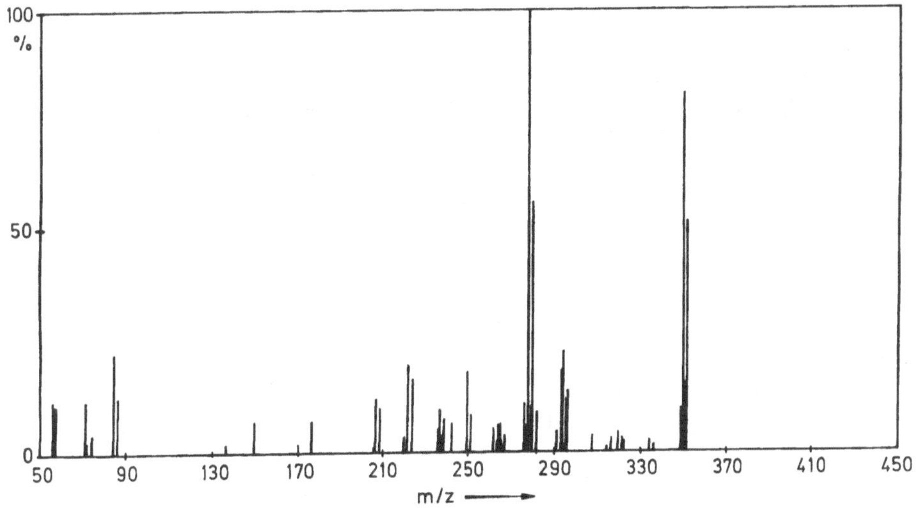

Fig. 44. 70 eV electron impact mass spectrum of DISPIRO3

One main fragmentation route involves successive losses of 1 CH_2, 2 CH_2, 3 CH_2, 3 CH_2 + 1 NH and 3 CH_2 + 2 NH (associated with H-transfers) to give maximal peaks at m/z 336 (1.3%), 321 (2.9%), 307 (3.4%), 291 (4.3%) and 276 (10.3%). Further consecutive losses of chlorine atoms give peaks at m/z 242 (6.0%) and 206 (11.3%). The remaining loop of the m/z 276 entity is fragmented in the same way as above: the successive loss of 1 CH_2, 2 CH_2, 3 CH_2, 3 CH_2 + 1 NH and 3 CH_2 + 2 NH (also associated with H-transfers) gives maximal peaks at m/z 262 (5.1%), 249 (17.0%), 236 (9.1%), 221 (18.6%) and 206 (11.3%). Further consecutive losses of chlorine atoms give minor peaks at m/z 170 (1.8%) and 136 (1.8%).

Another fragmentation route is observed with peaks at m/z 314 (8.8%) and 278 (100%), corresponding to the sucessive loss of one and two chlorine atoms (again associated with H-transfers) from the molecular ion. These two fragments lose in sequence the CH_2 and NH parts of their two loops in the same way as described above for M^+.

It is noteworthy that there are some peak overlaps provided by the different routes of fragmentation which result from the similarity of mass for two chlorine atoms (70) and for the [HN$-(CH_2)_3-$NH minus H] block (71).

In the low-mass range, peaks are observed at m/z 57 (10.1%), 71 (11.1%) and 149 (6.6%), which correspond to the HN$-(CH_2)_3$, HN$-(CH_2)_3-$N and N_3P_3N fragments. Moreover, two rather intense peaks are detected at m/z 86 (11.7%) and 84 (21.1%), which may correspond to the molecular ion of piperazine (provided by a rearrangement of four CH_2 groups and of two NH_2 groups of the DISPIRO 3 molecule) and to the [piperazine minus 2 H] fragment.

The 70 eV electron impact mass spectrum of DISPIRO 4 (M = 378.18) is shown in Fig. 45. The molecular ion M^+ was observed at m/z 377 (100.0%) with a set of satellites at m/z 378 (39.1%), 379 (66.7%) and 380 (17.4%). Fragmentation routes are quite similar to the ones detailed above for DISPIRO 3, except that the [M^+-2Cl] peak (which was the base peak (m/z 278) for DISPIRO 3) is not observed at all here.

Peaks of Fig. 45 are assigned as follows:

m/z 364 (2.4%): $M^+-1\ CH_2$; m/z 350 (3.1%): $M^+-2\ CH_2$; m/z 336 (4.5%): $M^+-3\ CH_2$; m/z 322 (11.1%): $M^+-4\ CH_2$; m/z 292 (64.6%): $M^+-4\ CH_2-2$ NH $= M_1$; m/z 278 (15.7%): $M_1-1\ CH_2$; m/z 264 (14.3%): $M_1-2\ CH_2$; m/z 250 (4.1%): $M_1-3\ CH_2$; m/z 238 (25.8%): $M_1-4\ CH_2$; m/z 221 (23.0%): $M_1-4\ CH_2-1$ NH; m/z 206 (24.1%): $M_1-4\ CH_2-2$ NH $= M_2$; m/z 172 (3.4%): M_2-1 Cl; m/z 135 (3.8%): M_2-2 Cl $= N_3P_3$.

Residual peaks can also be assigned:

m/z 342 (7.3%): M^+-1 Cl $= M_3$; m/z 287 (6.2%): $M_3-4\ CH_2$; m/z 271 (8.0%): $M_3-4\ CH_2-1$ NH; m/z 256 (6.9%): $M_3-4\ CH_2-2$ NH.

Anyhow, fragmentation routes for DISPIRO 3 and DISPIRO 4 are very similar to the ones previously described for SPIRO 3, i.e. $N_3P_3Cl_4$[HN$-(CH_2)_3-$NH] [30]

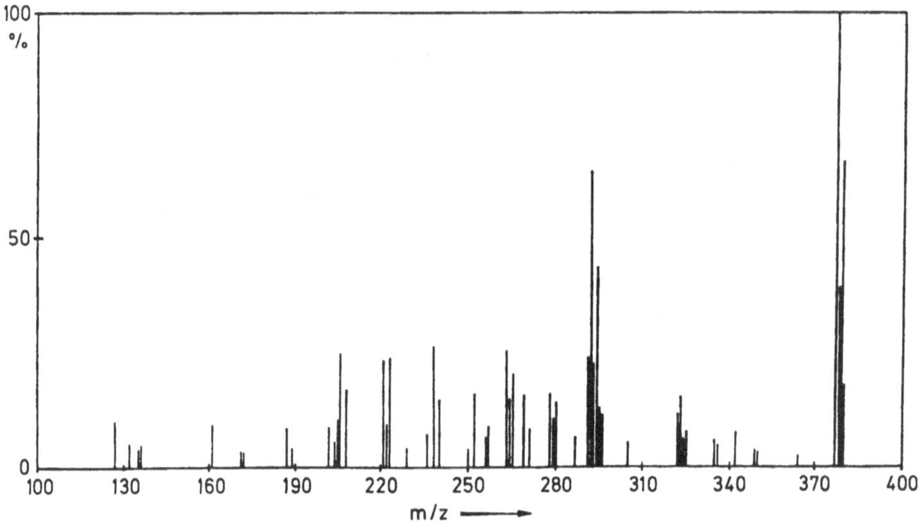

Fig. 45. 70 eV electron impact mass spectrum of DISPIRO4

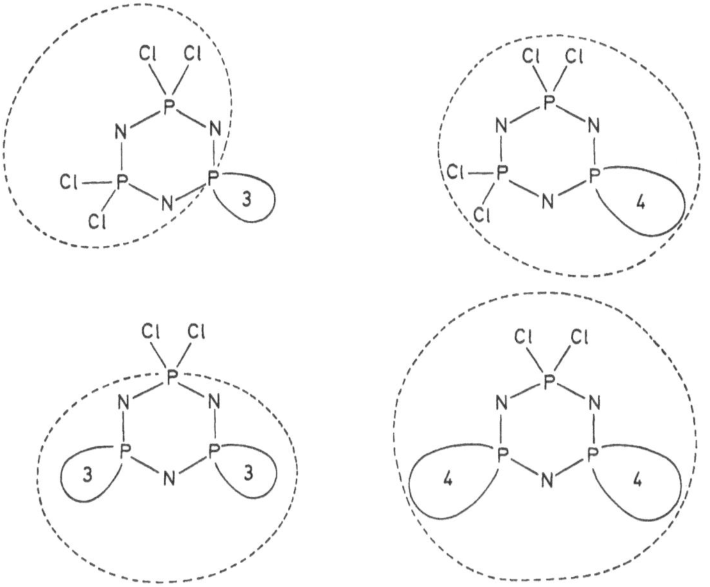

Fig. 46. Nature of the fragments corresponding to the base peaks in SPIRO3, SPIRO4, DISPIRO3 and DISPIRO4

and SPIRO 4, i.e. $N_3P_3Cl_4[HN{-}(CH_2)_4{-}NH]$[40]. However, one has to pay attention to the different kinds of base peaks (I = 100%). For these four structures we observe (Fig. 46) that base peaks for SPIRO 4 and DISPIRO 4 correspond to their molecular ions M^+, whilst the base peaks for SPIRO 3 and DISPIRO 3 correspond to M^+-loop and $M^+ - 2$ Cl, respectively. In other words, SPIRO 4 and DISPIRO 4 appear to be more stable upon electron impact than their $[HN{-}(CH_2)_3{-}NH]$-bearing relatives. These observations can presumably be related to the relative stability of the four chemicals vs. air moisture: DISPIRO 3 is indeed the only one of the series which is slowly hydrolyzed at room temperature, OH groups then replacing the "labile" Cl atoms of the structure.

NMR Spectroscopy

The ^{31}P NMR spectra of DISPIRO 3 and DISPIRO 4 were recorded on a Brucker WH 90 instrument (solvent: CD_2Cl_2; 85% H_3PO_4 as standard). The doublets at 13.6, 11.91 ppm and 17.19, 15.90 ppm, respectively, correspond to the loop-bearing phosphorus atoms and the triplets at 22.53, 21.39, 20.25 ppm and 25.66, 24.33, 22.92 ppm are attributed to the PCl_2 moieties. Table 18 shows ^{31}P NMR data for SPIRO and DISPIRO relatives. J_{P-P} values are found to be identical for SPIRO 4 and DISPIRO 4, whereas they differ when passing from SPIRO 3 to DISPIRO 3, the NMR behaviour of DISPIRO 3 being closer to that of SPIRO 4 and DISPIRO 4 than to that of SPIRO 3. In other words, the NMR behaviour of SPIRO 3 stands apart within the series and a reasonable explanation for it could be provided on the basis of its X-ray structure [30,33]: in SPIRO 3, the four phosphorus-nitrogen bonds associated with the P (loop) atom are practically equal, as well as the corresponding NPN

Table 18. ^{31}P chemical shifts and J(P—P) coupling constants (WH 90) for SPIRO3, SPIRO4, DISPIRO3 and DISPIRO4

Compound	Chemical shift (solvent CH_2Cl_2)		Δ[P(Cl_2)-P(loop)] ppm	$\lvert J_{P-P} \rvert$ Hz	Ref.
	PCl_2	P(loop)			
spiro-$N_3P_3Cl_4$[HN—$(CH_2)_3$—NH] (SPIRO 3)	21.39 20.09	8.87 7.58 6.37	13.16	41.68	[30]
dispiro-$N_3P_3Cl_2$[HN—$(CH_2)_3$—NH]$_2$ (DISPIRO 3)	22.53 21.39 20.25	13.06 11.91	8.91	47.06	This study
spiro-$N_3P_3Cl_4$[HN—$(CH_2)_4$—NH] (SPIRO 4)	21.47 20.18	14.36 13.07 11.78	7.76	47.06	[40]
dispiro-$N_3P_3Cl_2$[HN—$(CH_2)_4$—NH]$_2$ (DISPIRO 4)	25.66 24.33 22.92	17.19 15.90	7.75	47.06	This study

endo- and exocyclic bond angles. Such a T_d-like environment for a phosphorus atom belonging to a trimeric ring is rather unique in cyclophosphazenes and may explain the peculiar NMR behaviour mainfested by P (loop) in SPIRO 3. The X-ray structure of SPIRO 4 [40] indicates that the environment of P (loop) is as in other cyclophosphazenes (PN_{exo} bonds longer than PN_{endo} bonds, NPN_{endo} angles larger than NPN_{exo} angles). Thus, we could expect the X-ray structures of DISPIRO 3 and DISPIRO 4 to support this explanation of such structure-dependent NMR behaviour. Unfortunately, single crystals of both chemicals are mined very rapidly under X-ray exposure. We shall discuss this failure later.

5.1.2 The TRISPIRO 3 Derivative

Synthesis

The synthesis of TRISPIRO 3 was carried out in the same mixture of solvents as for the DISPIRO products. Three routes were followed:

1. the one-step route in which a mixture of $N_3P_3Cl_6$ and of 1,3-diaminopropane (1:6) is allowed to react at room temperature for 4 hours (Chivers' technique [61]);
2. the one-step route in which the DISPIRO 3 is allowed to react with a suitable excess of diamine;
3. the two-step route in which the SPIRO 3 derivative is obtained and purified in the first step and then allowed to react with a suitable excess of diamine to lead to the expected TRISPIRO chemical.

In every case, "suitable excess" means that the diamine is not only used for grafting the SPIRO loops on the N_3P_3 ring but also for trapping HCl. It is noteworthy that the yield of all these reactions drops dramatically either

i) when any other HCl-trapping amine is used or
ii) as soon as the ratio for the mixture of solvents differs from the exact (7:3) value.

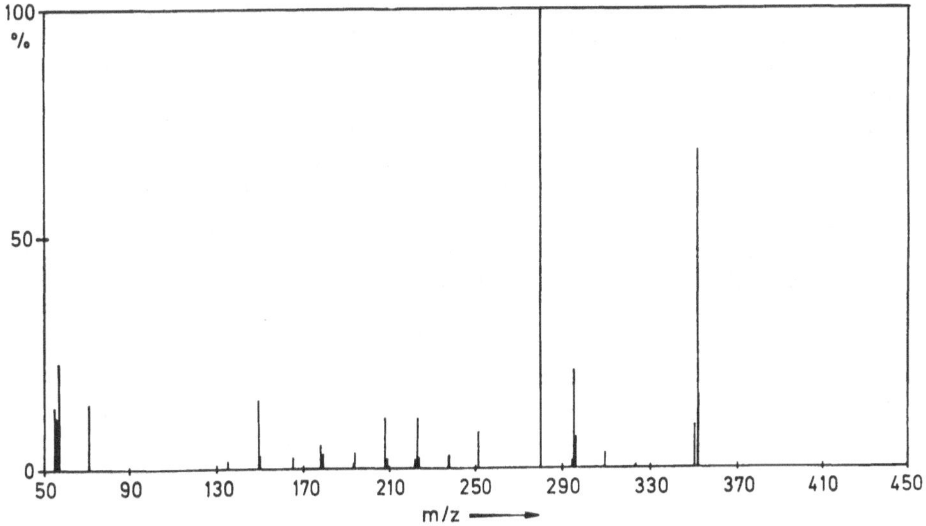

Fig. 47. 70 eV electron impact mass spectrum of TRISPIRO3

Mass Spectrometry of TRISPIRO 3

The 70 eV electron impact mass spectrum of TRISPIRO 3 (M = 351.28) is shown in Fig. 47. The molecular ion M^+ was observed as expected at m/z 351 (68.6%) with satellites at m/z 350 (8.9%) and 352 (15.1%).

There exists here a unique fragmentation route, firstly involving losses of 1 CH_2, 2 CH_2, 3 CH_2, 3 CH_2 + 1 NH and 3 CH_2 + 2 NH (associated with H-transfers) to give maximal peaks at m/z 337 (0.3%), 321 (4.6%), 309 (3.1%), 295 (20.5%) and 280 (100.0%). Thus, the loss of one SPIRO loop leads to the base peak m/z 280 (corresponding fragment coded as M1).

The two remaining loops in M1 are fragmented successively as above: the successive loss from M1 of 1 CH_2, 2 CH_2, 3 CH_2, 3 CH_2 + 1 NH and 3 CH_2 + 2 NH gives maximal peaks at m/z 266 (1.9%), 251 (7.7%), 237 (2.7%), 223 (10.4%) and 208 (10.4%). Fragmentation of the last loop from the m/z 208 fragment gives peaks at m/z 194 (3.1%), 178 (5.0%) 166 (4.6%), 149 (14.3%) and 135 (1.5%). Incidentally, the m/z 135 peak, corresponding to the N_3P_3 ring fragment, is observed here in contrast with DISPIRO 3 situation.

In the low mass range, peaks are observed (as in DISPIRO 3) at m/z 57 (22.4%) and 71 (13.5%), which correspond to the HN—$(CH_2)_3$ and HN—$(CH_2)_3$—N fragments.

X-Ray Structure of TRISPIRO 3

X-ray investigation of the TRISPIRO 3 structure displayed some interesting serendipitous features. TRISPIRO 3 crystallizes spontaneously at the end of the synthetic process, giving rise to single crystals suitable for X-ray analysis (solvent: 7:3 mixture of n-hexane and methylene chloride). However, such crystals are ruined very rapidly upon X-ray exposure and it was impossible to get enough reflections for structure refinement, even by using several crystals in sequence. A decrease of temperature from 300 K to 133 K did not help.

In order to improve the quality and the size of single crystals, we recrystallized TRISPIRO 3 in many solvents and/or mixtures of solvents and large single crystals were obtained in this way from methanol. Such crystals, inserted in a Lindemann capillary, were photographically investigated using a Weissenberg camera and Ni-filtered CuK_α radiation; the space group obtained here appeared to differ from the one which had been determined for the previous single crystals. Thus, the following problem had to be solved: either the two kinds of single crystals belong to two allotropic varieties of TRISPIRO 3 (like SOAz [62]) or crystals from methanol have clathrated some molecules of solvent (like MYKO 63 [42,46,47,63] or Allcock's clathrate [64]).

One single crystal from methanol was transferred to a SYNTEX $P2_1$ computer-controlled diffractometer. 25 reflections were used to orientate the crystal and to refine the cell dimensions. Data could be collected straightforwardly. The structure was determined using direct methods included in the Multan 80 program [49]. The refinement was performed taking into account anisotropic vibration for all atoms, except hydrogen. Fourier difference syntheses phased by the P, N and C atoms revealed not only the H atoms but also some extra electron densities located within intermolecular spaces which could be assigned to oxygen atoms of water (?)

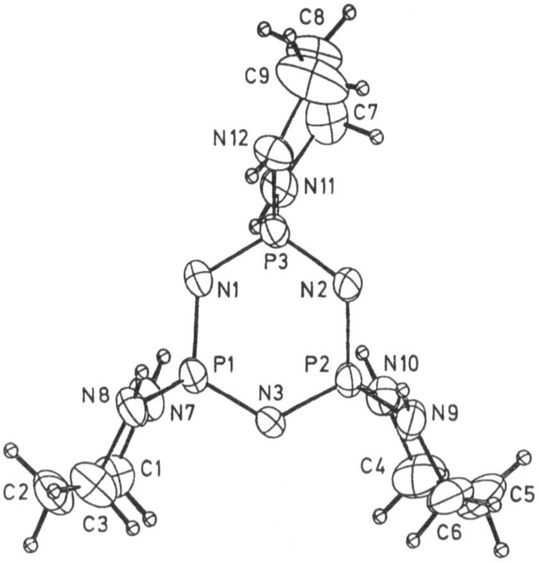

Fig. 48. Perspective view of the TRISPIRO 3A form

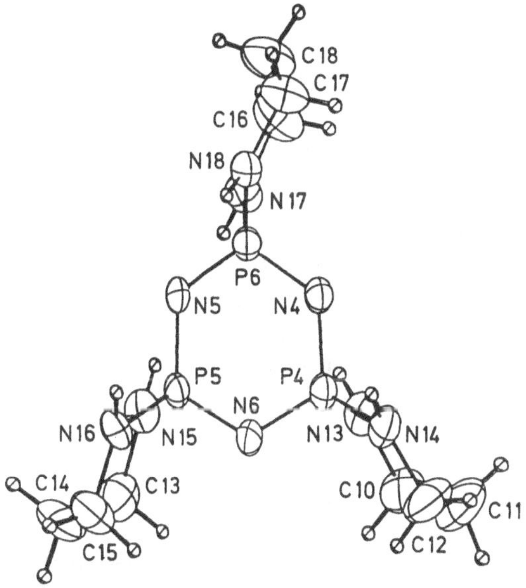

Fig. 49. Perspective view of the TRISPIRO 3B form

molecules clathrated to the TRISPIRO 3 molecules. Refinement of this clathrate was then achieved (final R = 0.094): the crystal structure is built from 4 TRISPIRO $\cdot$ 2 H_2O moieties, triclinic system, space group $P\bar{1}$, cell parameters a = 14.645(2), b = 11.518(4), c = 11.381(3) Å, α = 86.56(4), β = 103.58(2), γ = 103.92(4)°, V = 1811.27(5) $Å^3$.

Thus, single crystals of TRISPIRO 3 are not broken down by X-ray beams when water molecules are clathrated in the unit cell. These water molecules come from the wet commercial methanol which had been used for re-crystallization. Methanol

Table 19. Main bond lengths (Å) and bond angles (°) in the TRISPIRO 3A and TRISPIRO 3B forms

TRISPIRO 3A		TRISPIO 3B	
P1—N1	1.598(4) Å	P4—N4	1.602(5) Å
P1—N3	1.594(3)	P4—N6	1.610(6)
P2—N2	1.596(5)	P5—N5	1.592(5)
P2—N3	1.602(7)	P5—N6	1.600(4)
P3—N1	1.593(5)	P6—N4	1.599(3)
P3—N2	1.605(7)	P6—N5	1.611(6)
P1—N7	1.673(5)	P4—N13	1.658(6)
P1—N8	1.661(4)	P4—N14	1.647(5)
P2—N9	1.654(5)	P5—N15	1.666(4)
P2—N10	1.655(4)	P5—N16	1.664(4)
P3—N11	1.640(3)	P6—N17	1.655(5)
P3—N12	1.650(5)	P6—N18	1.652(4)
N7—C1	1.446(12)	N13—C10	1.454(15)
N8—C3	1.452(12)	N14—C12	1.460(14)
C1—C2	1.525(16)	C10—C11	1.486(17)
C2—C3	1.493(13)	C11—C12	1.485(21)
N9—C6	1.454(10)	N15—C13	1.462(11)
N10—C4	1.474(13)	N16—C15	1.488(11)
C4—C5	1.493(16)	C13—C14	1.544(13)
C5—C6	1.521(15)	C14—C15	1.442(17)
N11—C7	1.448(14)	N17—C16	1.464(11)
N12—C8	1.428(19)	N18—C17	1.440(12)
C7—C8	1.453(18)	C16—C18	1.490(15)
C8—C9	1.467(18)	C18—C17	1.486(20)
P1—N1—P3	123.6(2) (°)	P4—N4—P6	123.8(4) (°)
P1—N3—P2	123.3(2)	P4—N6—P5	121.1(3)
P2—N2—P3	120.5(4)	P5—N5—P6	122.0(4)
N1—P1—N3	114.5(3)	N4—P4—N6	114.8(4)
N1—P3—N2	115.3(4)	N5—P5—N6	116.9(3)
N2—P2—N3	115.0(3)	N4—P6—N5	114.5(4)
N7—P1—N9	99.3(2)	N13—P4—N14	102.6(3)
N9—P2—N10	104.0(3)	N15—P5—N16	101.5(3)
N11—P3—N12	103.3(4)	N17—P6—N18	100.4(5)

itself does not clathrate TRISPIRO 3; no suitable single crystals have been obtained from solutions in dry methanol.

Two crystallographically independent molecules, TRISPIRO 3A and TRISPIRO 3B, exist in the unit cell. Perspective views of these two forms (with the numbering of atoms) are represented in Figs. 48 and 49. The two molecular structures differ essentially in the conformation of one of the three loops, the two others being rather superimposable. It is noteworthy that neither the 3A nor the 3B form admits any symmetry element, i.e., neither two-fold nor three-fold axis. In other words, the three SPIRO loops in both 3A and 3B are not arranged in the "propeller-like" pattern which would have been expected if a three-fold axis had existed in the molecules. Main bond lengths and bond angles for 3A and 3B are listed in Table 19. Fractional atomic coordinates are given in Table 20.

Table 20. Fractional atomic coordinates with e.s.d.'s in parentheses for TRISPIRO3A, TRISPIRO3B and oxygens

Atom	x/a	y/b	z/c	Atom	x/a	y/b	z/c
P1	0.1929(1)	0.0518(2)	0.5742(2)	P4	0.5423(1)	0.3969(2)	0.8255(2)
P2	0.0682(1)	−0.0796(2)	0.7147(2)	P5	0.4024(1)	0.2832(2)	0.9514(2)
P3	0.0729(1)	0.1618(2)	0.6674(2)	P6	0.3675(1)	0.4724(2)	0.7956(2)
N1	0.1464(5)	0.1614(6)	0.5842(7)	N4	0.3425(5)	0.3811(6)	0.9017(6)
N2	0.0194(5)	0.0315(6)	0.7080(6)	N5	0.4762(5)	0.4897(6)	0.7784(6)
N3	0.1398(5)	−0.0707(6)	0.6260(6)	N6	0.5098(5)	0.3089(6)	0.9308(6)
N7	0.3116(5)	0.0925(7)	0.6327(6)	N13	0.5475(5)	0.3216(7)	0.7089(6)
N8	0.1950(5)	0.0379(7)	0.4312(6)	N14	0.6546(5)	0.4732(7)	0.8717(6)
N9	−0.0199(5)	−0.2009(6)	0.6893(6)	N15	0.4038(5)	0.2617(6)	1.0972(6)
N10	0.1247(5)	−0.0941(6)	0.8564(6)	N16	0.3363(5)	0.1508(7)	0.8925(6)
N11	0.1296(5)	0.2437(6)	0.7872(7)	N17	0.3484(5)	0.6024(6)	0.8229(6)
N12	−0.0097(5)	0.2337(7)	0.5988(6)	N18	0.2853(5)	0.4352(7)	0.6699(6)
O1	0.3759(7)	0.0274(9)	0.6033(9)	O10	0.6323(9)	0.2739(9)	0.7163(11)
O2	0.3654(7)	0.0273(9)	0.4668(10)	O11	0.7365(8)	0.4236(11)	0.8672(12)
O3	0.2634(7)	−0.0217(9)	0.4016(9)	O12	0.7252(9)	0.3646(14)	0.7501(13)
O4	0.1378(9)	−0.2150(10)	0.8954(9)	O13	0.4314(8)	0.1526(9)	1.1510(9)
O5	0.0460(10)	−0.3089(9)	0.8712(10)	O14	0.3662(9)	0.0400(9)	1.0819(11)
O6	−0.0056(7)	−0.3169(8)	0.7382(10)	O15	0.3694(7)	0.0464(9)	0.9561(11)
O7	0.0735(7)	0.2918(9)	0.8527(9)	O16	0.3257(8)	0.6853(9)	0.7234(9)
O8	0.0089(8)	0.3578(10)	0.7753(11)	O17	0.2442(9)	0.6276(11)	0.6247(12)
O9	−0.0535(11)	0.2967(15)	0.6658(13)	O18	0.2598(9)	0.5185(11)	0.5760(9)
				O1	0.0835(6)	0.0098(7)	0.0939(7)
				O2	0.4693(6)	0.3890(8)	0.4341(7)
				O3	0.2047(6)	0.2412(9)	0.1394(8)
				O4	0.2646(6)	0.3141(7)	0.3800(8)

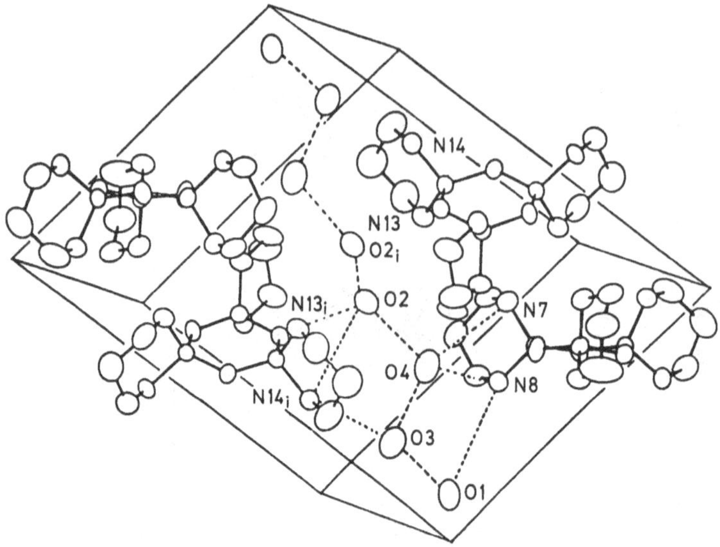

Fig. 50. Perspective view of the unit cell showing the "water-course" of water molecules and the pattern of hydrogen bonds

The perspective view of the unit cell (the 3A and 3B forms being drawn with white and black bonds for the N_3P_3 ring, respectively), as shown in Fig. 50, explains the remarkable stability of the crystal when irradiated. Two striking features are conspicuous:

i) the stacking in the crystal is due to strong intermolecular hydrogen bonds between oxygen atoms and suitable couples of N*H* groups of the loops (O4—N7: 3.79 Å;O4—N8: 3.16 Å; O2—N13i: 3.65 Å; O2—N14i: 3.96 Å; O1—N8: 3.80 Å and O3—N14i: 3.44 Å);

ii) even more interesting is the manner in which the eight oxygen atoms are arranged in the unit cell: they form a kind of water-course (indicated by dotted lines in Fig. 50) running between laminae of cyclophosphazenes, distances such as O1—O3, O3—O4 and O4—O2 being about 2.80 Å, that is comparable to the 2.75 Å value in ice. Thus, the two kinds of hydrogen bonds which exist in the unit cell confer a remarkable stability to the crystal, roughly similar to that in a crystal of ice. This is the explanation of the fact that crystals of this TRISPIRO dihydrate are so stable upon X-ray irradiation.

We tried to make use of this "dihydrate-trick" for reaching molecular structures of DISPIRO 3 and DISPIRO 4. Unfortunately, we have failed so far in obtaining suitable single crystals of such DISPIRO dihydrate from solutions in wet methanol: in every case, single crystals of the genuine chemicals are obtained which break down under X-ray exposure. The fact that DISPIRO compounds are not able to take up water from wet methanol may be related to the lower acidic character of N*H* hydrogens as measured by IR spectroscopy [65].

5.1.3 Conclusion

The reaction of $N_3P_3Cl_6$ with 1,3-diaminopropane and 1,4-diaminobutane (1:2 and 1:3) in suitable non-polar solvents allows the synthesis in a very high yield of the PNHR-containing DISPIRO and TRISPIRO derivatives without any significant side-polymerization. Another example of a DISPIRO structure was recently reported [66] upon the reaction of $N_3P_3Cl_6$ with 1,3-dihydroxypropane together with its SPIRO-ANSA isomer.

Thus, the graft on a N_3P_3 ring of two SPIRO loops or of one SPIRO loop and one ANSA arch seems to be trivial from now and we shall see in the two next sections how this conclusion is general.

5.2 The Polyspirodicyclotriphosphazenes [67]

One of the keys of the BASIC system lies in the fact that reaction of 1,3-diaminopropane (DAP) on a PCl_2-containing cyclophosphazene leads quantitatively to a purely SPIRO derivative.

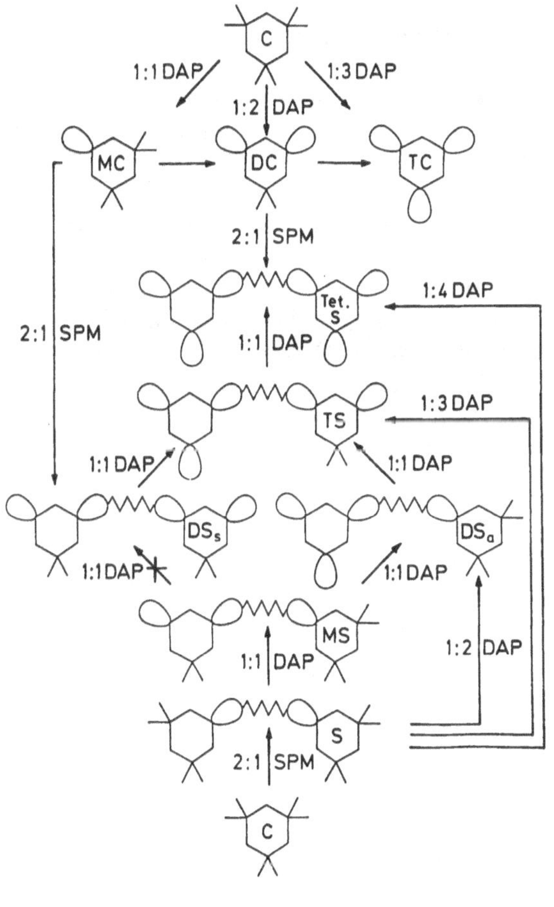

Fig. 51. Reaction pathways for the synthesis of SPIRO derivatives of SPM

Thus, we turned this structural peculiarity to account for preparing the SPIRO derivatives of SPM (see above). These new chemicals will be coded as *MS*, *DS*, *TS* and *TetS* for the MONOSPIRO, DISPIRO, TRISPIRO and TETRASPIRO configurations.

Figure 51 summarizes the various reaction pathways for preparing these derivatives and we shall detail now the direct (from SPM, *S*, itself) and the reverse (from $N_3P_3Cl_6$, *C*, itself) synthetic methods.

5.2.1 Synthesis of Polyspirodicyclotriphosphazenes

Direct Method

The starting material in this case, coded as *S*, proceeds from the reaction of $N_3P_3Cl_6$ (*C*) with spermine in 2:1 stoichiometric conditions[45,50]. A perspective view of *S* was shown in Fig. 36: the two N_3P_3 rings are cross-linked by two [HN—$(CH_2)_3$—N] SPIRO loops bridged through a BINO $(CH_2)_4$ methylenic chain. Then, four PCl_2 moieties stay available to react with 1,3-diaminopropane (DAP), leading in this way to new mono- and polySPIRO derivatives of *S*.

Reaction of DAP on *S* in (1:1), (2:1), (3:1) and (4:1) stoichiometric conditions were carried out in a 3:7 mixture of methylene chloride and 60–80 °C light petroleum as the solvent in the presence of the amount of triethylamine suitable to pick hydrogen chloride up. Reactions are achieved after two days. The crude final products are treated with a large excess of n-heptane to prevent the formation in situ of traces of clathrates given by such dicyclotriphosphazenes with CH_2Cl_2 (Traces of clathration by CH_2Cl_2 make the final products sticky and sometimes waxy. Treatments with normal heptane must be repeated till white microcrystalline powders will be obtained). Solvent is then removed in vacuo and the solids obtained were recrystallized from methylene chloride-light petroleum (60–80 °C) (3:7).

MONOSPIRO (*MS*), asym-DISPIRO (DS_a), TRISPIRO (*TS*) and TETRASPIRO (*TetS*) derivatives of *S* were prepared in such a way (Fig. 51). Molecular structures of *MS*, *TS* and *TetS* are univocal. The asymmetrical character of DS_a will be supported by NMR and mass spectrometry (see below).

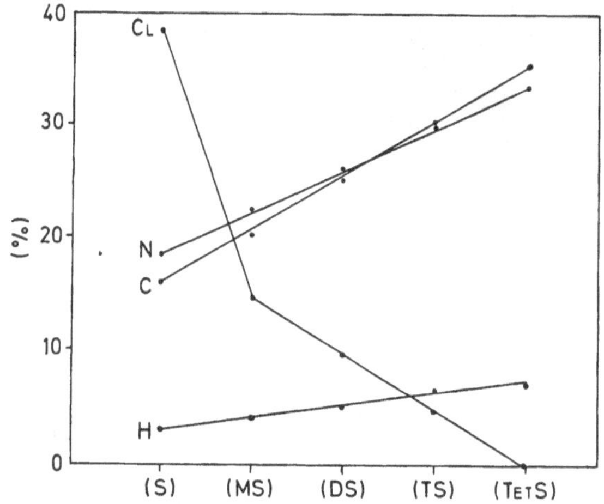

Fig. 52. Theoretical variations of C, H, N and Cl grades amongst the series

Table 21. Elemental analysis data for *S*, *MS*, DS_s, DS_a and *TS* samples

Chemical	Molecular Weight	Theoretical Values (%)	Experimental Data (%)
(S)	751.82	C: 15.97	16.01 ± 0.04
		H: 2.95	2.91 ± 0.02
		N: 18.63	18.34 ± 0.12
		Cl: 37.72	37.56 ± 0.14
(MS)	753.03	C: 20.75	20.66 ± 0.05
		H: 4.02	3.99 ± 0.02
		N: 22.34	22.40 ± 0.09
		Cl: 28.27	28.19 ± 0.13
(DS_s)	754.24	C: 25.48	25.37 ± 0.04
		H: 5.08	5.03 ± 0.02
		N: 26.00	25.88 ± 0.10
		Cl: 18.80	18.67 ± 0.14
(DS_a)	754.24	C: 25.48	25.50 ± 0.05
		H: 5.08	5.00 ± 0.02
		N: 26.00	25.78 ± 0.10
		Cl: 18.80	18.57 ± 0.11
(TS)	755.44	C: 30.21	30.14 ± 0.04
		H: 6.14	6.09 ± 0.02
		N: 29.66	29.56 ± 0.08
		Cl: 9.38	9.17 ± 0.11
		C: 34.92	34.91 ± 0.04
		H: 7.19	7.21 ± 0.02
		N: 33.32	33.19 ± 0.10

Elemental (C, H, N, Cl) analysis is a powerful tool for testing the purity of every term amongst the series. Fig. 52 shows indeed significant variations of C, H, N and Cl calculated grades when passing from *S* to *TetS*.
Experimental analytical data are gathered in Table 21.

Reverse Method

The starting material in this case are the SPIRO-$N_3P_3Cl_4[HN-(CH_2)_3-NH]$ (*MC*) and the DISPIRO-$N_3P_3Cl_2[HN-(CH_2)_3-NH]_2$ (*DC*) derivatives as provided upon reaction of DAP with *C* in (1:1) and in (2:1) stoichiometric conditions.

MC and *DC* structures containing two and one PCl_2 moieties respectively, reaction of these chemicals with spermine in (2:1) conditions may generate the symmetrical isomer DS_s of DS_a and *TetS* respectively.

Reactions are achieved in the same manner as described above for direct syntheses. The *TetS* sample obtained here is quite identical in any aspect to the one prepared from *S*. In contrast, melting point, mass spectrum and ^{31}P NMR data of DS_s differ, as expected, from DS_a characteristics.

In other words, the reverse technique allows to prepare the symmetrical DISPIRO derivative of *S* which could not be synthesized by the direct route from *S*.

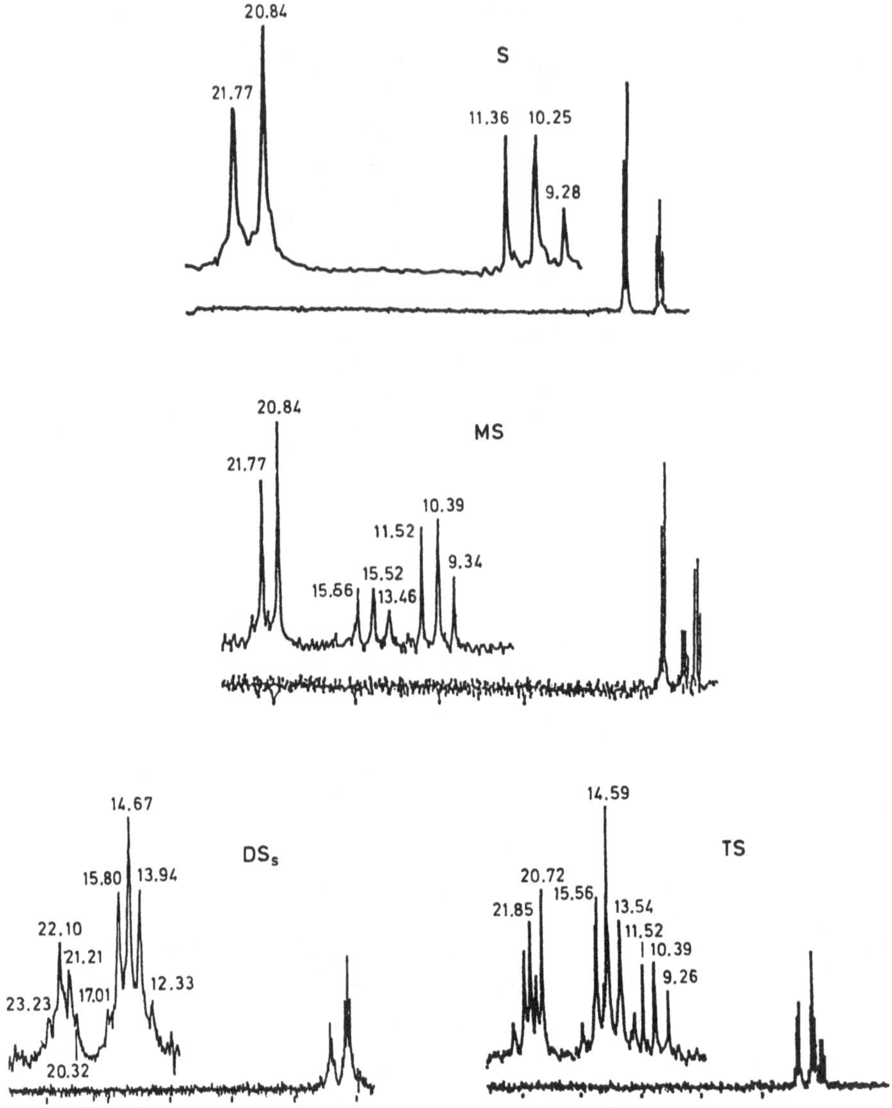

Fig. 53. ^{31}P NMR spectra (WH 90) of *S*, *MS*, *DS$_s$* and *TS*

5.2.2 ^{31}P NMR Spectroscopy

^{31}P NMR spectra (Brucker WH 90 in CD_2Cl_2 with H_3PO_4 as a standard) of *S*, *MS*, *DS$_s$* and *TS* are gathered in Fig. 53.

Spectrum of *S* is, as expected, of A_2B type, the doublet at 20.84 and 21.77 ppm being related to the PCl_2 moieties when the triplet at 9.28, 10.25 and 11.36 ppm is related to the P(SPIROBINO) entities.

Let us now consider the spectrum of *MS*. This spectrum is expected to be of ABC type, the molecule containing 3 P(Cl_2), 2 P(SPIROBINO) and 1 P(SPIRO). Signals related to $P(Cl_2)$ atoms are a doublet at 20.84 and 21.77 ppm, i.e., at exactly the same position as in *S*. The triplet at 9.34, 10.39 and 11.52 ppm corresponds to the P(SPIRO) atom. The triplet at 13.46, 14.52 and 15.56 ppm may be attributed to the P(SPIROBINO) atoms.

Spectrum of *TS* is roughly similar to the *MS* one except that patterns of the $P(Cl_2)$ doublet and of the P(SPIROBINO) triplet reveal sub-structures whose accurate first order analysis would require to record corresponding spectra at 101.27 MHz (Brucker WH 250) or even at 162.08 MHz (Brucker WH 400). However, it is noteworthy that the triplet at low-field related to P(SPIRO) moieties in *TS* keeps its "pure triplet" character as in *S* and *MS*, this triplet being superimposable (from a δ point of view) to the one in *MS*.

The spectrum of DS_s exhibits two multiplets centered on 14.67 and 22.10 ppm respectively. Such a pattern differs from the ABC-type ones in *MS* and *TS* by the fact that two of the three expected quadruplets overlap here. This assumption is supported by the record of the DS_s spectrum at 101.27 MHz: three quadruplets are then observed which are centered on 13, 15 and 22 ppm respectively (spectrum not reported).

When considering the $P(Cl_2)$ area of DS_s and *TS* spectra, a quadruplet at 23.23, 22.10, 21.21 and 20.32 ppm (DS_s) and 22.34, 21.85, 21.29 and 20.72 ppm (*TS*) is revealed.

The P(SPIROBINO) "false triplet" around 14.5 ppm stays relatively pure even if some shoulders make a certain sub-structure probable: P(SPIROBINO) triplets in DS_s and *TS* are at 13.94, 14.67 and 15.80 ppm, 13.54, 14.59 and 15.56 ppm respectively.

We have previously demonstrated that the value of δ P(SPIRO) and δ P(SPIROBINO) depends on the plus or minus Td-like neighbourhood of the loop-bearing P atom: closer to Td symmetry this neighbourhood, higher the low field shift of δ. As an example, δ is equal to 7.58 ppm for a quasi perfect Td environment as in the SPIRO$-N_3P_3Cl_4[HN-(CH_2)_3-NH]$ derivative, equal to 12–13 ppm in the DISPIRO$-N_3P_3Cl_2[HN-(CH_2)_3-NH]_2$ chemical (where significant distorsion from the Td local symmetry is observed) and equal at last to 18.6 ppm in the TRISPIRO-$N_3P_3[HN-(CH_2)_3-NH]_3$ where the neighbourhood of the three P atoms is far from a tetragonal symmetry. The evolution of δ P(SPIRO) amongst this series may be understood in terms of classical mechanistics: one loop induces a maximal stretching (from ternary symmetry) of the N_3P_3 ring (leading to a quasi-Td symmetry for the loop bearing P atom), two loops give less stretching at the total as a consequence of their resultant in the mathematical sense and three loops do not induce any stretching for symmetry reasons.

Thus, the neighbourhood of P(SPIROBINO) atoms is probably far from the Td symmetry both in *MS*, DS_s and *TS*. On the contrary, the local symmetry for P(SPIROBINO) atoms in *S* can be predicted as very close to the Td one because of the low-field shift (10.25 ppm to be compared with 14.5 ppm) of the corresponding triplet.

The environment of the six P atoms in *TetS* is the same, i.e. very far from the Td symmetry, owing to the fact that *TetS* spectrum reveals surprisingly one line (singlet) at 18.2 ppm (to be compared with the 18.6 ppm value mentioned above for the TRISPIRO derivative).

At last, comparison of spectra for the two DS_s and DS_a isomers shows that
i) the two patterns are quite identical (doublet-triplet-triplet)
ii) but DS_a chemical shifts are systematically higher than DS_s ones; values for DS_a are indeed 9.29, 10.34, 11.47 ppm for the P(SPIROBINO) moieties, 13.49, 14.62, 15.59 ppm for the P(SPIRO) atoms and 20.67, 21.80 ppm for the $P(Cl_2)$ entities (spectrum not represented). Incidentally, the triplet corresponding to the P(SPIROBINO) moieties in DS_a is centered on 10.4 ppm, conferring to the corresponding P atoms an environment similar to the ones in *S*, *MS*, *TS* and *TetS*. Parallely, the P(SPIRO) triplet in DS_a is low-field shifted versus DS_s by about 0.16 ppm when the $P(Cl_2)$ doublet in DS_a is high-field shifted versus DS_s by about 0.4 ppm. In other words, spectra of DS_s and DS_a are different enough to show that we have actually synthesize two different isomers. DS_s having been obtained by the direct method which must lead to a symmetrical compound, DS_a can be predicted as having an asymmetrical structure.

As a conclusion of this NMR study, we may notice that ^{31}P NMR looks suitable for discriminating the "symmetrical or not" character of two isomers such as DS_a and DS_s. We shall see below that mass spectrometry in this case, contrarily to what happens generally in cyclophosphazenic chemistry, is not very convenient for this purpose.

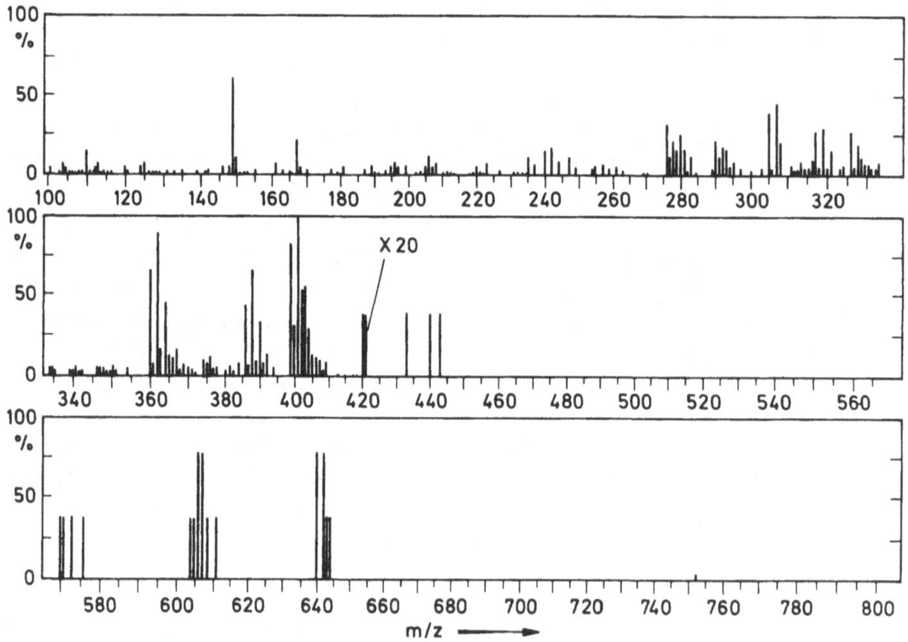

Fig. 54. 70 eV electron impact mass spectrum of *S*

5.2.3 Mass Spectrometry

Having proved on several occasions that electron impact mass spectrometry is a powerful tool for identifying and testing the purity of cyclophosphazenes, we used in a first step this technique for identification of polyspirodicyclotriphazenes described here.

As an example, the EI spectrum of *S* is visualized in Fig. 54. This figure emphasizes the hard nut to be cracked when using EI technique in cyclophosphazenes when their mass becomes higher than 700: 70 eV are generally not enough to reveal molecular ions, whatever the quality of the signal amplification is. Amplifier and multiplier have to work in saturated conditions to make intensity of M^+ ion observed but, as a consequence, several base peaks then appear in the spectra whose analysis becomes meaningless.

Thus, the EI technique is not suitable for the identification (through position and isotopic distribution around the M^+ m/z value) of compounds described in this section.

In order to reveal molecular ions, we moved to desorption/chemical ionization mass spectrometry. This "direct chemical ionization" (DCI), initially described by Mc Lafferty in 1973 [68], was widely used and improved recently with the aim of decreasing sharply thermal degradation of samples under vaporization and to extend mass spectrometry capability both to poorly volatile and to very fragile compounds having high molecular masses [69–74].

Real mechanism of DCI is still unknown. It seems however that all concur in the belief that DCI spectra reveal two kinds of ions: those provided by ionization of the

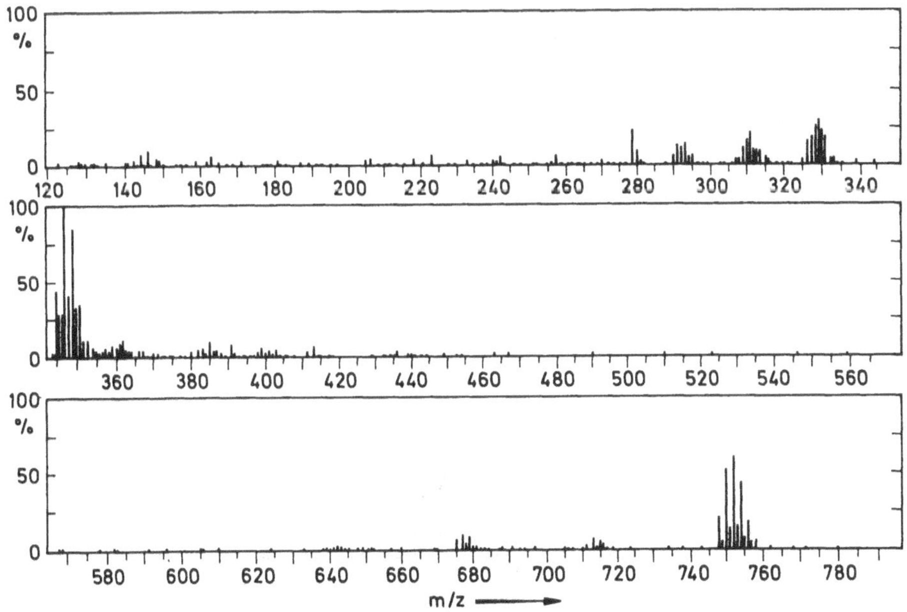

Fig. 55. DCI mass spectrum of *S*

molecule itself with ionic decomposition in following of M^+ and those provided by pyrolitic process with ionization in sequence of desorbed neutral fragments [75].

DCI spectra were recorded on the Ribermag R1010 mass spectrometer. NH_3 is used as ionizing vector. The sample is laid down on a tungsten spiral wire as transmitter. Filament is continuously heated, intensity varying from 0 to 600 mA with a speed of about 5 $\mu A \cdot sec^{-1}$. The scan speed was as high as possible, corresponding to an integrator speed of 1 msec.

The DCI spectrum of *S* is presented Fig. 55. The MH^+ molecular ion is clearly revealed at normal amplification but the $(M + NH_4)^+$ related peak is not observed in that case. The superfine substructure of the MH^+ ion is visualized in Fig. 56 whose computational analysis by classical techniques supports the existence of eight Cl atoms in the molecule.

It is noteworthy that fragmentation of *S* by DCI has nothing to do with the one by EI: the base peak in DCI is observed at m/z 347 when it is observed at m/z 401 in EI. The m/z 401 peak corresponds to $N_3P_3Cl_4[HN—(CH_2)_3—N](CH_2)_4$ when the m/z 347 corresponds to a sub-fragment of the previous one, namely $N_3P_3Cl_4[HN—(CH_2)_3—N]$ (taking into account hydrogen transfers).

Needless to say that such a well-known discrepancy between EI and DCI spectra, which was commonly observed in other fields of chemistry [75–77], makes analysis of real fragmentation routes quite intricate. Actually, the DCI technique appears to be really useful in our case from two points of view only:

i) for assigning the mass of the chemical (through MH^+ and $(M + NH_4)^+$ peaks) and

ii) for comparing fragmentation modes of two isomers like DS_a and DS_s. In that case indeed, the comparison of the two DCI spectra is meaningful *under the express*

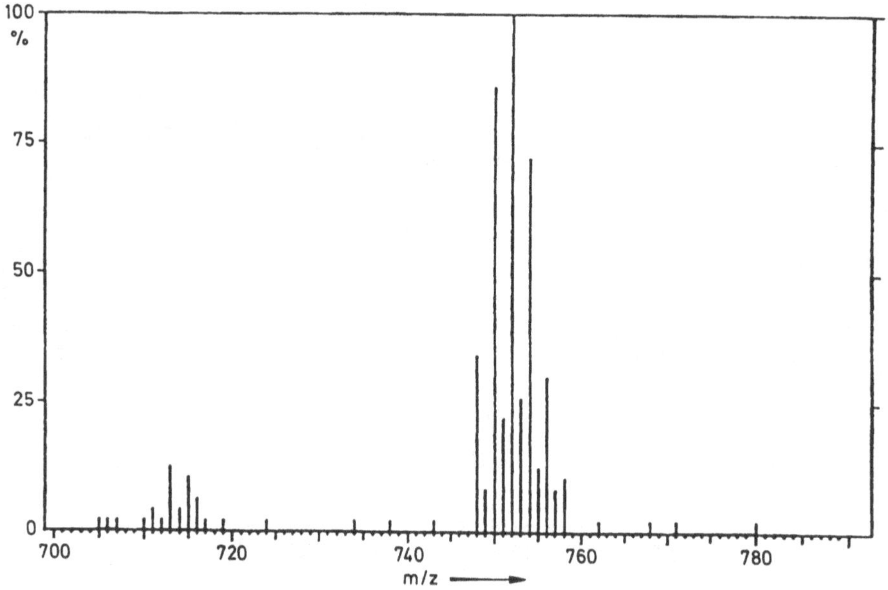

Fig. 56. Superfine substructure of the MH^+ ion from *S* (DCI technique)

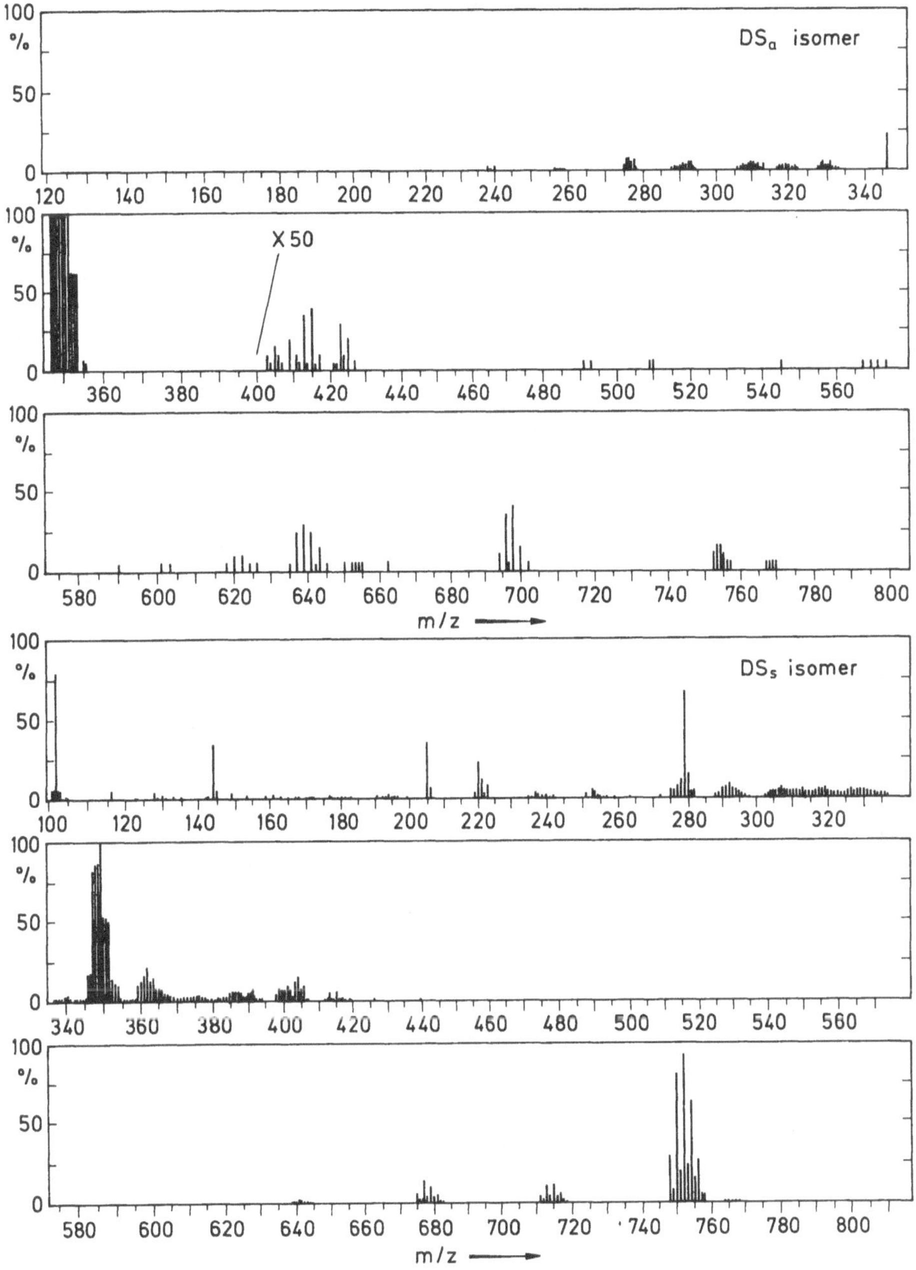

Fig. 57. DCI spectra of DS_a and DS_s isomers

condition that the two spectra are recorded in sequence on the same apparatus by the same operator [75].

Figure 57 gathers DCI spectra of DS_a and DS_s cautiously recorded in this way. The two spectra look quite different, mainly in the "over 700" and "around 280"

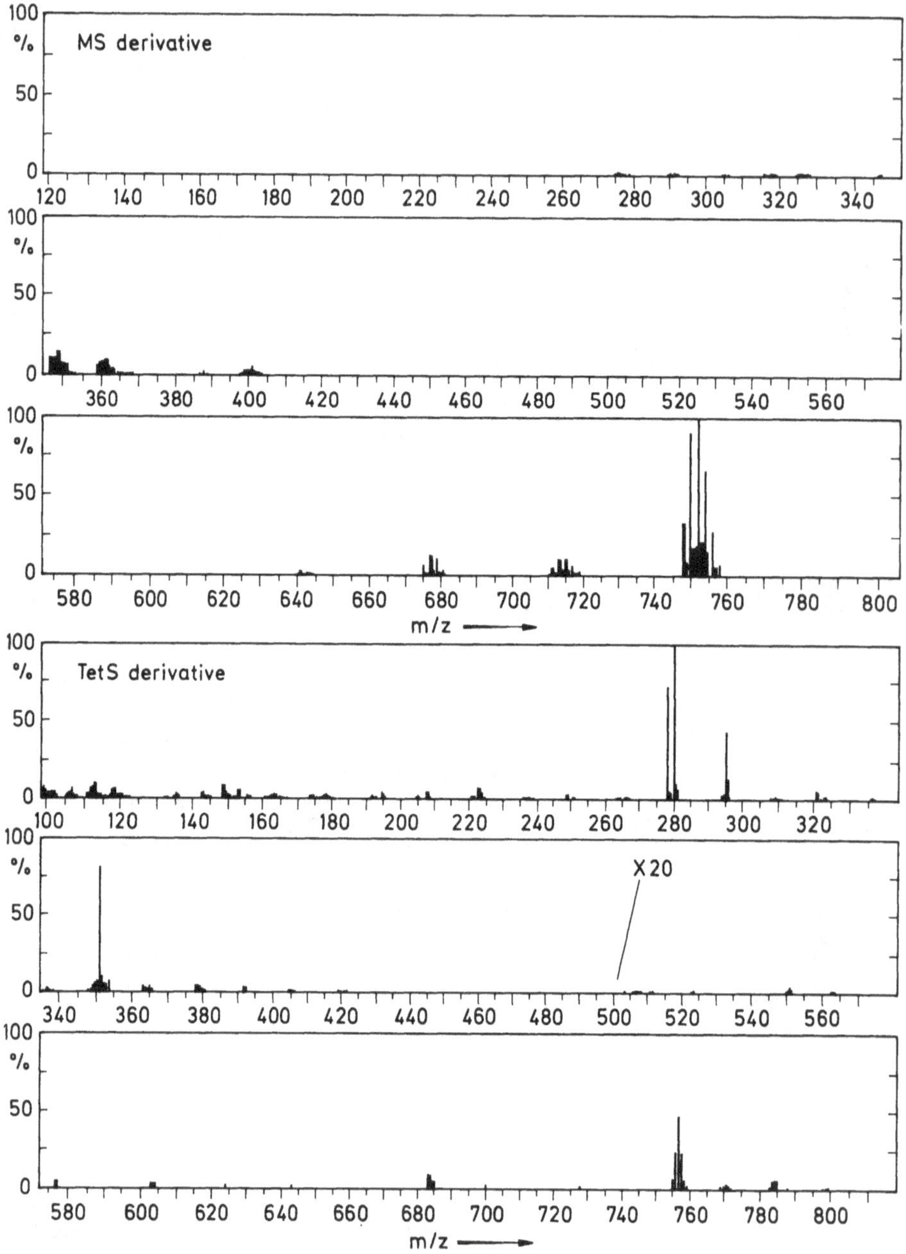

Fig. 58. DCI spectra of *MS* and *TetS*

areas. In contrast, base peak is observed around m/z 350 in both cases. Moreover, the $(M + NH_4)^+$ peak is detected for DS_s when it is not for DS_a despite of the high intensity of the corresponding MH^+ peak. It seems once more that the relative magnitude of MH^+ and $(M + NH_4)^+$ peaks can be considered as random even when spectra are cautiously registered.

Anyhow, the fact that mass patterns for DS_a and DS_s look so different prompt us to think that the two DISPIRO derivatives of *S* prepared by direct and reverse method respectively are not the same.

At last, DCI spectra of *MS* and *TetS* are indicated in Fig. 58. The former displays essentially a MH^+ peak (which is the base peak) without significant $(M + NH_4)^+$ satellite when the latter displays MH^+ *and* $(M + NH_4)^+$ peaks, its base peak being observed at m/z 278 which corresponds to the $N_3P_3[HN{-}(CH_2)_3{-}NH]_2$ fragment (taking into account H-transfers).

Our concluding remarks about the use of the DCI technique in our case will be the following

i) DCI allows to reveal in a gentle manner molecular ions (through MH^+ and $(M + NH_4)^+$ peaks) and to identify unambiguously chemicals

ii) comparison of fragmentation routes through a series are in contrast much more tricky. Thus, the DCI technique, which is so powerful for investigations of *mixtures* of compounds (namely in the field of biological metabolites), does not seem so helpful for studying fragmentation patterns. The EI technique in that case appears more powerful from this view point, at least in cyclophosphazenes, whatever the technical problems inherent in high molecular masses.

5.2.4 Conclusion

The synthesis of the first polyspirodicyclotriphosphazenes was achieved

i) upon reaction of 1,3-diaminopropane on the product of the reaction of $N_3P_3Cl_6$ with spermine (direct route) and

ii) upon reaction of spermine on the SPIRO-$N_3P_3Cl_4[HN{-}(CH_2)_3{-}NH]$ and the DISPIRO-$N_3P_3Cl_2[HN{-}(CH_2)_3{-}NH]_2$ derivatives. The trick for getting these chemicals in the monomeric state with a high yield is to use a sharp (3:7) mixture of methylene chloride and 60–80 °C light petroleum as the solvent. Molecular structures were ascribed by using in concert ^{31}P NMR spectra and EI/DCI mass spectrometry techniques. Relative merits of EI and DCI spectrometries are discussed: DCI is the suitable tool for revealing molecular ions of compounds whose molecular mass is higher than 700 but EI is more convenient for analysis of fragmentation routes and of molecular bonds relative fragility.

5.3 The Second SPIRO-ANSA Cyclotriphosphazene and the first DISPIRO-DIANSA-BINO Cyclotriphosphazene as Examples of the BASIC System [78]

This section reports on the synthesis and characterization of the second SPIRO-ANSA structure (for the first one, see Ref. [66]) and of the first DISPIRO-DIANSA-BINO cyclophosphazene.

The ^{31}P spectrum (Brucker WH 90) of the starting material, e.g., the ANSA-$N_3P_3Cl_3(CH_3)[HN{-}(CH_2)_3{-}O]$, is presented in Fig. 59. We added 1,3-diaminopropane (1:2) (half of the diamine being used for trapping HCl) in the NMR tube containing 58 mg of the ANSA material in $CDCl_3$ and 1 hour later a new ^{31}P spectrum was recorded which is presented in Fig. 60.

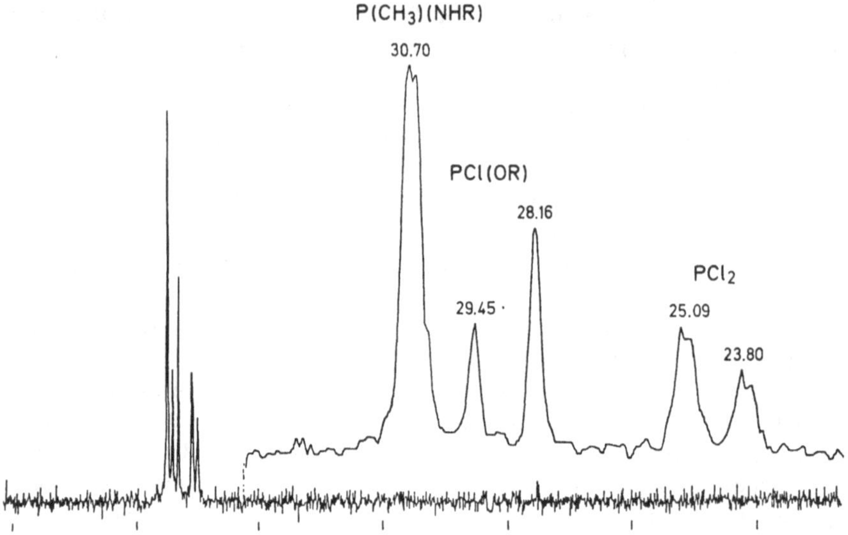

Fig. 59. ^{31}P NMR spectrum (WH 90) of the ANSA—$N_3P_3Cl_3(CH_3)[HN—(CH_2)_3—O]$

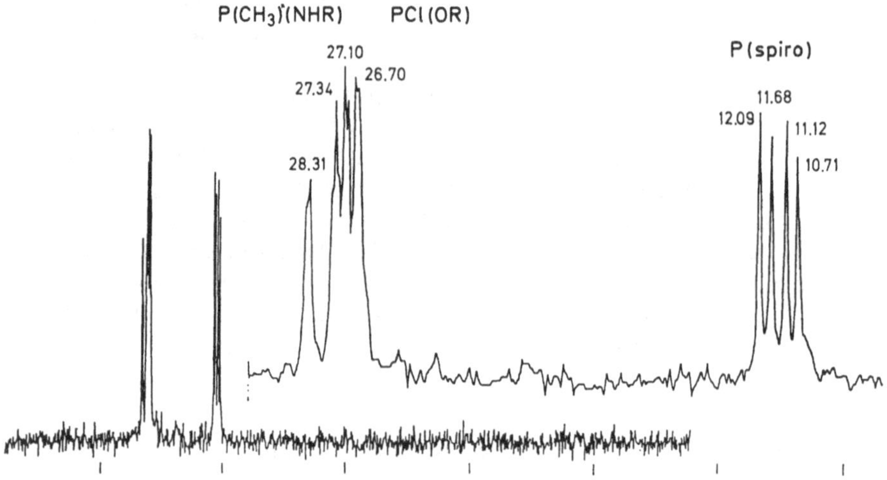

Fig. 60. ^{31}P NMR spectrum (WH 90) of the SPIRO—ANSA—$N_3P_3Cl(CH_3)[HN—(CH_2)_3—NH]$ $[HN—(CH_2)_3—O]$

Comparison of Figs 59 and 60 shows that the PCl_2 doublet of doublets around 24.5 ppm of the ANSA starting material has vanished and that the expected doublet of sharp doublets around 11.4 ppm is revealed which may be assigned unambiguously to a SPIRO loop-bearing phosphorus atom in an ABC-type cyclotriphosphazene. Infrared spectrum of this new compound (I) revealing a unique P—Cl band at 555 cm^{-1}, we can assign to (I) the SPIRO-ANSA structure which is visualized in Fig. 61.

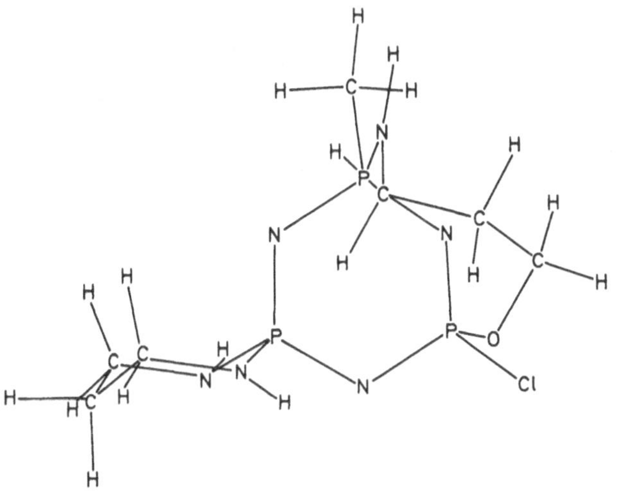

Fig. 61. The SPIRO—ANSA structure assumed for (I) from IR spectroscopy and ^{31}P NMR

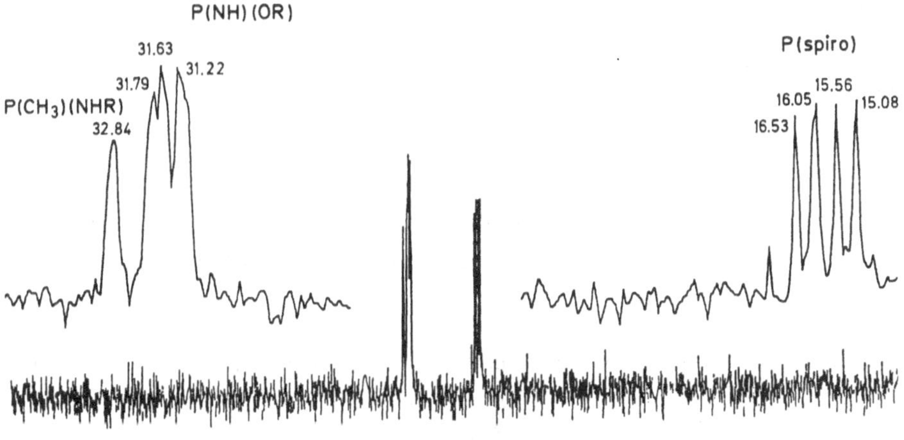

Fig. 62. ^{31}P NMR spectrum (WH 90) of the DISPIRO—DIANSA—BINO—$\{N_3P_3(CH_3)[HN-(CH_2)_3-NH][HN-(CH_2)_3-O](CH_2)_2\}_2$

This structure contains one remaining labile Cl atom. 43 mg of (I) in $CDCl_3$ were treated as previously in a NMR tube with 15.1 mg of 1,6-diaminohexane, i.e. in (2:1 + 1) conditions, half of the diamine being used for trapping hydrogen chloride.

A new ^{31}P NMR spectrum was registered after 2 hours which is presented in Fig. 62. The doublet of sharp doublets has shifted from 11.4 to 15.8 ppm. Such a high-field shift of about 4 ppm is a characteristics of the building somewhere in the molecule of a BINO bridge. This new compound (II) is also an ABC-type cyclophosphazene whose molecular weight from EI and DCI mass spectrometry is equal to 674. Infrared spectrum of (II) does not reveal P—Cl stretching frequency anymore. Thus, we may reasonably assign to (II) the DISPIRO-DIANSA-BINO structure of Fig. 63. X-ray structures of (I) and (II) are now in progress in our Laboratory.

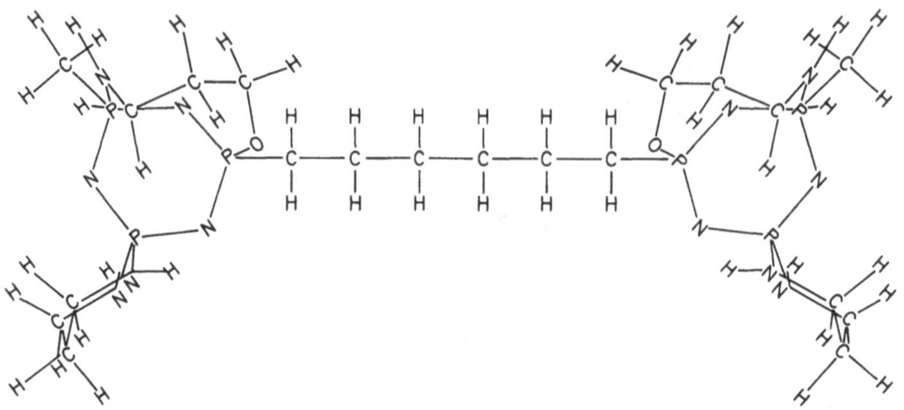

Fig. 63. The DISPIRO—DIANSA—BINO structure assumed for (II) from IR spectroscopy and ^{31}P NMR

As a conclusion, we must emphasize once more the high stereospecificity and stereoselectivity of the two reactions described here. It seems from now that any merged chemical containing several SPIRO and/or ANSA and/or BINO moieties may be commonly synthesized in a quantitative manner. This is at least a nice game for the chemist and we propose to label this game as "*BASIC*", i.e. "*B*INO*A*NSA-*S*PIRO *I*N *C*YCLOPHOSPHAZENES.

6 Attempts at the Production of more Selective Antitumorals from the Polyamines-Linked Chlorinated Cyclophosphazenes Described above as Precursors [19,79,80]

The reaction of $N_3P_3Cl_6$ with natural polyamines as tumor finders leads then stereoselectively and regioselectively to well-defined specific configurations, mainly SPIRO, ANSA, BINO, DISPIRO, TRISPIRO, SPIRO-ANSA, SPIRO-BINO, DISPIRO-BINO, DISPIRO-DIANSA-BINO.

The corresponding polyamines-linked chlorinated cyclophosphazenes constitute potential precursors of antitumor agents when their structures contain three chlorine atoms at least [81,82].

As a first step in exploring the potential of this approach, this section describes the synthesis, chemical characteristics and the initial pharmacological data of the two novel antitumor agents obtained through peraziridinylation of the SPIRO-$N_3P_3Cl_4$ [HN—$(CH_2)_3$—NH] and $N_3P_3Cl_4$[HN—$(CH_2)_4$—NH] derivatives.

6.1 Experimental

6.1.1 The SPIRO-$N_3P_3Az_4$[HN—$(CH_2)_3$—NH] Vectorized drug

Synthesis

48.5 mmole of aziridine in 35 ml of a 2:1 mixture of petroleum ether 60–80 °C and CH_2Cl_2 are added dropwise in 2 hours to a mixture of 11.0 mmole of SPIRO-$N_3P_3Cl_4$

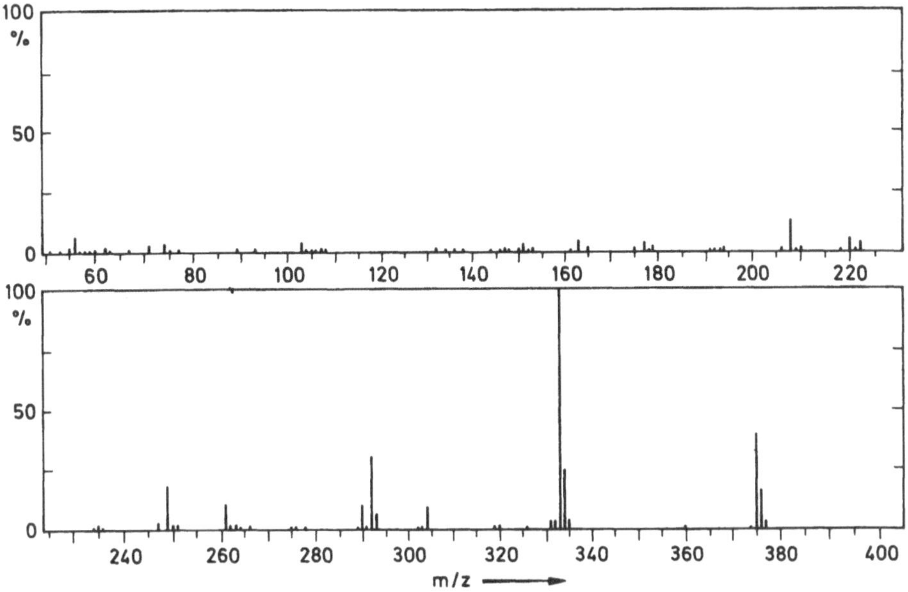

Fig. 64. 70 eV electron impact mass spectrum of (II)

[HN$-$(CH_2)$_3$$-$NH] (I) and of 44.5 mmole of NEt_3 in 120 ml of the same solvent The medium is stirred under argon pressure in an icebath. The reaction takes 48 hours and is considered complete when the IR stretching frequencies (525 and 580 cm^{-1}) of the P$-$Cl bonds in (I) have disappeared. Hydrochloride is then filtered off, solvent is removed in vacuo at 30 °C to give a residue which, upon recrystallization from dry CCl_4, yields the title compound (II) (4 g, 87%), m.p. 134 °C, t.l.c. Rf = 0.68 with CH_3OH as eluant. The analytical data obtained (calculated values in parentheses): C% 34.98 (35.20), H% 6.40 (6.44), N% 33.38 (33.59) and P% 24.59 (24.76) are highly consistent with the title structure.

Mass Spectrometry of (II)

The spectrum was recorded (Fig. 64) on a R1010 Ribermag quadrupole mass spectrometer using a direct inlet system.

The molecular ion M^+ is observed at m/z 375 (38.2%) with two satellites at m/z 376 (16.2%) and 377 (2.8%) due to nitrogen isotopic distribution. One major fragmentation route is detected involving the successive loss of 1–4 Az groups (associated with H-transfers) to give maximal peaks at m/z 333 (100.0%), 292 (29.2%), 249 (17.9%) and 208 (12.9%). Another fragmentation route for M^+ is observed in which the diamino loop is firstly expelled as a whole (m/z 304, 9.5%), the resulting fragment then successively losing 1–4 Az moieties to give maximal peaks at m/z 261 (10.6%), 220 (5.0%), 177 (3.3%) and 135 (1.1%).

NMR Spectroscopy of (II)

The ^{31}P NMR spectrum, recorded on a Brucker WH 90 instrument, exhibits a doublet at 38.34 and 37.45 ppm (PAz_2 entities) and a triplet at 19.69, 18.80 and 17.91 ppm

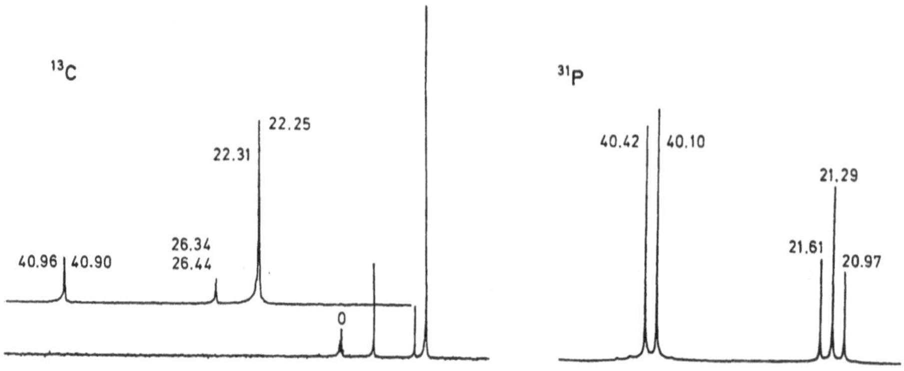

Fig. 65. ^{13}C and ^{31}P NMR spectra of (II) (WH 250)

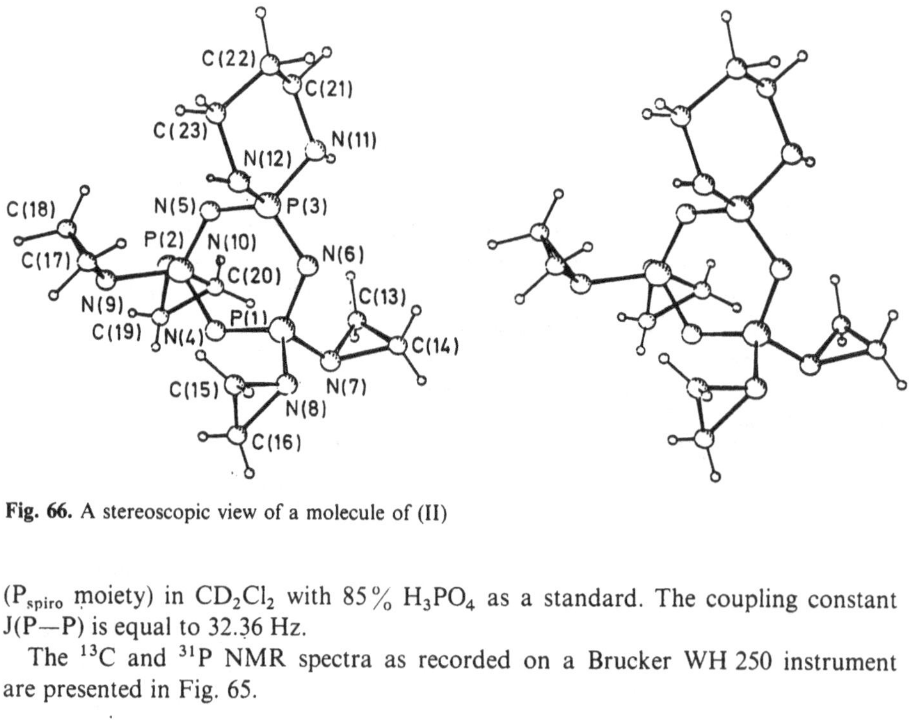

Fig. 66. A stereoscopic view of a molecule of (II)

(P_{spiro} moiety) in CD_2Cl_2 with 85% H_3PO_4 as a standard. The coupling constant J(P—P) is equal to 32.36 Hz.

The ^{13}C and ^{31}P NMR spectra as recorded on a Brucker WH 250 instrument are presented in Fig. 65.

X-ray Crystal and Molecular Structure of (II)

Single crystals, suitable for X-ray crystallography, were obtained directly from the synthesis of the compound (solvent CCl_4 or CH_2Cl_2). The selected crystal was inserted in a Lindemann capillary to prevent a possible attack by moisture. Photographic investigations were conducted using a Weissenberg camera and a Ni-filtered Cu K_α radiation.

The chosen single crystal was a colourless small prism, 0,60 mm long. It was readily mounted on an automatic SYNTEX $P2_1$ diffractometer. The cell parameters were

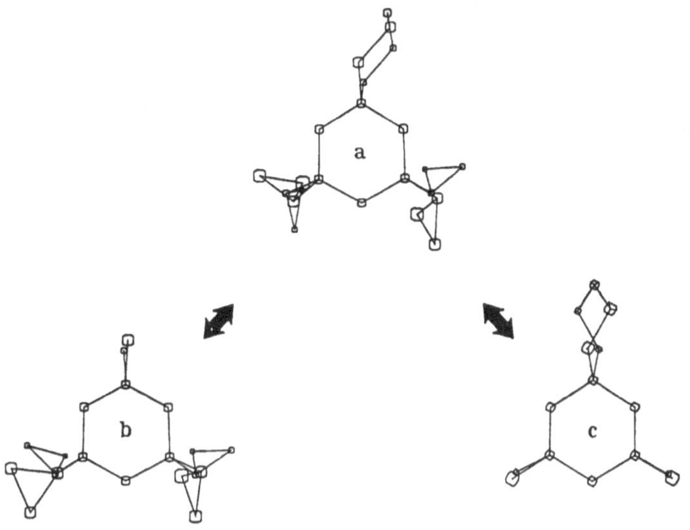

Fig. 67. Az conformations and loop configuration in (II), (a), versus gem—$N_3P_3Az_4Cl_2$, (b), and (I), (c)

determined at room temperature by optimizing the settings of 25 reflections hkl spread out in the reflecting sphere (MoK_α radiation).

The title compound crystallizes in the monoclinic system, space group $P2_1/a$, with cell parameters a = 18.369(3), b = 9.715(1), c = 9.722(2) Å, β = 94.79(1)°, V = 1728.9(5) Å^3, Z = 4, final R = 0.0470.

A stereoscopic view is shown in Fig. 66 together with the numbering of the atoms. This view clearly illustrates the SPIRO structure, the six-membered phosphazene ring being strictly planar. A second view (with the N_3P_3 ring in the plane) reveals considerable puckering of the SPIRO loop (Fig. 67). This figure also presents the same views for

i) gem-$N_3P_3Az_4Cl_2$ and

ii) SPIRO-$N_3P_3Cl_4$[HN—$(CH_2)_3$—NH] (I)

and alterations in both the Az conformations and in the loop configuration when

i) substitution of Cl atoms by the loop and

ii) substitution of Cl atoms by Az groups.

Main bond lengths and bond angles are listed in Table 22. Exocyclic nitrogen atoms display a pronounced pyramidal character as previously observed in genuine MYKO 63 [63], in MYKO 63.$3CCl_4$ anticlathrate [42], in 2 MYKO 63.C_6H_6 clathrate [47], in the three allotropic varieties of SOAz [62], in SOF and SOPHi [83] and in MYCLAz itself [57]. Their distance from the corresponding PCC planes is remarkably constant and equal to 0.69 ± 0.01 Å.

The distorsion of the diamino loop may be estimated by distances between the C(21) and C(23) atoms and the N(11)—P(3)—N(12) plane (which contains C(22)): these distances are equal to 0.296(3) and 0.311(2) Å, respectively.

Some interesting features are observed when comparing the neighbourhoods of atom P(3) in the title compound and in the starting chlorinated material (I): the

Table 22. Bond lengths and angles (with e.s.d.'s in parentheses) in (II)

P(1)—N(4)	1.594(3)	N(7)—C(13)	1.461(6)
P(1)—N(6)	1.575(2)	N(7)—C(14)	1.469(6)
P(1)—N(7)	1.686(4)	C(13)—C(14)	1.455(8)
P(1)—N(8)	1.682(3)	N(8)—C(15)	1.476(5)
P(2)—N(4)	1.599(3)	N(8)—C(16)	1.479(6)
P(2)—N(5)	1.571(4)	C(15)—C(16)	1.475(8)
P(2)—N(9)	1.689(3)	N(9)—C(17)	1.464(6)
P(2)—N(10)	1.690(4)	N(9)—C(18)	1.474(7)
P(3)—N(5)	1.611(2)	C(17)—C(18)	1.457(8)
P(3)—N(6)	1.599(3)	N(10)—C(19)	1.454(6)
P(3)—N(11)	1.678(5)	N(10)—C(20)	1.461(5)
P(3)—N(12)	1.644(3)	C(19)—C(20)	1.459(7)
		N(11)—C(21)	1.482(6)
		N(12)—C(23)	1.492(6)
		C(21)—C(22)	1.508(7)
		C(22)—C(23)	1.519(8)
N(4)—P(1)—N(6)	117.9(1)	P(1)—N(7)—C(13)	117.1(3)
N(4)—P(1)—N(7)	111.2(3)	P(1)—N(7)—C(14)	119.1(1)
N(6)—P(1)—N(7)	107.1(5)	C(13)—N(7)—C(14)	59.5(5)
N(4)—P(1)—N(8)	107.0(7)	P(1)—N(8)—C(15)	116.4(8)
N(6)—P(1)—N(8)	112.6(6)	P(1)—N(8)—C(16)	118.1(2)
N(7)—P(1)—N(8)	99.2(7)	C(15)—N(8)—C(16)	59.8(6)
P(1)—N(4)—P(2)	121.7(1)	P(2)—N(9)—C(17)	118.1(8)
N(4)—P(2)—N(5)	117.2(5)	P(2)—N(9)—C(18)	116.4(3)
N(4)—P(2)—N(9)	112,5(4)	C(17)—N(9)—C(18)	59.4(6)
N(5)—P(2)—N(9)	106.7(6)	P(2)—N(10)—C(19)	119.1(1)
N(4)—P(2)—N(10)	105.4(1)	P(2)—N(10)—C(20)	117.5(5)
N(5)—P(2)—N(10)	114.1(5)	C(19)—N(10)—C(20)	60.0(4)
N(9)—P(2)—N(10)	99.4(3)	P(3)—N(11)—C(21)	114.6(8)
P(2)—N(5)—P(3)	122.0(7)	P(3)—N(12)—C(23)	117.3(5)
N(5)—P(3)—N(6)	115.7(4)	N(11)—C(21)—C(22)	112.5(4)
N(5)—P(3)—N(11)	107.9(7)	N(12)—C(23)—C(22)	109.9(8)
N(6)—P(3)—N(11)	108.7(3)	C(21)—C(22)—C(23)	112,3(4)
N(5)—P(3)—N(12)	111.6(3)		
N(6)—P(3)—N(12)	109.0(1)		
N(11)—P(3)—N(12)	102.9(4)		
P(3)—N(6)—P(1)	122.1(8)		

presence of the loop pulls P(3) away from N(4) along the two-fold axis here as in (I) [33], the endocyclic P(3)—N(5) and P(3)—N(6) bonds being much longer (1.599 and 1.611 Å) than the average value (~1.58 Å) normally expected for a trimeric cyclophosphazene ring. However, the endocyclic N(5)—P(3)—N(6) angle is less pinched here (115.74°) than in (I) (111.5°), and exocyclic P(3)—N(11) and P(3)—N(12) bond lengths are much longer here (1.678 and 1.644 Å) than in (I) (1.618 Å). In other words, the peculiar pseudo-tetrahedral situation which had been observed for P(3) in (I) disappears when the four chlorine atoms are replaced by four Az groups and, consequently, the ^{31}P signal for the loop-bearing P(3) atom, which was at 7.58 ppm [33] for (I), shifts to the normal value, i.e. within the 18–22 ppm range [29].

6.1.2 The SPIRO-$N_3P_3Az_4$[HN—$(CH_2)_4$—NH] Vectorized Drug

Synthesis

79.39 mmole of aziridine in 50 ml of a 4:1 mixture of petroleum ether 60–80 °C and CH_2Cl_2 is added dropwise in 2 hours to a mixture of 12.23 mmole of SPIRO $N_3P_3Cl_4$[HN—$(CH_2)_4$—NH] (III) and of 58.2 mmoles of NEt_3 in 300 ml of the same solvent. The medium is stirred under argon pressure in an ice-bath during 2 days and the reaction is considered complete when the IR stretching frequencies (530 and 585 cm^{-1}) of the P—Cl bonds in (III) have disappeared. Hydrochloride is then filtered off, solvent is removed in vacuo at 40 °C to give a white powder which, upon recrystallization from dry CCl_4, yields the title compound (IV) (3.5 g, 68%, m.p. 171 °C, t.l.c. Rf = 0.41 with CH_3OH as eluant). Analytical data (calculated values in parentheses) were: C% 36.93 (37.02), H% 6.75 (6.75), N% 32.08 (32.38) and P% 23.75 (23.87).

Mass Spectrometry of (IV)

The mass spectrum of the title compound is shown in Fig. 68. The molecular ion M^+ is observed at m/z 389 (83.3%). One major fragmentation route is detected involving the successive loss of 1–4 Az groups (associated with H-transfers) to give maximal peaks at m/z 347 (96.8%), 304 (100.0%), 263 (30.9%) and 220 (20.6%). Each of these sub-fragments of M^+ may lose 1 CH_2, 2 CH_2, 3 CH_2, 4 CH_2, 4 CH_2 + 1 NH and 4 CH_2 + 2 NH from the loop to give peaks of minor importance, again associated with hydrogen transfers. The peak at m/z 135, corresponding to the N_3P_3 ring alone, is observed here with high intensity (3.9%) in comparison to that observed for other cyclotriphosphazenes.

It is noteworthy that the nature and magnitude of the peaks given by fragmentation

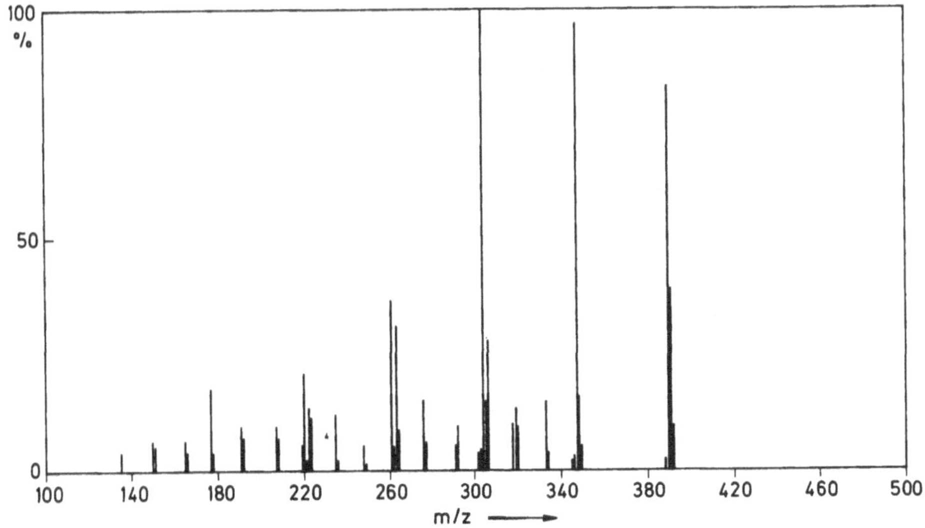

Fig. 68. 70 eV electron impact mass spectrum of (IV)

of M^+ are very similar to those detailed above for SPIRO-$N_3P_3Az_4$[HN—$(CH_2)_3$—NH] (Fig. 64). Both mass spectra are indeed superimposable, except for the m/z 135 peak. This similarity supports the assumption of a SPIRO structure for the title compound, in agreement with the analytical data. X-ray investigations are now in progress with the aim of providing a definite proof of such a SPIRO structure.

NMR Spectroscopy of (IV)

The ^{31}P NMR spectrum, recorded on a Brucker WH 90 instrument, displays a doublet at 38.74 and 37.77 ppm (PAz_2 entities) and a triplet at 23.65, 22.68 and 21.71 ppm which may be reasonably assigned to the loop-bearing P atom. J(P—P) coupling constant is equal to 35.29 Hz (CD_2Cl_2, 85% H_3PO_4).

The ^{13}C and ^{31}P NMR spectra, as recorded on a Brucker WH 250 instrument, are shown in Fig. 69.

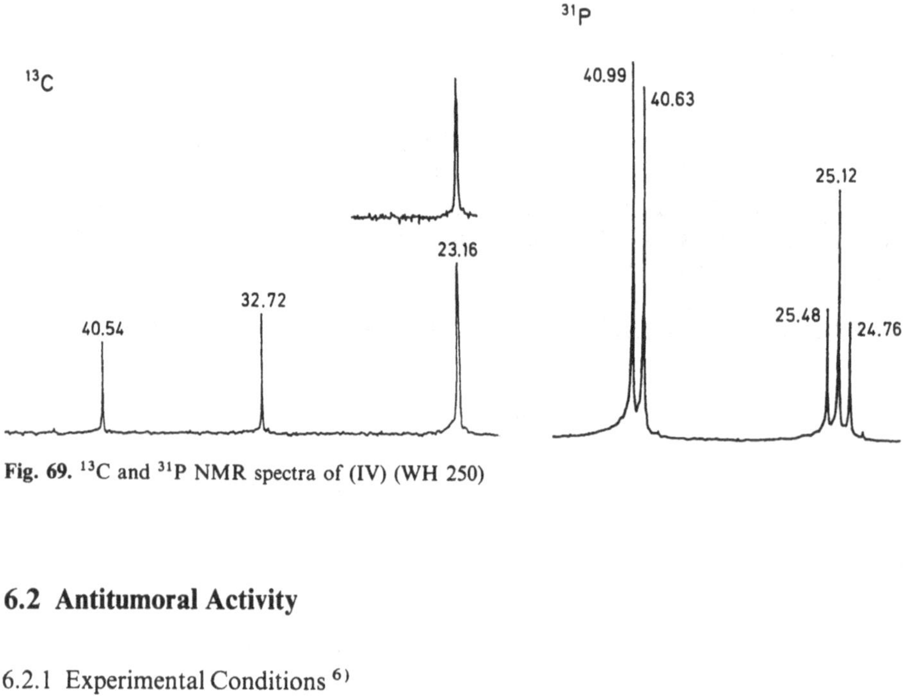

Fig. 69. ^{13}C and ^{31}P NMR spectra of (IV) (WH 250)

6.2 Antitumoral Activity

6.2.1 Experimental Conditions [6)]

For the tests on the P388 and L1210 leukaemias, 10^6 and 10^5 cells respectively were transplanted intraperitoneally (i.p.) on day 0 in compatible CD2F1 male mice (20 ± 2 g), the drug treatment being initiated 24 hours later. Male CD2F1 mice were also used as recipients for the compatible P815 mastocytoma (10^6 cells i.p.). The drugs were always dissolved in sterile saline (owing to their high water solubility, larger than 20 gl^{-1}) and administrated i.p., ten animals per group at least being used. Results presented are representative of at least 3 experiments performed. The percent increase in median lifespan over untreated controls (T/C%) was calculated as an evaluation criterion of effectiveness [84)].

Table 23. Comparative antitumor activity of DIAM3, DIAM4 and MYCLAz against P388, L1210 and P815 tumors

	DIAM3 (LD_0 = 75 MG/KG)						DIAM4 (LD_0 = 125 MG/KG)						MYCLAZ (LD_0 = 25 MG/KG)	
	Q1Dx1		Q1Dx5		Q1Dx9		Q1Dx1		Q1Dx5		Q1Dx9		Q1Dx1	
	MG/KG	T/C%	MG/KG	T/C%	MG/KG	T/C%	MG/KG	T/C%	MG/KG	T/C%	MG/KG	T/C%	MG/KG	T/C%
P 388	50	200	40	304	20	260	50	194	40	304	20	225	24	168
	25	172	20	250	10	240	25	161	20	181	10	200	18	158
	12,5	155	10	181	5	175	12.5	150	10	163	5	180	12	152
			5	159					5	150			6	130
			2.5	145					2.5	145				
L 1210	50	127	40	150	40	188	100	150	40	187	40	181	24	140
	25	122	20	168	20	181	50	122	20	193	20	163	18	146
	12.5	111	10	168	10	150	25	116	10	150	10	150	12	115
			5	125	5	131	12.5	105	5	175	5	125		
			2.5	125	2.5	118			2.5	131	2.5	118		
P 815	50	250	40	168	20	383	50	311	40	172	20	316	24	158
	25	170	20	154	10	255	25	165	20	145	10	244	20	153
	12.5	155	10	145	5	205	12.5	145	10	136	5	177	18	144
			5	113	2.5	155			5	127	2.5	172	12	150
			2.5	136	1.25	129			2.5	109	1.25	150	10	127

6.2.2 Results

The first experiments were aimed at determining in mice the highest non-lethal dose (LDo) of these compounds upon single i.p. injection. Observation time was 30 days. In these conditions it was found that normal CD2F1 mice could consistently resist without gross adverse signs or symptoms single i.p. doses of 75 and 125 mg kg^{-1} for DIAM3 (II) and DIAM4 (IV), respectively. As similar results were also seen in Swiss mice, these doses were taken as LDo. When higher doses were employed, dose-dependent lethality was seen, death occuring within 5–6 days. The LDo dose for MYCLAz in the same murine strains was found to be 25 mg kg^{-1}.

Antitumoral activity was evaluated by using both single-injection treatments on day 1 (Q1DX1) and repeated injection protocols involving drug administrations on days 1–5 (Q1DX5) or on days 1–9 (Q1DX9). As shown in Table 23 by representative data, a significant, dose-dependent antitumoral effect was observed in mice transplanted with the P388, L1210 and P815 neoplasms with single parenteral administrations of DIAM3 and DIAM4. For instance, T/C % values in the 200–300 % range were obtained in the P388 and P815 systems using DIAM3 and DIAM4 at single 50 mg kg^{-1} doses, whereas the L1210 tumor proved less responsive to both chemicals. This Table also shows that in each of these tumor models both agents produced significantly better results in terms of lifespan prolongation when repeated rather than single injection treatments were employed, although no attempts to identify optimal treatment schedules were made. It can be seen additionally that both DIAM3 and DIAM4 were clearly superior to MYCLAz. In fact, although essentially similar T/C % values were seen when equal doses of MYCLAz and of DIAMs were injected in single or repeated treatments (data not shown), the lower toxicity displayed by the novel compounds permitted their use in higher doses and thus the obtainment of higher levels of antineoplastic acitivity. Although in none of the three experiments performed did the differences between the T/C % values seen with DIAM3 and DIAM4 reach statistical significance, it should be noted that in 2 experiments the administration of the latter was associated with higher T/C% throughout the dose range tested than was seen with DIAM3. Furthermore, in murine systems, DIAM4 possesses a wider margin than DIAM3 between the antitumorally most effective dose and the LDo, a finding which suggests a lower general toxicity for the former agent. In this connection, it is of interest that putrescine possesses a higher affinity for malignant cells than 1,3-diaminopropane [18].

6.3 Conclusion

Benefits from MYCLAz targeting through its linkage to 1,3-diaminopropane and 1,4-diaminobutane (putrescine) as tumor finders are as follows:

i) (II) and (IV) are 3 and 5 times less toxic (from LDo values) than MYCLAz. Such a decrease of acute toxicity allows to inoculate higher doses of the drug (i.e. MYCLAz) when linked to diamines (i.e. when vectorized) than when in the genuine state, these higher doses leading to higher effectivenesses;

ii) moreover, it is noteworthy from Table 23 that, contrarily to what happens with MYCLAz itself, no significant cumulative toxicity occurs when (II) and (IV) are injected within heavy chronical Q1D schedules: for example, 9 injections Q1D with 20 mg kg^{-1} daily for (II) on P388 correspond to a total dose on 9 days of 180 mg kg^{-1}, that is 2.4 times the LDo value of (II); similarly, 5 injections Q1D at 40 mg kg^{-1} in the same drug/tumor system, correspond to a total dose of 200 mg kg^{-1}, that is 2.7 times the LDo value;

iii) more important yet is the fact that therapeutic indexes (T.I.) of (II) and (IV), defined as the ratio of the LDo value over the minimal active dose, are very high when compared to the MYCLAz one: TI values are indeed from 6 to 10 for (II) and (IV) within monoinjection protocol and they reach the 60 and 100 values respectively within polyinjection protocols (to be compared with 4 for MYCLAz). Such high TI values prove that MYCLAz reaches more selectively the tumor when vectorized through diamines as tumor finders;

iv) finally, it seems from Table 23 that (IV) (from putrescine) would display a slightly but significantly higher activity at the total than (II). This observation can be related to the higher affinity for malignant cells exhibited in vivo by putrescine when compared to 1,3-diaminopropane [18]. If such a relationship between the affinity of polyamines and their capability to vectorize the drug towards the tumor was right, more promizing activities yet could be expected from targetings through spermidine and spermine which are considered as having the highest affinities in vivo for malignant cells.

Attempts in this way are now in progress in our laboratory in collaboration with EORTC Screening and Pharmacology Group at Mario Negri Institute, Milano, Italy. This approach looks so much the more exciting that we have discriminated recently the probable prior target of our drugs, both when genuine and vectorized, that is cell membrane [85] and that the same drugs, designed initially as anticancer agents, also display promising immuno-modulating properties against polyclonal activation of lymphocytes B and glomerulonephritis [86–88].

"The author would like to express his sincere thanks to his co-workers whose names appear in the References: they work indeed as convicts for the privilege of their stroke who is greatly indebted to them".

7 References

1. Cf for example: Kulkarni, P. N., Blair, A. H., Ghose, T. I.: Cancer Res. *41*, 2700 (1981)
2. Cf for example: Porter, C. W., Bergeron, R. J., Stolowich, N. J.: ibid. *42*, 4072 (1982)
3. Labarre,J.-F.: Topics in Current Chemistry *102*, 1 (1982)
4. Cernov, V. A., Litkina, V. B., Sergievskaya, S. I., Kropacheva, A. A., Pershina, V. A., Sventsitkaya, L. E.: Farmakol. Toksikol. (Moscow) *22*, 365 (1959)
5. Labarre, J.-F., Cros, S., Faucher, J.-P., Francois, G., Levy, G., Paoletti, C., Sournies, F.: 2nd Int. Symp. Inorganic Ring Systems (IRIS), Göttingen, Gesellschaft Deutscher Chemiker, Proceedings p. 44 (1978)
6. Labarre, J.-F., Faucher, J.-P., Levy, G., Sournies, F., Cros, S., Francois, G.: Europ. J. Cancer *15*, 637 (1979)
7. Malfiore, C., Marmonti, L., Filippeschi, S., Labarre, J.-F., Spreafico, F.: Anticancer Research, *3*, 425 (1983)
8. Labarre, J.-F., Lahana, R., Sournies, F., Cros, S., Francois, G.: J. Chim. Phys. Fr. *77*, 85 (1980)
9. Sournies, F.: Ph. D. Thesis, Paul Sabatier Univ., Toulouse, France 1981
10. Labarre, J.-F., Sournies, F., Cros, S., Francois, G., van de Grampel, J. C., van der Huizen, A. A.: Cancer Lett. *12*, 245 (1981)
11. Labarre, J.-F., Sournies, F., van de Grampel, J. C., van der Huizen, A. A.: CNRS Fr. Pat. n° 79–17336, July 4, 1979; World Extension, July 4, 1980

12. Kitazato, K., Takeda, S., Unemi, N.: J. Pharm. Dyn. *5*, 803 (1982)
13. Yajima, N., Kondo, K., Morita, K.: 10th Meet. Jpn. Environ. Mutagen Soc., Tokyo, December 3, Proceedings p. 648 (1981)
14. Nasca, S., Jezekova, D., Coninx, P., Garbe, E., Carpentier, Y., Cattan, A.: Cancer Treatment Rep. *66*, 2039 (1982)
15. Guerch, G., Labarre, J.-F., Sournies, F., Manfait, M., Spreafico, F., Filippeschi, S.: Inorg. Chim. Acta *66*, 175 (1982)
16. Enjalbert, R., Guerch, G., Sournies, F., Labarre, J.-F., Galy, J.: Z. Kristallogr. Mineral. *164*, 1 (1983)
17. Pohjanpelto, P.: J. Cell. Biol. *68*, 512 (1976)
18. Tabor, C. W., Tabor, H.: Ann. Biochem. *45*, 285 (1976)
19. Guerch, G.: Ph. D. Thesis, Paul Sabatier Univ., Toulouse, France 1983
20. Allock, H. R.: Phosphorus-Nitrogen Compounds, Academic Press, New York 1972
21. Shaw, R. A.: Z. Naturforsch. *31b* 641 (1976)
22. Krishnamurthy, S. S., Sau, A. C., Woods, M.: Adv. Inorg. Chem. Radiochem. *21*, 41 (1978)
23. Das, S. K., Keat, R., Shaw, R. A., Smith, B. C.: J. Chem. Soc. 5032 (1965)
24. Das, S. K., Keat, R., Shaw, R. A., Smith, B. C., Woods, M.: J. Chem. Soc., Dalton Trans., 709 (1973)
25. Lingley, D. J., Shaw, R. A., Woods, M., Krishnamurthy, S. S.: Phosphorus and Sulfur *4*, 379 (1974)
26. Becke-Goehring, M., Boppel, B.: Z. Anorg. Allg. Chem. *322*, 239 (1963)
27. Krishnamurthy, S. S., Ramachandran, K., Vasudeva Murthy, A. R., Shaw, R. A., Woods, M.: Inorg. Nucl. Chem. Lett. *13*, 407 (1977)
28. Babu, Y. S., Manohar, H., Ramachandran, K., Krishnamurthy, S. S.: Z. Naturforsch. *33b*, 588 (1978)
29. Krishnamurthy, S. S., Ramachandran, K., Vasudeva Murthy, A. R., Shaw, R. A., Woods, M.: J. Chem. Soc., Dalton Trans., 840 (1980)
30. Guerch, G., Graffeuil, M., Labarre, J.-F., Enjalbert, R., Lahana, R., Sournies, F.: J. Mol. Struct. *95*, 237 (1982)
31. Monsarrat, B., Prome, J. C., Labarre, J.-F., Sournies, F., van de Grampel, J. C.: Biomed. Mass Spectrom. *7*, 405 (1980)
32. Parkes, H. G., Shaw, R. A.: 3rd Int. Symp. Inorganic Ring Systems (IRIS), Graz, Proceedings p. 40 (1981)
33. Enjalbert, R., Guerch, G., Labarre, J.-F., Galy, J.: Z. Kristallogr. Mineral. *160*, 249 (1982)
34. Faucher, J.-P.: Ph. D. Thesis, Paul Sabatier Univ., Toulouse, France 1975 and references therein
35. Allcock, H. R.: Acc. Chem. Res. *11*, 81 (1978)
36. Pople, J. A., Beveridge, D. L.: Approximate Molecular Orbital Theory, McGraw-Hill, New York 1970
37. Labarre, J.-F.: Structure and Bonding *35*, 1 (1978) and references therein
38. Wiberg, K. A.: Tetrahedron *24*, 1083 (1968)
39. Guerch, G., Faucher, J.-P., Graffeuil, M., Levy, G., Labarre, J.-F.: J. Mol. Struct., Theochem. *88*, 317 (1982)
40. Guerch, G., Labarre, J.-F., Roques, R., Sournies, F.: J. Mol. Struct. *96*, 113 (1982)
41. Krishnamurthy, S. S., Sau, A. C., Vasudeva Murthy, A. R., Keat, R., Shaw, R. A., Woods, M.: J. Chem. Soc., Dalton Trans. 1405 (1976) and 1980 (1977)
42. Galy, J., Enjalbert, R., Labarre, J.-F.: Acta Cryst. *B36*, 392 (1980)
43. Mahmoun, A.: Ph. D. Thesis, Paul Sabatier Univ., Toulouse, France, 1984
44. Castera, P., Lahana, R., Enjalbert, R.: to be published
45. Labarre, J.-F., Guerch, G., Sournies, F., Lahana, R., Enjalbert, R., Galy, J.: J. Mol. Struct. *116*, 75 (1984)
46. Cameron, T. S., Chan, C., Labarre, J.-F., Graffeuil, M.: Z. Naturforsch. *35b*, 784 (1980)
47. Cameron, T. S., Labarre, J.-F., Graffeuil, M.: Acta Cryst. *B38*, 168 (1982)
48. Guerch, G., Labarre, J.-F., Lahana, R., Roques, R., Sournies, F.: J. Mol. Struct. *99*, 275 (1983)
49. Main, P., Fiske, S. J., Hull, S. E., Lessinger, L., Woolfson, M. M., Germain, G., Declercq, J.-P.: MULTAN 80, a System of Computer Programs for the Automatic Solution of Crystal

Structures from X-ray Diffraction Data, Universities of York (Gt. Britain) and Louvain-la-Neuve (Belgium), 1980
50. Guerch, G., Labarre, J.-F., Lahana, R., Sournies, F., Enjalbert, R., Galy, J., Declercq, J.-P.: Inorg. Chim. Acta *83*, L33 (1984)
51. Lahana, R., Crasnier, F., Labarre, J.-F.: ibid. *90*, L65 (1984)
52. Harris, P. J., Williams, K. B.: ICPC Meet., Abstract (posters) 169, Nice, France, September 1983
53. Harris, P. J., Williams, K. B.: Inorg. Chem. *23*, 1495 (1984)
54. Allcock, H. R., Harris, P. J.: ibid. *20*, 2844 (1981)
55. Enjalbert, R., Galy, J., Harris, P. J., Williams, K. B., Lahana, R., Labarre, J.-F.: J. Amer. Chem. Soc. (1985) in press
56. Microanalytical Data
Compound II. Calc. for $C_4H_{11}N_4OP_3Cl_4$: C, 13.11; H, 3.00; N, 15.30.
Found: C, 13.20; H, 2.94; N, 15.18
Compound III. Calc. for $C_4H_{10}N_4OP_3Cl_3$: C, 14.57; H, 3.03; N, 17.00
Found: C, 14.76; H, 3.06; N, 16.87
57. Enjalbert, R., Guerch, G., Sournies, F., Labarre, J.-F., Galy, J.: Z. Kristallogr. Mineral. *164*, 137 (1983)
58. Cromer, D. T., Waber, J. T.: Internat. Tables for X-Ray Crystallography, Vol. 4, Table 2.2A, Birmingham, Kynoch Press 1974
59. Ritchie, R. J., Harris, P. J., Allcock, H. R.: Inorg. Chem. *19*, 2483 (1980)
60. El Murr, N., Lahana, R., Labarre, J.-F., Declercq, J.-P.: J. Mol. Struct. *117*, 73 (1984)
61. Chivers, T., Hedgeland, R.: Can. J. Chem. *50*, 1017 (1972)
62. Galy, J., Enjalbert, R., Van der Huizen, A. A., van de Grampel, J. C., Labarre, J.-F.: Acta Cryst. *B37*, 2205 (1981)
63. Cameron, T. S., Labarre, J.-F., Graffeuil, M.: ibid *B38*, 2000 (1982)
64. Allcock, H. R., Siegel, L. A.: J. Amer. Chem. Soc. *86*, 5140 (1964)
65. Labarre, J.-F., Guerch, G., Lahana, R., Mahmoun, A., Sournies, F., Mathis, R., Willson, M., Mathis, F.: Spectrochim. Acta (1985) in press
66. Contractor, S. R., Hursthouse, M. B., Parks, H. G., Shaw, L. S., Shaw, R. A., Yilmaz, Y.: J. Chem. Soc., Chem. Comm., 675 (1984)
67. Sournies, F., Lahana, R., Labarre, J.-F.: J. Mol. Struct. (1984) in press
68. Baldwin, M. A., McLafferty, F. W.: Org. Mass Spectrom. *7*, 1353 (1973)
69. Dell, A., Williams, D. H., Morris, H. R., Smith, G. A., Feeney, J., Roberts, G. C. K.: J. Amer. Chem. Soc. *97*, 2497 (1975)
70. Hansen, G., Munson, B.: Anal. Chem. *50*, 1130 (1978)
71. Cotter, R. J.: ibid. *51*, 317 (1979)
72. Cotter, R. J., Fenselau, C.: Biomed. Mass. Spectrom. *6*, 287 (1979)
73. Issachar, D., Ynon, J.: Anal. Chem. *52*, 49 (1980)
74. Bruins, A. P.: ibid. *52*, 605 (1980)
75. Cf. for example: Patouraux, D.: Ph. D. Thesis, Dijon Univ., France, 1980 and references therein
76. Beckey, H. D., Rollgen, F. W.: Org. Mass Spectrom. *14*, 188 (1979)
77. Beckey, H. D.: ibid. *14*, 292 (1979)
78. Sournies, F., Labarre, J.-F., Harris, P. J., Williams, K. B.: Inorg. Chim. Acta *90*, L61 (1984)
79. Labarre, J.-F., Guerch, G., Levy, G., Sournies, F.: CNRS Fr. Pat. n^0 82–19768, Nov. 25, 1982; World Extension, October 25, 1983
80. Labarre, J.-F., Guerch, G., Sournies, F., Spreafico, F., Filippeschi, S.: J. Mol. Struct. *117*, 59 (1984)
81. Guerch, G., Sournies, F., Labarre, J.-F., Manfait, M., Spreafico, S., Filippeschi, S.: Bioinorg. Chim. Acta *66*, 175 (1982)
82. Guerch, G., Labarre, J.-F., Oiry, J., Imbach, J.-L.: Inorg. Chim. Acta *67*, L5 (1982)
83. Cameron, T. S., Labarre, J.-F., Sournies, F., van de Grampel, J. C., van der Huizen, A. A.: J. Mol. Struct. (1985) in press
84. Geran, R. I., Greenberg, N. H., MacDonald, M. M., Schumacher, A. M., Abbott, B. J.: Cancer Chemother. Rep. *3*, 1 (1972)

85. Trombe, M. C., Beaubestre, C., Sautereau, A.-M., Labarre, J.-F., Laneelle, G., Tocanne, J.-F.: Biochem.Pharmacology *33*, 2749 (1984)
86. Dueymes, M. D., Fournie, G. J., Carentz, F., Mignon-Conte, M. A., Labarre, J.-F., Conte, J. J.: Europ. Soc. for Clinical Investigation, Annual Meet., April 17–19, 1984, Milan, Italy (Abstract)
87. Fournie, G. J., Dueymes, M. D., Carentz, F., Mignon-Conte, M. A., Labarre, J.-F., Conte, J. J.: IXth Internat. Congr. of Nephrology, June 11–16, 1984, Los Angeles, USA (Abstract)
88. Dueymes, M. D., Fournie, G. J., Carentz, F., Mignon-Conte, M. A., Labarre, J.-F., Conte, J. J.: Clinical Experimental Immunology, *59*, 169 (1985)

Author Index Volumes 101–129

Contents of Vols. 50–100 see Vol. 100
Author and Subject Index Vols. 26–50 see Vol. 50

The volume numbers are printed in italics